The Cell in Mitosis

The Cell in Mitosis

Proceedings of the
First Annual Symposium
held under the provisions of
The Wayne State Fund
Research Recognition Award

Edited by

LAURENCE LEVINE

Department of Biology
Wayne State University
Detroit, Michigan

1963

ACADEMIC PRESS • NEW YORK AND LONDON

ACADEMIC PRESS INC.
111 Fifth Avenue, New York, New York 10003

United Kingdom Edition published by
ACADEMIC PRESS INC. (LONDON) LTD.
Berkeley Square House, London W.1

LIBRARY OF CONGRESS CATALOG CARD NUMBER: 63-14754

Second Printing, 1968

PRINTED IN THE UNITED STATES OF AMERICA

Contributors

DAVID P. BLOCH, *Botany Department and The Plant Research Institute, University of Texas, Austin, Texas.*

ROBERT C. BUCK, *Department of Microscopic Anatomy, University of Western Ontario, London, Canada.*

L. R. CLEVELAND, *Department of Zoology, University of Georgia, Athens, Georgia.*

ALFRED M. ELLIOTT, *Department of Zoology, University of Michigan, Ann Arbor, Michigan.*

LAURENCE LEVINE, *Department of Biology, Wayne State University, Detroit, Michigan.*

LIONEL I. REBHUN, *Department of Biology, Princeton University, Princeton, New Jersey and Marine Biological Laboratory, Woods Hole, Massachusetts.*

O. H. SCHERBAUM, *Department of Zoology, University of California, Los Angeles, California.*

ANDREW G. SZENT-GYÖRGYI, *Institute for Muscle Research at the Marine Biological Laboratory, Woods Hole, Massachusetts.*

G. B. WILSON, *Department of Botany and Plant Pathology and Biology Research Center, Michigan State University, East Lansing, Michigan.*

ARTHUR M. ZIMMERMAN, *Department of Pharmacology, State University of New York, Downstate Medical Center, Brooklyn, New York.*

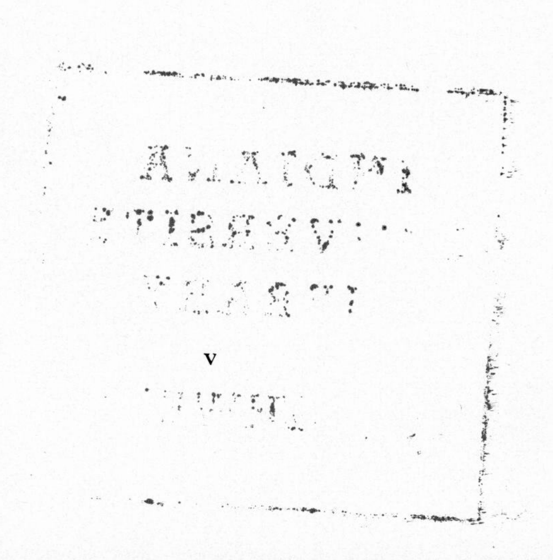

Preface

The symposium whose proceedings are recorded here was organized in accordance with the provisions of the Wayne State Fund Research Recognition Award, and was held November 6-8, 1961. This award, made annually to an assistant professor on the basis of a research proposal, was established in 1960 by a gift from the Wayne State Alumni Fund.

The central theme of the symposium, the cell in mitosis, has inherent and sometimes inexorable complexities. It is doubtful that the ordinary mortal, endowed with one approach, technique, or point of view, could hope to completely resolve these. However, a group of creative scientists with diverse backgrounds, communicating with each other in an atmosphere favoring informality and the free flow of discussion, may one day do so. The resolution of the complexities posed by the cell in division would have to depend upon pulling the threads of the molecular and submolecular from the fabric of macromolecular design that drives the cell toward the consummation of its life cycle. We need the insight and capacity to attack all facets of these relationships at every amenable level.

In this symposium, physiologists, electron microscopists, light microscopists, and biochemists explored various aspects of cell division. Their syntheses of data and ideas should serve a broad spectrum of cell biologists. The readers of these proceedings will find some of the chapters unique either because of the data presented or the way that the material is organized for the first time in one place. They will also find that the issues were not completely resolved but were sharpened to the point of departure into further experimentation.

The discussion is of extreme importance in a symposium for the proper balance and assessment of contributions. To make this as effective as possible a specifically invited discussant was assigned the duty of relating to the assemblage a distillate of his co-contribution wherein experiments could be suggested, avenues of argument explored and embellished, and different points of view presented. In short, there was provided a platform for the interchange of ideas and stimulation of further discussion which could be continued into the summation.

A particular debt of gratitude is owed to the chairmen who ably served to maintain the dignity of the sessions and the sequence of comments. They were Donald Costello, University of North Carolina; Maurice Bernstein, Wayne State University; J. G. Carlson, University of Tennessee; Hans Went, Washington State University; and Werner G. Heim, Wayne State University.

LAURENCE LEVINE

January 1963

Contents

Function of Flagellate and Other Centrioles in Cell Reproduction

L. R. Cleveland

The Central Spindle and the Cleavage Furrow

Robert C. Buck

Saltatory Particle Movements and Their Relation to the Mitotic Apparatus

Lionel I. Rebhun

The Fine Structure of *Tetrahymena pyriformis* during Mitosis

Alfred M. Elliott

Chemical Prerequisites for Cell Division

O. H. Scherbaum

Chemical Aspects of the Isolated Mitotic Apparatus

Arthur M. Zimmerman

Studies on the Disruption of the Mitotic Cycle

G. B. Wilson

The Histones: Syntheses, Transitions, and Functions

David P. Bloch

Contractility

Andrew G. Szent-Györgyi

Introduction

Laurence Levine

Department of Biology, Wayne State University, Detroit, Michigan

I MUST ADMIT that I could have been tempted to plagiarize some introduction to a hypothetical symposium of half a century ago. It would, of course, have been necessary to make a few changes. For instance, one would delete the word protoplasm and insert, in appropriate places, phase and electron microscope, synchronized cell populations, isolated spindles, micrurgy, tracer techniques, deoxyribonucleic acid (DNA) double helices, and perhaps adenosine triphosphate (ATP). The basic issues, however, would still remain at the core. The basic essential mechanisms of mitogeneses which underlie the emergence of the elegant geometries of mitosis, the congression of chromosomes onto metaphase configuration and progression through telophase remain enigmatic to this day.

I do not mean to imply that no progress has been made. The issues have yielded to the contemporary biologist and his tools. For example, recent advances in fixation and sectioning technique have made it possible to resolve the special morphology of the dividing cell with the electron microscope so that the identity of spindle fibers and centrioles would be apparent. Furthermore, the contribution of specific particles toward the scheme of cell division and the behavior of the endoplasmic reticulum during mitosis could be assessed. The level of looking not only permits resolution of fine structure but advances in optical design have made possible the transformation of phase changes into intensity changes. Thus artifacts may be assessed, the reality of the spindle fibers directly demonstrated within certain living cells, and the potential promise inherent in observing the living cell in mitosis realized. Furthermore, the skill of the micrurgist is abetted by ultrarefinements and such important features of subcellular morphology as nucleoli directly manipulated and transplanted. Populations of a variety of species dividing in synchrony may be treated as reagents and subjected to biochemical assay and the hope of identifying the impact of division on the chemistry of the cell is manifest.

One of the most elaborate machines of the diving cell, the spindle, may be isolated intact and disassembled so that its subunits could be identified and characterized. The contemporary biologist has a model for genetic replication in DNA double helicies which behave consistently with experiment. There is a glimmering that the translation of chemical potential into motility and contractility could be rationalized as some interaction between ATP and specialized cellular polyelectrolyte. There is the hope that it may be understood how chemical potential could be directed toward those specific orientations of subunits which predicate the spindle or how the collapse of such potential and the ensuing randomness means death.

Thus, it can be seen that I really could not plagiarize because with all the qualifications for progress, and the foregoing are but tastes, I would have to have written an introduction anyway. I submit that the ingredients for progress are all around us. As further evidence advances have been such that two symposia will have convened in about 2 years. The problems underlying mechanisms of cell division have attracted a widening circle of workers and there is much hope in this area – hope for the blending of the ingredients of progress and the syntheses that would permit the fine structure of the living dividing cell to be resolved as a fabric of meaningful chemical change. It is further hoped that this and other similar conferences would provide the catalyst.

Functions of Flagellate and Other Centrioles in Cell Reproduction

L. R. Cleveland

Department of Zoology, University of Georgia, Athens, Georgia

I. Introduction and General Considerations

Few organelles have been referred to by so many different names as the centriole (Boveri, 1895, 1901). Once in the development of cytology it almost seemed that each cytologist had to have his own name for the centriole. Of late, however, centriole is used much more than any of the other terms; yet, even today, there are still some who think centriole and centrosome (Boveri, 1888) are synonyms, and so use them.

Many of the early names such as centrum, centrosphere, sphere, attraction sphere (Strasburger, 1895) and idiozome (Meves, 1898) were used sometimes to refer to a centriole; at other times to refer to a centrosome. Later, terms such as central body and central apparatus sometimes have referred to a centriole, sometimes to a centrosome, and sometimes to both centriole and centrosome together. Centroplast, as often used in the Heliozoa, in most instances refers to a centriole which, during some stage of the life cycle of the cell, is usually surrounded by a centrosome.

The term blepharoplast (Webber, 1897), once almost universally employed in the description of flagellate protozoa as the organelle from which the flagella arise, is an unfortunate one. It was first used by Webber (1897) in a study of the pollen cells of *Ginko* to distinguish it from what Hirasé (1894) had already referred to as the attraction spheres of *Ginko*. Since organs of locomotion arose from the bodies at the ends of the central spindle Webber thought they were not centrioles (or centrosomes) even though their appearance and behavior were typical of these organelles. There are other instances, notably in the pteridophytes *Equisetum* and *Marsilia* (Sharp, 1912, 1914), in the collar cells of sponges, in flagellated spermatozoa, and in flagellates, which show clearly that the term blepharoplast as used by Webber, and still sometimes employed in flagellates, as the organelle from which flagella arise, refers to a body which is ontogenetically, phylogenetically, and functionally a centriole. In other words, since centrioles, at some time in their life cycle in so many organisms, are capable of producing flagella, in addition to their more usual function of producing the achromatic figure, there is no justification for giving those centrioles a special name such as blepharoplast. To do so only serves to create confusion. In fact, it is quite likely that in the course of time more examples will be found like that of Wolbach (1928) who believes he has seen, in two human tumors of striated muscle, the origin of myofibrils from centrioles. And Costello (1961) believes centrioles play a role in determining the cleavage plains in early embryonic development.

Another reason for not using blepharoplast is the fact that it is also sometimes used synonymously with basal body or basal granule. Dual terminologies should be reserved for the exclusive use of those who prefer confusion to clarity.

It now seems desirable to define centriole and centrosome briefly. Centrosome as first used by Boveri (1888) referred to the large, hyaline body lying at the astral center. Later Boveri (1895, 1901) redefined the term centrosome and used it only for the hyaline body which surrounds the small, deeply staining body which he termed the centriole. Failure of subsequent investigators to use these terms according to Boveri's second or revised definition has resulted in much needless confusion. Further, the fact that no centrosome is present in some cells has also been a cause for confusion, as well as the fact that the centriole is sometimes near the limit of visibility, or stains faintly, if at all, with most stains. Studies of living cells with phase contrast have helped to clear up some doubtful cases.

But there are still certain plant and animal cells in which centrioles have not been seen. To consider this question and argue for their existence in all cells because of their functions in many cells where they may be seen would be going beyond the scope of this article. It is sufficient to consider here the roles of centrioles in those cells where they are large and may be seen plainly, both in living and fixed and stained material.

However, before doing so it seems desirable to mention briefly the fact that for a period in the 1920's, following the large number of papers by well-known authors describing centrioles and their function so clearly that almost no one doubted their existence, there was a period when the very existence of centrioles was seriously challenged. In fact, the idea that centrioles were artifacts produced by fixation and staining was almost universally accepted, even though the promoters of this idea really had little logic on their side of the argument. The time was ripe for discarding many well-established concepts rather easily. But the artifact notion did not survive for long, although we do find Wilson as late as 1931 pleading for careful, continuous observations on centrioles. He actually used the term central bodies because he had been attacked so vigorously for using centrosome and centriole that, in self defense, he coined the terms central bodies and central apparatus. However, looking back on the argument, it is now clear that these new terms served in no way to lift the confusion regarding centriole and centrosome, nor to allay the contention that they were artifacts. Careful studies of centrioles and centrosomes in many types of cells finally turned the tide of thinking and forced an acceptance of the early classical views of Van Beneden, Hermann, Flemming, Boveri, Heidenhain, Meves, Wilson, Conklin, and others.

The early conceptions of the centriole by Boveri and others have been extended considerably, especially in the last 30 years. The centriole, in addition to being that small heavily staining dot surrounded by a centrosome, seen in the early days of cytology, may also be rod-shaped or v-shaped, and it may or may not be surrounded by a centrosome. In certain genera of flagellates it is extremely long; for example, in the larger species of *Barbulanympha* it is 28–35 microns in length and 4–5 in width (Cleveland, 1934). In *Pseudotrichonympha* (Cleveland, 1934, 1935) the centrioles, although more narrow than those of *Barbulanympha,* are considerably longer; in a few species of *Pseudotrichonympha* they are rather tightly coiled and, if uncoiled, their total length would be approximately 500 microns (Cleveland, 1957a).

In the case of these very long centrioles of hypermastigote flagellates only their distal ends are surrounded by a centrosome. For most of their length, they lie free in the cytoplasm. Many genera of flagellates have such centrioles (for a list see Cleveland, 1957a); but not all hypermastigote flagellates have long centrioles; nor do all of these flagellates possess centrosomes. In some, centrosomes are completely absent and, when such organisms are studied in conjunction with those that possess centrosomes, it is possible to learn precisely the slight role a centrosome plays in cell reproduction and the several important roles played by the centriole. The centrosome merely serves to make the central spindle cylindrical (or nearly so) instead of flat. It plays no role whatever in the production of the achromatic figure or any other organelle; it is not autonomous. The centriole, on the other hand, is autonomous, and the centrosome, when present, is produced anew by the centriole in each cell generation.

For many reasons the term achromatic figure is preferable to the one used most, the spindle, although I doubt if it will ever supplant spindle. Some cytologists use spindle as a synonym of achromatic figure; others use it in the same sense as Hermann (1891) originally used the term central spindle, namely, for the fibers which extend from one pole to the other, the so-called continuous fibers. Hence, it is often difficult in reading a paper to determine in which sense the author is using the term spindle. No such double meaning applies to the usage of achromatic figure, and, as recent studies of living cells show, it is really achromatic.

The achromatic figure is composed entirely of astral rays. Some of the rays arising from one centriole meet and join those arising from the other;

they grow along one another, and thus form the central spindle (connecting fibers, interzonal fibers). Other rays join centromeres and become chromosomal fibers (mantle fibers, half fibers, traction fibers). Still other rays remain free, and play no role in cell reproduction (Cleveland, 1934, 1935, 1938, 1954b, 1955, 1956, 1957a,b).

Mention should be made of the fact that in the protozoa several unfortunate terms have been used for the central spindle. Alexieff (1913) used blepharoplastodesmose, and Kofoid and Swezy (1915) object to this term because of its length and coin the term paradesmose. When asked why they did not use central spindle, these authors could only answer that the structure which they called paradesmose lay outside the nuclear membrane. Since the nuclear membrane remains intact during nuclear reproduction, the central spindle must of necessity lie outside this membrane (in most flagellates it does depress the nuclear membrane and thus assumes a central position). The difference, in this respect, between higher organisms and that which occurs in protozoa lies not in the nature and function of the central spindle but in the fact that the nuclear membrane does not break down. Therefore, the term central spindle is just as applicable in the protozoa as in any cell, and the sooner the usage of paradesmose is discarded, the better.

Unlike paradesmose, which has been used a great deal by protozoologists for central spindle, centrodesmose (Heidenhain, 1894), netrum (Boveri, 1901), and desmose (Janicki, 1915) have been used so seldom to refer to the central spindle that they need not be considered here.

In studying the astral rays, the chromosomal fibers, and the central spindle of higher organisms in living cells with phase contrast, one is not able to trace these fibers beyond the centrosome (in several genera of flagellates they may be traced easily through centrosome to centriole). They seem to stop there. The same is true of most fixed and stained preparations. This explains why, for so long, many cytologists gave the centrosome credit for producing these fibers. Some, of course, were cautiously content with saying the fibers converge at the periphery of the centrosome.

The Centrioles of Flagellates

Not only can the centrioles of several genera of flagellates be seen much better, owing to their size and other features, but their roles in cell reproduction are very much plainer than in any metazoan or plant cell. It therefore seems desirable to devote most of our time to them.

II. Functions, Types, and Life Cycles of Flagellate Centrioles

So far as our knowledge goes at present, the centrioles of flagellates may be placed into five groups, based on morphology and life cycle. All of these centrioles may be seen plainly in living cells in all stages of their life cycle. They never disappear, although in some genera they are much smaller in some phases of their life cycle than in others. They clearly fall into the class of so-called autonomous organelles; once a cell loses them, it can never regain them; only centrioles have the ability to reproduce centrioles. Of course, they are dependent on other types of cytoplasms for their ability to reproduce themselves and to carry out their roles in cell reproduction.

A. Life Cycle and Functions of the Centrioles of *Barbulanympha*

In the resting cell of *Barbulanympha* two elongate centrioles are present. They are about 30 microns in length and 4 to 5 in width in the larger species. A centrosome about 5 microns in diameter surrounds the distal end of each centriole (Fig. 1A). The distal or posterior ends of the centrioles are free and lie from 15 to 30 microns apart, while the anterior ones, so long as the cell is resting, are held close together by an interconnection.

In early prophase, astral rays arise from the distal end of each centriole, pass through the centrosome and extend beyond it (Fig. 1B). As these rays increase in length, those between the centrioles soon meet (Fig. 1C) and join one another to form the central spindle (Fig. 1D). Meanwhile, the anterior ends of the centrioles become farther separated and a small new centriole is formed from the anterior end of each old one (Fig. 1E). As the central spindle increases in length by continued elongation of the fibers composing it, each new centriole grows, increases in length, and extends posteriorly near the old one which has produced it (Fig. 1F). These new centrioles do not become very long before a small new centrosome begins to form on and surround the distal end of each (Fig. 1G). Finally, following nuclear duplication, the central spindle pulls in two (Fig. 1H). By this time the two old centrioles as well as the two new ones lie a considerable distance apart, and the new centrioles are about half as long as the old ones. Cytokinesis follows shortly, and the new centriole of each daughter cell is now two-thirds grown (Fig. 1I). Within an hour or two the new centriole becomes as long as the old one, and the cell returns to the resting stage (Fig. 1A).

Shortly after each new centriole is formed, and sometime before it begins to elongate, it begins the production of a whole new set of extranuclear organelles (flagella, parabasals, axostyles). These new organelles become fully grown by early anaphase or shortly thereafter. Thus, the parent reproducing cell is able at cytokinesis to deliver to each daughter two complete sets of extranuclear organelles, one set being new and having been produced by the new centriole and the other being old and having been produced by the other or old centriole. Hence, in cell reproduction of *Barbulanympha* each daughter obtains a new centriole and a new set of extranuclear organelles (Cleveland, 1954b).

But the centrioles of *Barbulanympha* do not always carry out this type of life cycle, nor do they always function in this manner in the production of new extranuclear organelles. In gametogenesis, which is haploid,

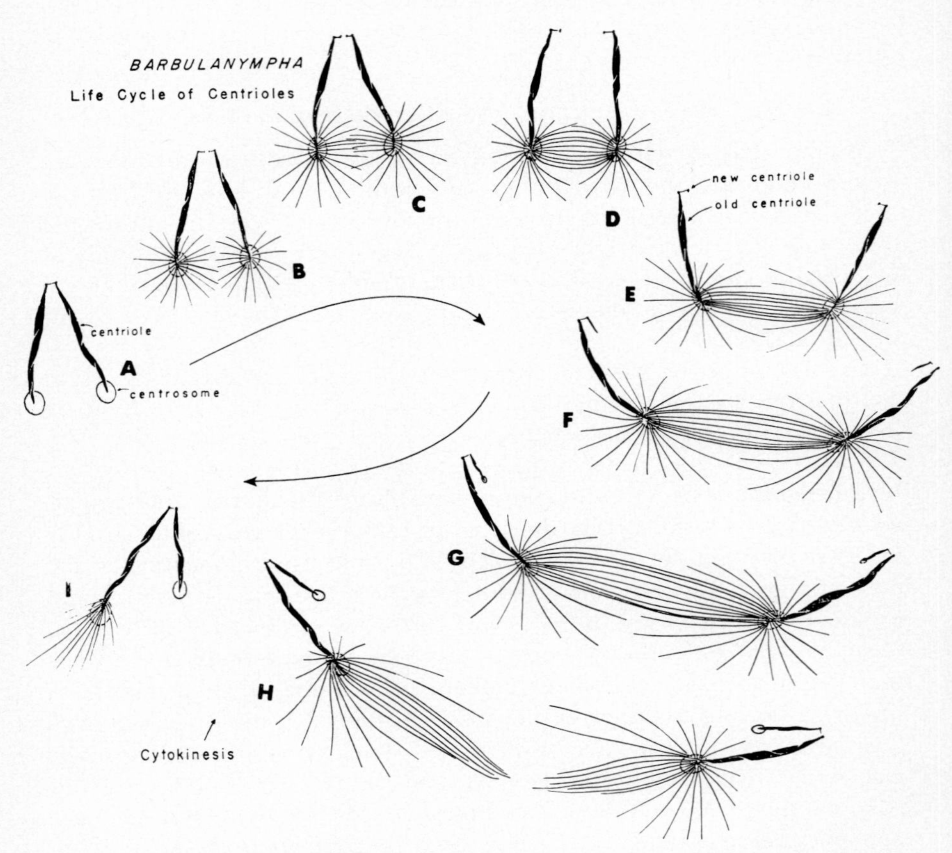

FIG. 1. *Barbulanympha,* life cycle of centrioles. A, resting cell. Two elongate centrioles interconnected anteriorly. A centrosome surrounds the distal or posterior end of each centriole. B, early prophase. Astral rays arise from distal end of each centriole; anterior ends of centrioles are separating. C,D, joining and overlapping of astral rays to form central spindle; centrioles move farther apart anteriorly; formation of a new centriole at the anterior end of each old one. E, elongation of central spindle; new centrioles farther apart. F, continued elongation of central spindle, and elongation of new centrioles posteriorly. G, continued growth of central spindle and the formation of a small new centrosome around the distal end of each new centriole. H, central spindle pulls in two following nuclear duplication. I, cytokinesis; new centriole (on right) nearly as large as old one; only remnant of achromatic figure remains.

in all asexual mitoses, and in meiosis which is zygotic, the behavior described above occurs; but in the zygote, which may be formed by fusion of gametes, autogamy, or endomitosis, the functions of the two remaining centrioles (the two male ones disintegrate) are altered greatly. First, there is reorganization of the centrioles, followed about 35 days later by discarding and renewal of all extranuclear organelles except the centrioles. Stated briefly, the centriole reorganization occurs as follows: All of each centriole except the very anterior, slightly enlarged end undergoes gradual, but complete disintegration; then these anterior ends, which still maintain their interconnection, become free and migrate posteriorly to the nucleus, 30 to 50 microns away. Here they lie in a resting condition for about 35 days. Each dotlike reorganized centriole, at the end of this resting period, produces new extranuclear organelles, such as flagella, parabasals, and axostyles, but not new centrioles. As these centrioles begin the formation of new organelles, they also begin to elongate posteriorly; and by the time each centriole has produced a fully grown new set of organelles it has also elongated to its full length and has developed a fully grown centrosome that surrounds its distal end. At this time the centrioles of the zygote present the same size and appearance as do those of a resting asexual cell, or a gamete. Meanwhile, the two old sets of flagella, axostyles, and parabasals have been gradually disintegrating, and by the time the two new sets have become fully grown, which requires about 5 days, they complete the disintegration process. It should be emphasized that, owing to the fact that the old and new organelles overlap one another for 5 days, one growing during this period and one disintegrating, makes it easy to interpret correctly the role played by the centrioles in the production of flagella, parabasals, and axostyles. A further fact that the new flagella are intracytoplasmic during their entire period of growth and development and for a day after the old ones, which are extracytoplasmic, have disintegrated, makes it easy to differentiate between new and old flagella (Cleveland, 1954a).

The reorganization process just described shows several things clearly: a definite relationship between hypermastigote centrioles and those of higher forms of life; the ability of a centriole at certain times to function more than once in the formation of flagella, axostyles, and parabasals, just as it is able to do at all times in the formation of the achromatic figure; the ability of centrioles to function in the production of extranuclear organelles without reproducing themselves, and also without accompanying nuclear or cytoplasmic reproduction; the inability of flagella, parabasals, and axostyles to reproduce themselves; and most important of all the fact that the anterior tip of these unusually large centrioles of flagellates is their reproducing portion. In the course of evolution of many of the hypermastigotes the centriole, unlike that of higher forms of life, developed so that its achromatic-figure reproducing region became widely separated from its own reproducing region. A long stalk connects the two regions, and thus puts the achromatic-figure reproducing region near the nucleus so that it may function in nuclear reproduction (as will be seen later, sometimes the nucleus lies out of reach of the achromatic figure and when it does the latter is unable to carry out its function; thus we learn much about the functions of centrioles).

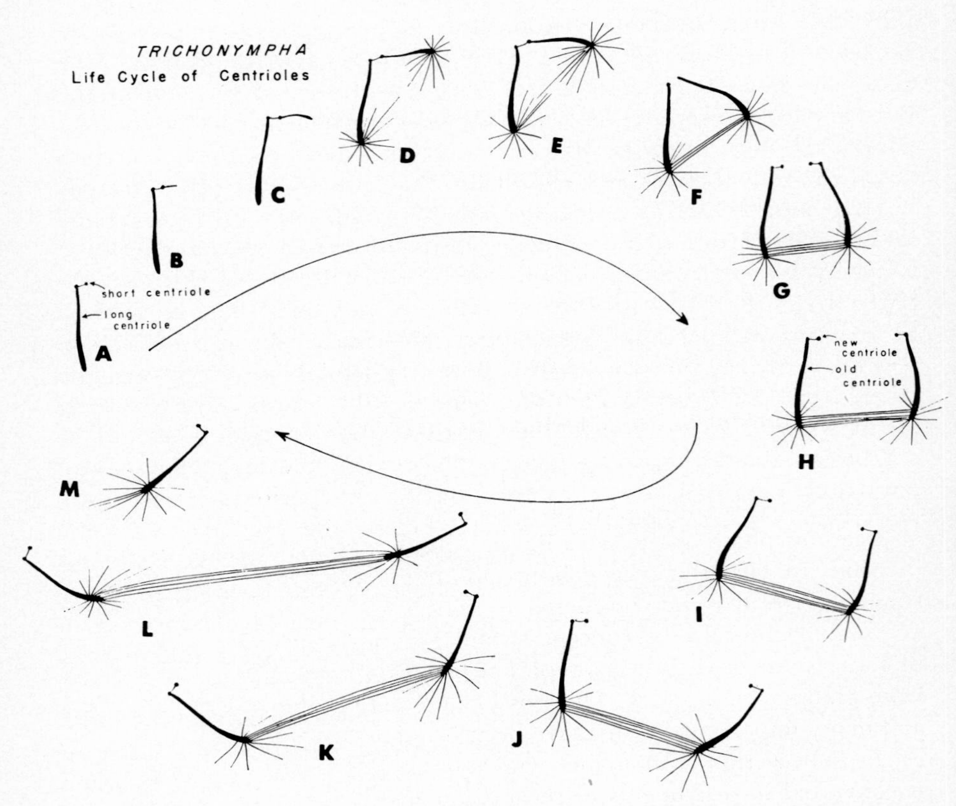

FIG. 2. *Trichonympha,* life cycle of centrioles. A, two resting centrioles, one long, one short. B,C, early prophase elongation of short centriole. D, elongation complete; astral rays grow out from distal end of each centriole. E,F, formation of central spindle by growth of astral rays along one another; anterior interconnection of centrioles lost. G,H, formation of a new centriole from anterior end of each old one. I–L, elongation of central spindle and separation of centrioles. M, cytokinesis, one long and one short centriole interconnected anteriorly. With loss of remnant of achromatic figure, centrioles return to resting stage.

B. Life Cycle and Functions of the Centrioles of *Trichonympha*

The centrioles of *Trichonympha* never have a centrosome surrounding any portion of them; in fact, there is no structure, in any of the many species of this genus, which has the slighest resemblance to a centrosome (Cleveland, 1949b, 1960).

In the resting cell two centrioles are present in the anterior end, some distance anterior to the nucleus; one is long and one is short (Fig. 2A). At the onset of cell reproduction, the short one gradually elongates (Fig. 2B, C). As soon as it becomes fully elongated, astral rays begin to grow out directly from the distal end of each centriole (Fig. 2D). These rays, like those of *Barbulanympha* and other genera of hypermastigote flagellates, are firmly attached to the centrioles; in fact, microdissection experiments carried out on *Barbulanympha* show that a cell has to be torn to shreds before they are disconnected from the centrioles. Often they still adhere to it after all cytoplasm has been torn away.

Just as already described in *Barbulanympha,* the rays arising from one centriole meet in the region between the centrioles, grow along one another, and thus form the central spindle (Fig. 2E, F). Meanwhile, the connection holding the anterior ends of the centrioles together is lost and these ends move apart. Shortly thereafter each elongate centriole, just as already described in *Barbulanympha,* produces a new, dotlike centriole at its anterior end, and each new centriole maintains connection with the elongate one which produced it (Fig. 2G, H). At the same time each new centriole produces a new set of flagella and parabasals (Cleveland, 1949b). The central spindle now elongates and, together with astral rays which become chromosomal fibers, carries out its function in nuclear reproduction (Fig. 2I–L); it finally pulls in two (Fig. 2M), cytokinesis occurs, and the central spindle, along with the free astral rays, disappears, leaving one long and one short centriole (Fig. 2A).

C. Life Cycle and Functions of the Centrioles of *Pseudotrichonympha*

In the resting cell of *Pseudotrichonympha* both centricles are short, lie close together in the anterior end, and appear as slightly curved rods (Fig. 3A). At the onset of cell reproduction, each centriole begins to elongate posteriorly (Fig. 3B), and shortly thereafter a small centrosome may be seen forming around the distal end of each (Fig. 3C). Slightly later astral rays begin to grow out from the distal end of each elongate centriole, and extend through the centrosomes (Fig. 3D). Just as in *Trichonympha* and *Barbulanympha,* the rays between the distal ends of the centrioles meet to form the early central spindle (Fig. 3E, F).

Meanwhile, the anterior ends of the centrioles move apart, and the central spindle being produced by the distal ends of the centrioles grows and increases in length (Fig. 3G). At the same time, each old centriole produces a new one from its anterior end, as described in *Barbulanympha* and *Trichonympha*. Now the central spindle elongates as a result of

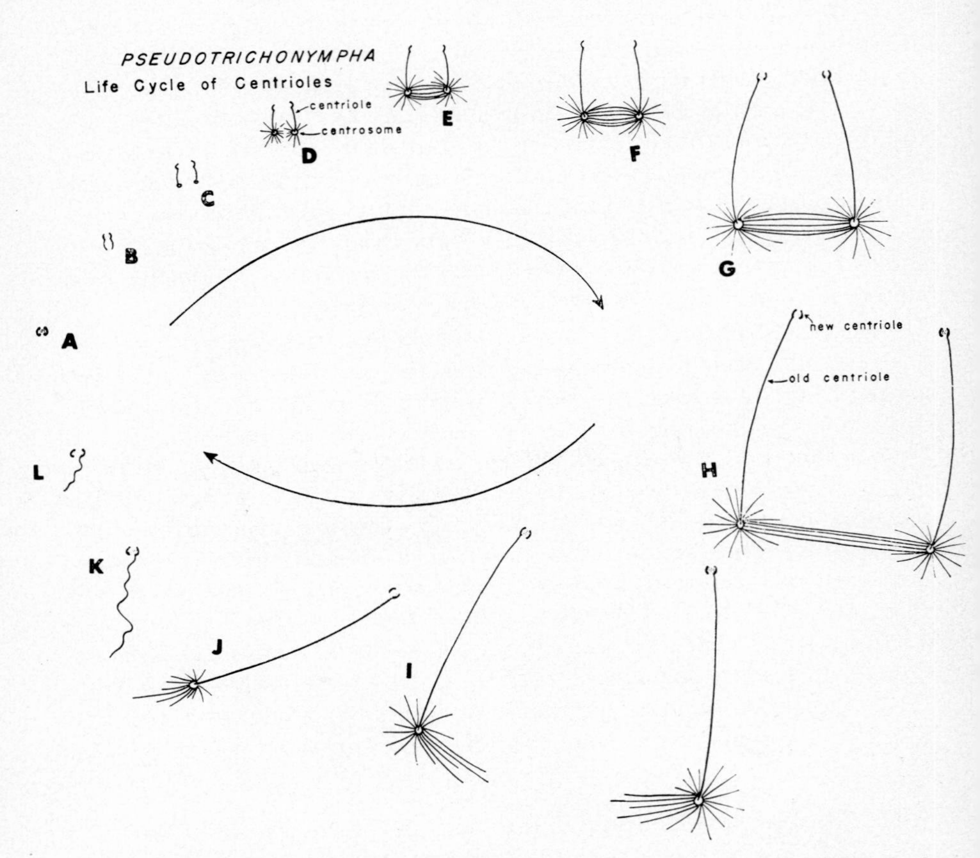

FIG. 3. *Pseudotrichonympha,* life cycle of centrioles. A, two short, curved, rod-shaped centrioles of resting cell. B, posterior elongation of centrioles at onset of cell reproduction. C, formation of small centrosome around distal end of each centriole. D, growth of astral rays from distal ends of centrioles. E,F, formation of central spindle by meeting and joining of astral rays; separation of centrioles anteriorly. G,H, increase in length of central spindle and of old centrioles; formation of a new centriole from anterior end of each old one. New centrioles do not elongate. I, central spindle pulls in two. J, cytokinesis, followed by degeneration of achromatic figure. K,L, progressive stages in degeneration of old, elongate centriole, beginning at its posterior end, and progressing, eventually, to the point where it is the same size and shape as the new one, thus returning to the resting condition of the centrioles shown in Fig. 3A.

growth and, together with astral rays which become chromosomal fibers, functions in nuclear reproduction (Fig. 3H). Finally the central spindle pulls in two (Fig. 3I), cytokinesis occurs (Fig. 3J), and the central spindle, together with the remaining astral rays, degenerates, but the long centriole which goes to each daughter cell at cytokinesis does not remain elongate, as is the case in *Barbulanympha* and *Trichonympha*. Degeneration which begins at its posterior end (Fig. 3K, L) extends anteriorly until the centriole, which once was quite elongate, is now the same size as the new one, from which it cannot be differentiated (Fig. 3A). Thus, the centrioles of *Pseudotrichonympha* during each cell generation are greatly reduced in size, just as occurs in those of *Barbulanympha* during the zygote stage (Cleveland, 1954a).

D. Life Cycle and Functions of the Centrioles of *Macrospironympha*

In the resting cell of *Macrospironympha* two rod-shaped centrioles lie close together in the anterior end some distance anterior to the nucleus (Cleveland, 1956). No centrosomes are associated with any portion of them, but both distal ends of the centrioles are surrounded by a common rostral body (Fig. 4A). At the onset of cell reproduction the nucleus becomes free and usually migrates to the posterior end of the cell. At the same time the distal portion of each centriole, together with the rostral body, becomes detached. These dotlike, detached, posterior ends of the centrioles and the rostral body within which they lie are free to move anywhere in the cell (Fig. 4B). However, in thousands of examples studied they always migrate to the nucleus, although it sometimes takes them longer to do so than at other times. By the time they reach the nucleus (and sometimes shortly before they do) the rostral body begins to disintegrate. Sometimes one can see a small achromatic figure forming within the rostral body as it begins to disintegrate, but usually the formation of an achromatic figure does not begin until this body is fairly well disintegrated, with remnants of it scattered here and there, at which time the achromatic figure lies partly across and outside the ever intact nuclear membrane (Fig. 4C). It is thus plain that the small detached piece of each centriole is the achromatic-figure producing portion of this organelle. The central spindle is formed by the cooperative functions of these pieces, just as it is from the distal ends of the intact centrioles of *Barbulanympha* and *Trichonympha*. Soon after these pieces begin to form astral rays, a centrosome develops around each piece, and at the same time each old centriole develops a new one which elongates, becomes fully grown and lies beside the old one; the achromatic figure grows and increases in length and takes up a central position with respect to the nucleus. As the central spindle elongates the nuclear mem-

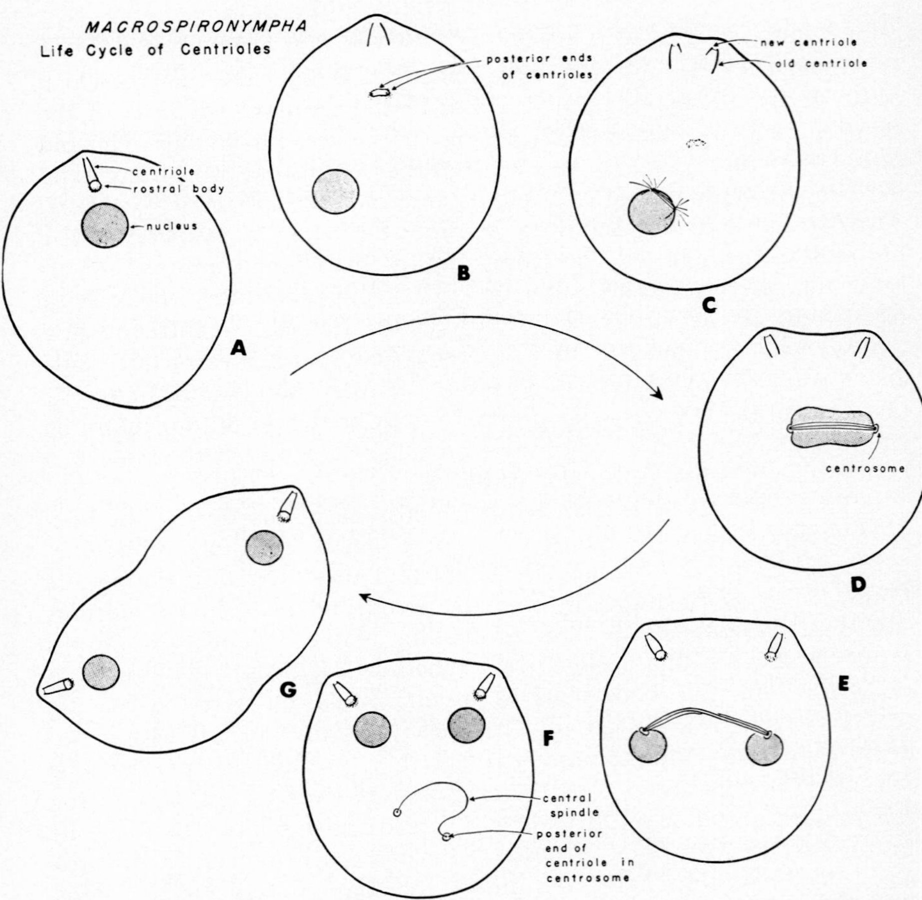

FIG. 4. *Macrospironympha,* life cycle of centrioles. A, two elongate centrioles lie in the anterior end of a resting cell. At their distal ends they are surrounded by a common rostral body. B, early prophase; nucleus has lost its connection with rostrum and has migrated posteriorly; at same time rostral body has lost its connection with elongate centrioles, has migrated posteriorly, and has carried with it a portion of the distal end of each centriole. C, rostral body is disintegrating and has freed the distal ends of the centrioles. These ends have migrated to the nucleus and have formed an early achromatic figure on the border of the nuclear membrane. Meanwhile, the two portions of the centrioles left in the anterior end of the cell have separated and the anterior end of each old portion has begun the development of a new centriole. D, an elongate central spindle lies across the nucleus, with a small dot-like portion of an elongate centriole at each end. A centrosome now surrounds each portion. New centrioles in anterior end are fully grown and one lies beside each old centriole. E, elongation of central spindle has separated nucleus into two daughters. In anterior end of cell two new rostral bodies have formed, one surrounding each group of centrioles, preparatory to cytokinesis. F, each daughter nucleus has migrated to a position posterior to a rostral body and its two elongate centrioles. Remains of greatly elongated central spindle still attached at each end to the migrating portion of a centriole. G, elongation of cell preparatory to cytokinesis. Migrating portions of centrioles, centrosomes, and central spindle have disintegrated.

brane does likewise (Fig. 4D). The nuclear membrane soon becomes dumbbell shaped and later pulls in two. The daughter nuclei now become well separated, owing to the continued elongation of the central spindle. Since a daughter nucleus is fastened to each centriole via chromosomal fibers and each end of the central spindle is also fastened to a centriole, it is thus plain how elongation of the central spindle pushes the nuclei apart (Fig. 4E). Finally, the chromosomal fibers break down, the connection between daughter nuclei and centrioles is lost, thus leaving the nuclei free to migrate. At the same time the four fully grown centrioles which are now present in two groups, one old one and one new one composing each group, move somewhat farther apart; and a new rostral body which has been forming meanwhile now surrounds the distal ends of each group. Now the free daughter nuclei leave the achromatic-figure producing portion of the centrioles to which they were firmly attached earlier and migrate to a position slightly posterior to the new rostral bodies, one nucleus migrating to the base of each rostral body. In thousands of examples I have studied not once have the daughter nuclei failed to follow this scheme of migration perfectly.

During all these events just described the two dotlike detached portions of the centrioles, which together formed a central spindle, remain very plain indeed. A very long, bent central spindle connects them, and each is still surrounded by a centrosome (Fig. 4F). It is usually some time after the daughter nuclei have migrated before degeneration begins in these detached portions of the centrioles, the central spindle, and the centrosomes, but all three eventually degenerate completely, slightly before the cell elongates and becomes constricted in the middle (Fig. 4G) preparatory to cytokinesis.

This unusual behavior on the part of centrioles and nuclei has been seen in only one other genus of hypermastigotes (Cleveland, 1951).

This rather detailed account of the life cycle of the centrioles of *Macrospironympha* seems justified because here we learn clearly and beyond question not only that the achromatic figure-producing portion of a centriole may be separate and distinct from the propagative or reproductive portion, but also that the two portions are able to carry out their respective functions when they lie 100 to 200 microns apart. In *Pseudotrichonympha* (Fig. 3) it has been shown that the achromatic figure-producing portion of a centriole degenerates and is renewed during each cell generation. Many genera of hypermastigotes behave in this manner (Cleveland, 1957a) and the centrioles of the polymastigote *Trichomonas* have been shown to have the same behavioral pattern (Cleveland, 1961). We also noted the same behavior in the zygote of *Barbulanympha* (see p. 9).

E. Life Cycle and Functions of the Centrioles of *Joenia*

In the resting cell of *Joenia* and many related genera (Cleveland, 1957b), the centrioles are broad rods with no centrosome surrounding any portion of them during any stage of their life cycle (Figs. 5A, 6A). Unlike the centrioles already considered, they never elongate. Another feature which differentiates them from all other flagellate centrioles studied except two [*Idionympha* (Cleveland, 1934) and *Holomastigotoides* (Cleveland, 1949a)] and also from most centrioles of higher forms of life, is the fact that a small, arrested central spindle lies between the centrioles in the resting cell (Fig. 6A). This small central spindle, which

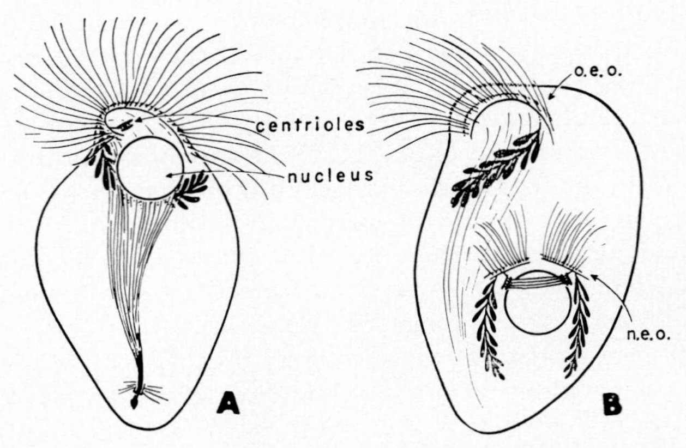

FIG. 5. *Joenia,* low power diagrams showing relationship between centrioles, extranuclear organelles, and nucleus during cell reproduction. A, resting cell, showing one set of extranuclear organelles (only one is ever present in resting cell) and small arrested central spindle between two rod-shaped centrioles (see arrow). B, early prophase. Posterior migration of nucleus and centrioles after both have been detached from old extranuclear organelles *(o.e.o.);* formation of central spindle across nucleus; development of two new centrioles and two daughter sets of new extranuclear organelles *(n.e.o.).* The old or parent set of extranuclear organelles is degenerating in the anterior end of cell.

was formed in the late prophase to telophase of the previous generation, arises from all of each centriole. Unlike *Barbulanympha, Trichonympha, Pseudotrichonympha,* and *Macrospironympha,* just described, *Joenia* in the resting cell possesses only one set of extranuclear organelles. The other genera have two.

Another striking difference between the centrioles of *Joenia* and those already considered, is the fact that at the onset of cell reproduction the centrioles, which at this time lie in the anterior end with one of them attached to the set of parent extranuclear organelles which it produced in the previous generation, lose their connection with these organelles at the same time that the nucleus loses its connection with them. Then both nucleus and centrioles migrate together to the posterior end of the cell (Fig. 5A, B). The parent extranuclear organelles, such as flagella, parabasals, and axostyle, which are left by themselves in the anterior end as a result of the migration of nucleus and centrioles, soon begin to degenerate; but, before much progress in degeneration is made, the cen-

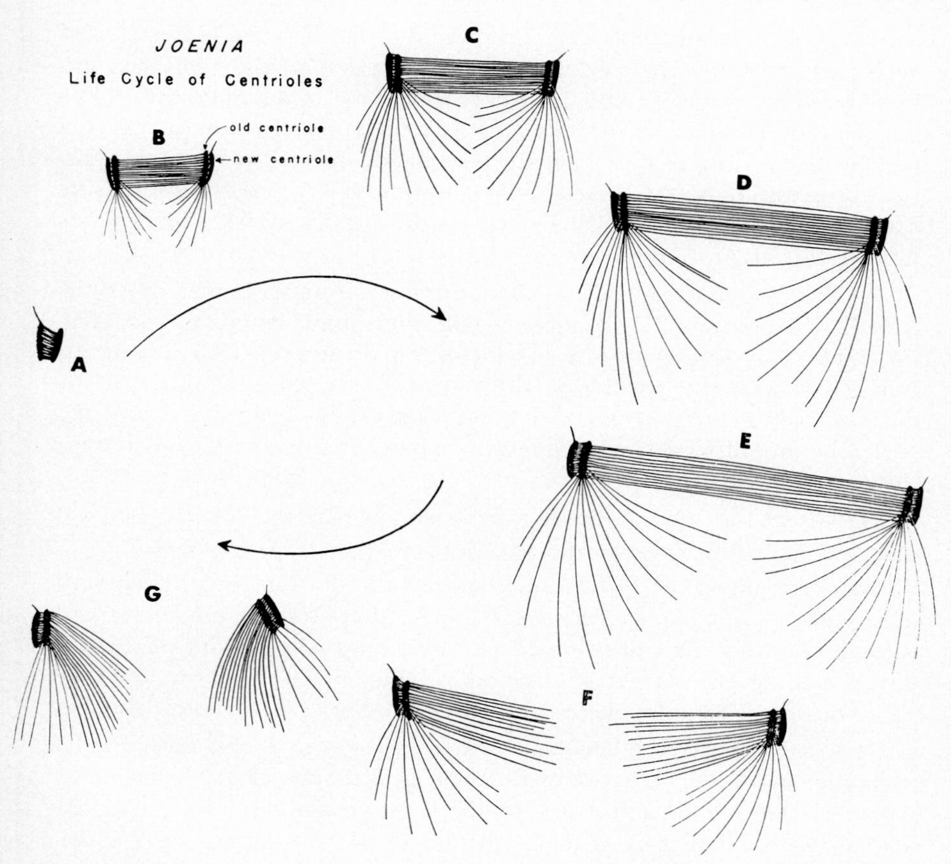

FIG. 6. *Joenia,* life cycle of centrioles. A, two rod-shaped resting centrioles connected by an arrested central spindle. B, prophase. Old centrioles have produced astral rays and new centrioles. The formerly arrested central spindle has increased in length by growth. A new set of extranuclear organelles takes its origin from a rod directed anteriorly from each new centriole. C, continued growth of central spindle and the new centrioles, which are now as large as the old ones. D,E, continued increase in length of central spindle. At each end of the old central spindle a new, arrested central spindle is formed, between the new and old centrioles, from astral rays arising from these centrioles. F,G, central spindle pulls in two; cytokinesis.

trioles which migrated to the posterior region of the cell have each produced a new centriole, and each new centriole is well along in the production of a new set of flagella, parabasals, and axostyles (Fig. 5B). At cytokinesis one of the two new sets of extranuclear organelles of the parent reproducing cell goes to each daughter. Thus, any resting cell has all new extranuclear organelles because no old ones are ever carried over to daughter cells.

Sometimes the arrested central spindle that lies between the centrioles begins to grow and increase in length shortly before the centrioles migrate; at other times growth does not begin until after migration. The central spindle does not increase in length very much before astral rays may be seen arising from each centriole (the old ones). At the same time, each centriole that was present in the resting cell lays down by its side (the outer side in each case) a new centriole which quickly becomes as large as its parent (Fig. 6B, C).

At this time a short rod extends anteriorly from each new centriole, and the new extranuclear organelles take their orgin from this rod (Fig. 5B). This rod, which must be considered a part of a centriole, maintains connection with the organelles which it produces until the onset of the next cell generation. It is the degeneration of this rod which frees the centrioles and allows them to migrate. The old centrioles do not possess such a rod (Fig. 6B–G).

Details of the production of a new centriole by the side of each old one are interesting, but cannot be dealt with here (Fig. 6B–G).

Each new centriole in conjunction with the old one which produced it produces a small new central spindle, so that from anaphase onward the cell has three central spindles, two new ones and the old one which is functioning in nuclear reproduction (Fig. 6D–G). From mid-telophase onward (Fig. 6E) there is no growth in either new central spindle.

In order to make the behavior of the centrioles in *Joenia* clear a brief statement must be made regarding the production of the achromatic figure by the two old centrioles, or those present in the resting cell. As already noted, they have between them a small arrested central spindle. At the onset of cell reproduction, this central spindle begins to increase in length as a result of the growth of the fibers composing it. In other words, the arrested fibers begin to grow again. From this point onward, the growth of the achromatic figure and its function in nuclear reproduction are both the same as in any cell (Fig. 6). It should be noted, however, that the central spindle, like that of *Trichonympha* and *Holomastigotoides* where no centrosomes are present (Fig. 2), is always flat (Fig. 6B–F), never cylindrical (Cleveland, 1949a,b, 1953).

III. Participation of Centrioles in Formation of Achromatic Figure

Since the usual behavior is for two centrioles to participate in the formation of the achromatic figure, this condition will be considered first.

A. Two Centrioles

No cell shows the role of the centrioles in achromatic figure production so plainly as *Barbulanympha*. For this reason only the centrioles of this organism will be considered. The role of the centrioles of other organisms has been dealt with elsewhere, and in this paper that of four other genera has been considered briefly and diagrammatically. Also, at the end of this lecture a motion picture will be shown, both in color and black and white, of the centrioles and central spindle of *Barbulanympha*. These scenes are taken from a longer picture, "Sexuality and other features of the flagellates of *Cryptocercus*," which may be purchased at cost of printing. If there is enough demand the scenes shown here may be printed and sold separately at cost.

The centrioles and achromatic figure of *Barbulanympha* may be seen in living material easily and clearly with an ordinary light microscope, even at low power. I well remember demonstrating the centrioles and central spindle of *Barbulanympha* at Woods Hole in 1931 to a fairly large group, including Wilson, Morgan, McClung, and Conklin. All of these men had interesting comments to make. Some of the comments were recorded. Those of Wilson and Conklin seem worthy of quoting. Wilson said, "I am vindicated; never did I think I would live to see centrioles and the achromatic figure so plainly." Conklin, who had difficulty at first seeing what he was supposed to see, remarked when he did see it, "Oh, I understand why I could not see it at first; it is so very plain; I was looking for something difficult to see." Then, after a moment, he remarked, "I could almost swing on that central spindle like a trapeze bar."

In the living, resting cell of *Barbulanympha* (Fig. 7A) the centrioles may be seen plainly extending to the center of the centrosome. Many still photographs, just as plain as the drawings of Fig. 7 have been made, but it seems useless to include them since a motion picture illustrating the same will be shown.

At the onset of cell reproduction astral rays grow out from the distal end of each centriole and pass through the centrosome which surrounds this end. As already described, the rays between the centrioles meet, join, and form the central spindle (Fig. 7B, C, D). As the growth of these fibers continues, the central spindle becomes longer and thus pushes the centrioles apart (Fig. 7E).

If, however, the distal ends of the two centrioles lie far apart, many astral rays are produced from each of these ends; they grow and become quite long, but no central spindle is formed because the rays from one centriole cannot meet those from the other (Fig. 9D).

At this point it should be remarked that *Barbulanympha* is a very accommodating organism (Cleveland, 1955). It will carry out experiments on itself while you observe the results. For example, one may observe cell reproduction when no nucleus at all is present, and many other unusual conditions, such as one centriole and one nucleus, two centrioles and two nuclei, and as many as forty nuclei with any number of centrioles ranging from three to fifty. Such combinations of centrioles and nuclei permit one to reach many valuable conclusions regarding functions of centrioles in the formation of the achromatic figure and its role in chromosomal movement. Only a few of these conclusions can be considered in this brief paper.

B. Three Centrioles

When three centrioles cooperate in the formation of the achromatic figure, the number and type of central spindles produced depends on the position of the distal ends of the centrioles with respect to one another. If two distal ends lie fairly close together, usually only two central spindles are formed (Fig. 8A). The length of the central spindles is determined by the distance between the distal ends of the centrioles (Fig. 8B, C). When the three distal ends are equidistant apart, as if at the corners of an equilateral triangle, all three central spindles are of equal length (Fig. 8B).

C. Four Centrioles

These multiple centrioles, it should be explained, are produced in two ways; irregularity in centriole production (Fig. 9A), and multiple fusion of gametes (Cleveland, 1955).

When four centrioles are present, in groups of two, such a situation usually results from the production, during the previous generation, of more than two new centrioles. In an example such as that shown by Fig. 9C, either two central spindles, more or less parallel to each other, are formed or there may be, as this figure shows, a crisscrossing of the fibers of the central spindle, resulting in the formation of four central

---

FIG. 7. *Barbulanympha.* Formation of the achromatic figure by two centrioles. A, resting cell with a large centrosome surrounding the distal end of each centriole. The centriole extends to the center of its centrosome. B, astral rays that grow out from the centrioles pass through the centrosomes. C,D, formation of central spindle by astral rays arising from one centriole meeting, joining, and growing along those arising from the other. E, continued growth of these astral rays increases length of central spindle and thus pushes centrioles apart.

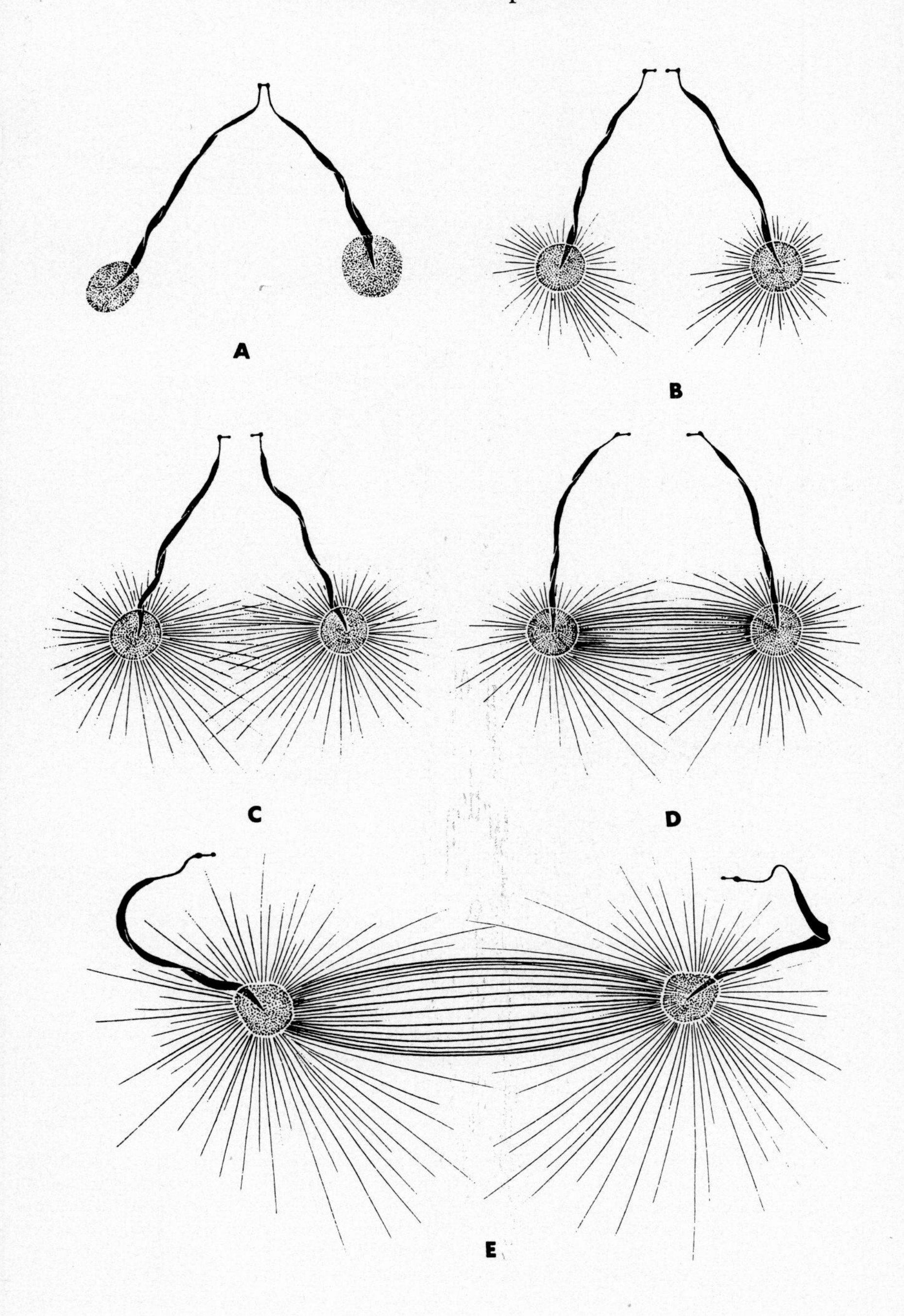
A
B
C
D
E

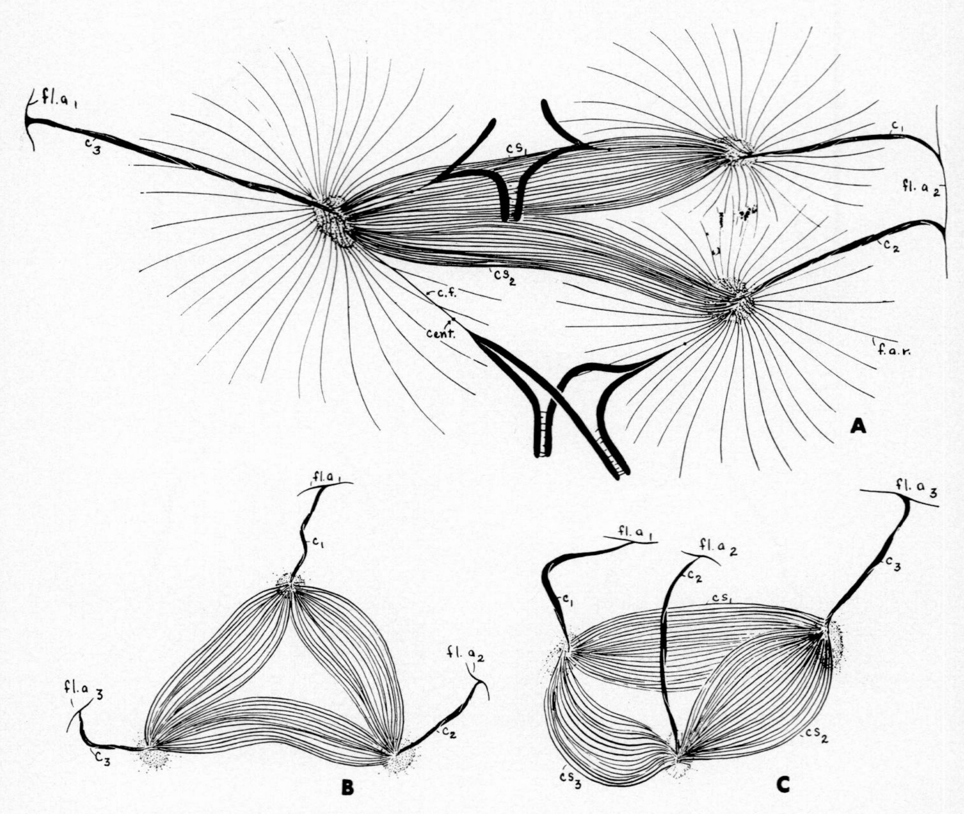

FIG. 8. *Barbulanympha.* Formation of the achromatic figure by three centrioles. Abbreviations: *c,* centriole; *cs,* central spindle; *cent.,* centromere; *fl.a.,* flagellated area; *f.a.r.,* free astral rays; *c.f.,* chromosomal fibers. Centrioles, central spindle, and flagellated areas are numbered. Astral rays and chromosomes are omitted in B and C. A, two of the centrioles lie close together and only two central spindles are formed, yet the chromosomes, by means of chromosomal fibers, are being pulled to three poles as the growth and pushing of the central spindles places tension on the chromosomal fibers. B, three centrioles with equidistant distal ends participating in formation of three central spindles of equal length, each centriole participating twice. C, three centrioles whose distal ends are not the same distance apart participating in the formation of three central spindles of unequal length.

FIG. 9. *Barbulanympha.* Formation of achromatic figure by multiple centrioles. A, early stage in irregularity of formation of four centrioles. B, formation of seven central spindles when the distal ends of the five centrioles participating lie more or less an equal distance apart. Note that the lower middle centriole, which is functioning in the production of four central spindles, has most of its astral rays invested in this function. Each of the two upper centrioles is functioning in the production of three central spindles, while the two on the lower ends are functioning in the

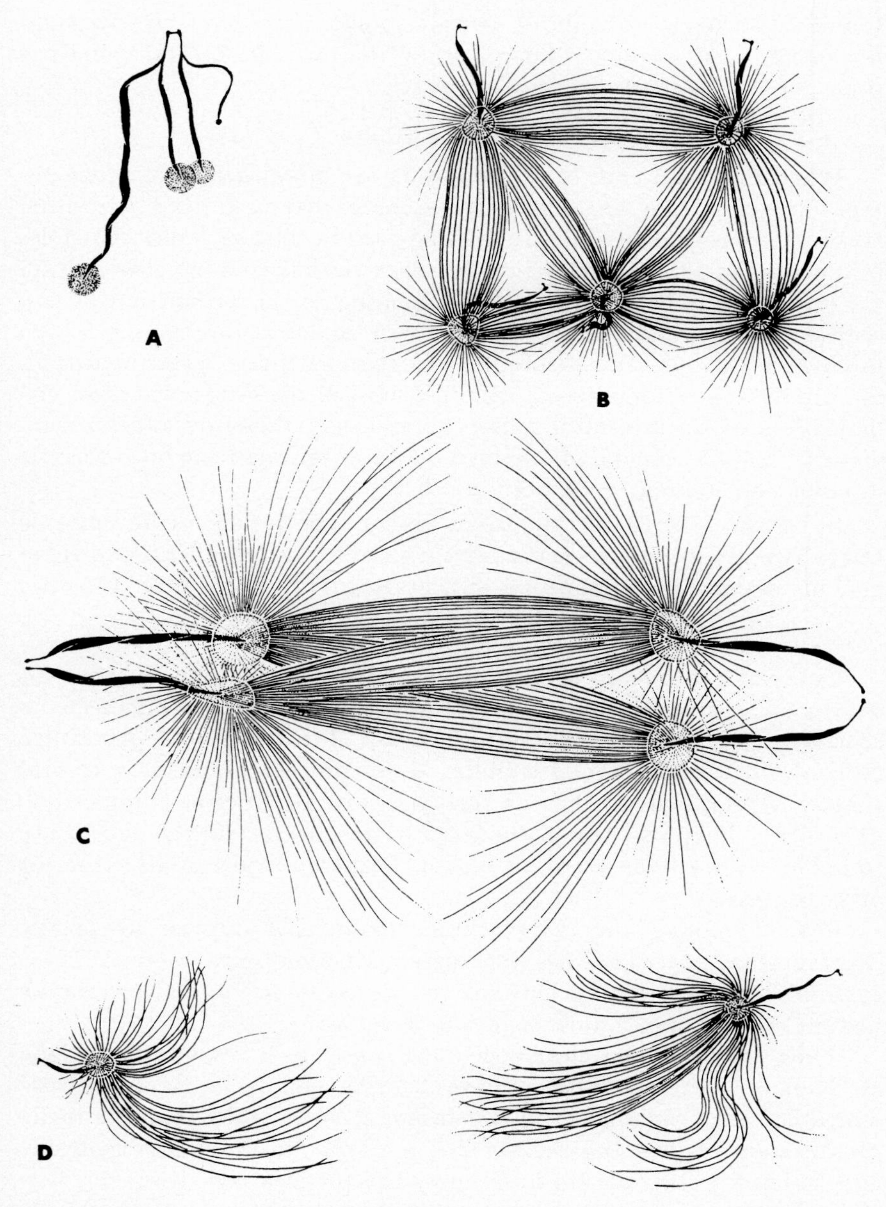

production of only two central spindles. C, the four centrioles whose distal ends lie in groups of two are functioning in the production of four central spindles. Note the crisscrossing of the fibers composing these central spindles. D, distal ends of the centrioles lie too far apart to cooperate in central spindle formation; therefore no such body is produced.

spindles. Such an example of central spindle formation clearly could not occur, if, as was once thought to be the case, the central spindle is spun out by two centrioles moving apart.

D. Five Centrioles

When the distal ends of the centrioles are more or less an equal distance apart, as many as seven central spindles may be formed (Fig. 9B). If this figure is studied, one will see that each of the two upper centrioles is functioning in the production of three central spindles, the two on the lower left and right are each functioning in the production of two central spindles, and the one in the lower center is functioning in the production of four central spindles. In the one that is participating in the production of four central spindles, most of the astral rays produced by it are used up in central spindle formation. Obviously, central spindles such as these can only be formed by the astral rays from one centriole meeting and joining those from another.

If, on the other hand, the distal ends of the centrioles lie more or less in two groups, not so many central spindles are produced, and there may or may not be crisscrossing of central spindle fibers (Fig. 10A, B).

E. One Centriole

Unipolar achromatic figures are sometimes fairly abundant in *Barbulanympha,* and they show beyond question that in this organism no central spindle is ever present (Fig. 11A, B). It is impossible for a single centriole to form a central spindle. Two must cooperate if a central spindle is produced. If they lie too far apart cooperation is impossible (Fig. 9D). They also cannot cooperate when they lie side by side. Very probably what is true for centrioles in *Barbulanympha* is also true for other organisms.

It should also be pointed out that in thousands of unipolar achromatic figures studied never once has nuclear reproduction been observed. However, in many cases the nucleus and the always intact nuclear membrane become considerably lobulated in late telophase.

Thus, we have seen clearly that the centriole is responsible for the production of the achromatic figure, flagella, parabasals, axostyles, and sometimes other organelles, but we do not know how the centriole really produces any of these organelles. They are firmly attached to a centriole, and without a centriole they are never present in a cell. However, it is not enough to say, as I have done, that these organelles grow out from a centriole. What is the nature of the syntheses, which are either carried out or controlled by a centriole that brings about the production of such a variety of organelles?

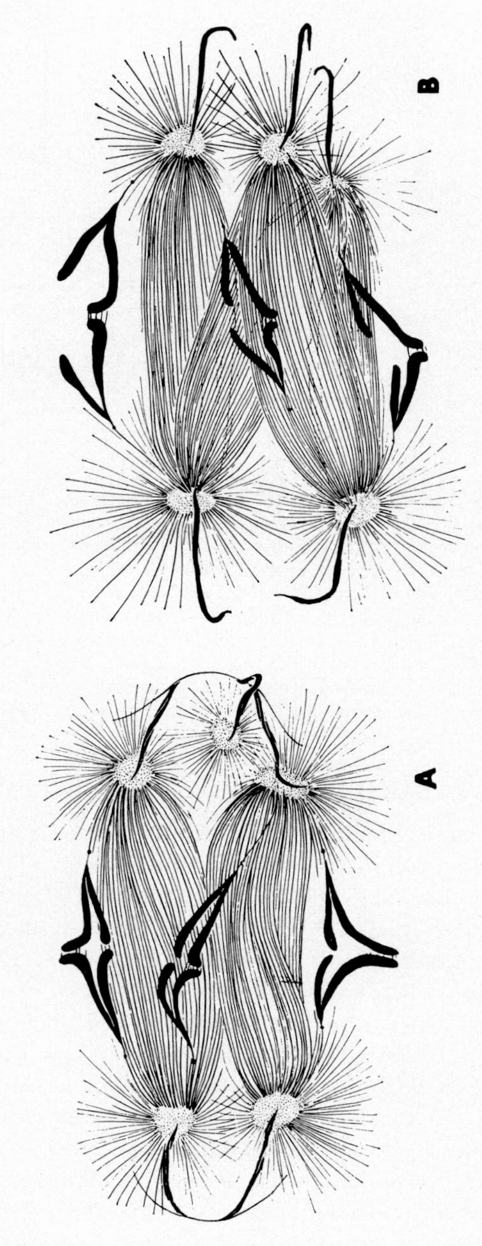

Fig. 10. *Barbulanympha*. Formation of achromatic figure by five centrioles whose distal ends lie in two groups. A, two central spindles with no crisscrossing of their fibers. B, four central spindles with crisscrossing of their fibers. Note that daughter chromosomes do not always move along the same central spindle.

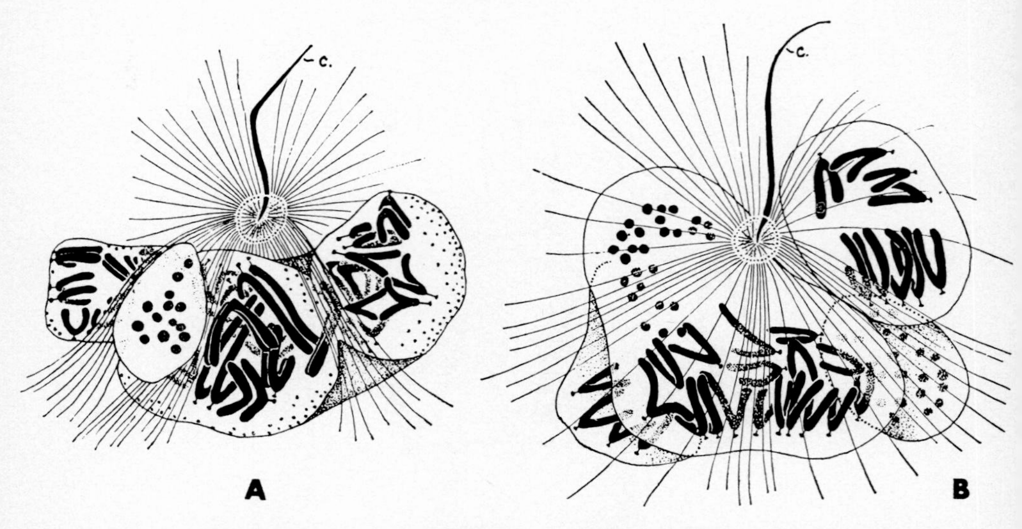

FIG. 11. *Barbulanympha.* One centriole attempting to form an achromatic figure. Abbreviation: *c,* centriole. A, the single centriole lies at the anterior border of the always intact nuclear membrane. Its fibers pass over the lobulated nuclear membrane at several points. No central spindle is formed, and the once paired chromosomes, which are now in early telophase, are falling apart without ever having been connected, via chromosomal fibers, with the distal end of the centriole or pole. B, the distal end of the single centriole lies over the nucleus. No central spindle is formed. The chromosomes, which are in late telophase and were once paired intimately, now show no evidence of pairing or of ever having been connected to a pole. No nuclear reproduction ever occurs when only one centriole is present. The chromosome number is doubled.

IV. Function of Achromatic Figure in Chromosomal Movement

It does not seem desirable to review here the various theories that have been advanced to explain chromosomal movement. This has been done well enough by Cornman (1944), Schrader (1951), Ris (1955), and Specht (1961). Cornman rightly concludes that all theories except the one involving expanding and contracting structures possess many untenable contradictions. Schrader concludes that, "If we can successfully determine the detailed structure of the spindle, we will have the solution of the mechanism of mitosis within our grasp." He says further, "The case of such protozoa as *Barbulanympha* may present highly specialized conditions, but in its demonstration that an extranuclear spindle may be capable of moving chromosomes which remain inside the nucleus throughout the entire division, it offers a challenge that cannot be ignored." Since Schrader wrote these words, studies of *Barbulanympha* and other related organisms have gone a long way toward supplying a simple yet satisfactory explanation of the mechanics of the movement of chromosomes to the poles or centrioles (Cleveland, 1953, 1954b, 1955, 1956, 1957a,b, 1958a,b,c, 1960, 1961).

Since other organisms, protozoan, metazoan, and plants, in which chromosomal movement has been studied, serve only to reinforce the conclusions supplied by *Barbulanympha,* it does not seem necessary to consider them in this short article. In fact, only a very small amount of the studies carried out on *Barbulanympha* can be reviewed in this resume.

Let's consider first two examples where five centrioles but only one nucleus are present in the cell. In these examples it is perfectly clear that the anaphase daughter chromosomes sometimes in their poleward movement do not move along the same central spindle. In other words, they do not move to opposite poles when they have a choice of more than two poles (Fig. 10). When only two poles are present they have no choice; they must move to one or the other, and in so doing must also travel along the central spindle. Hence, the notion that chromosomes in going poleward use the central spindle to travel along, must be abandoned.

The notion that chromosomal fibers contract and thus pull the chromosomes poleward must also be abandoned. The elongation of the central spindle as a result of growth puts tension on the chromosomal fibers and thus enables them to pull the chromosomes poleward. Chromosomal movement is brought about by a combination of pushing and pulling; the elongating central spindle pushes and the chromosomal fibers pull (Cleveland, 1954b). When the distal ends of the daughter chromosomes become free, tension on the chromosomal fibers is released and the chromosomes move very near the poles. This makes the chromosomal fibers appear to contract, although they lack the ability to do so.

Let's consider next an example where five centrioles and two nuclei are present in a cell, and one nucleus lies a considerable distance from any of the centrioles, while the other lies near all five of the centrioles (Fig. 12A). If we look at the nucleus which lies near the centrioles in detail, we see anaphase and telophase daughter chromosomes moving to each of the five poles (centrioles). And in this nucleus there are several examples of chromosomes that are not moving along any central spindle (Fig. 12B). Now if we look at the other nucleus in detail, we see no tendency whatever for the daughter chromosomes to move apart and thus form daughter nuclei (Fig. 12C). Instead, remnants of pairing may still be seen as the daughters fall apart. There are no poles to attract them. In later stages of this same type of cell — and many have been studied — one may gradually trace the behavior of the chromosomes, step by step, as they return to the resting stage, and the nucleus thus doubles its number of chromosomes.

The fact that the nuclear membrane always remains intact through-

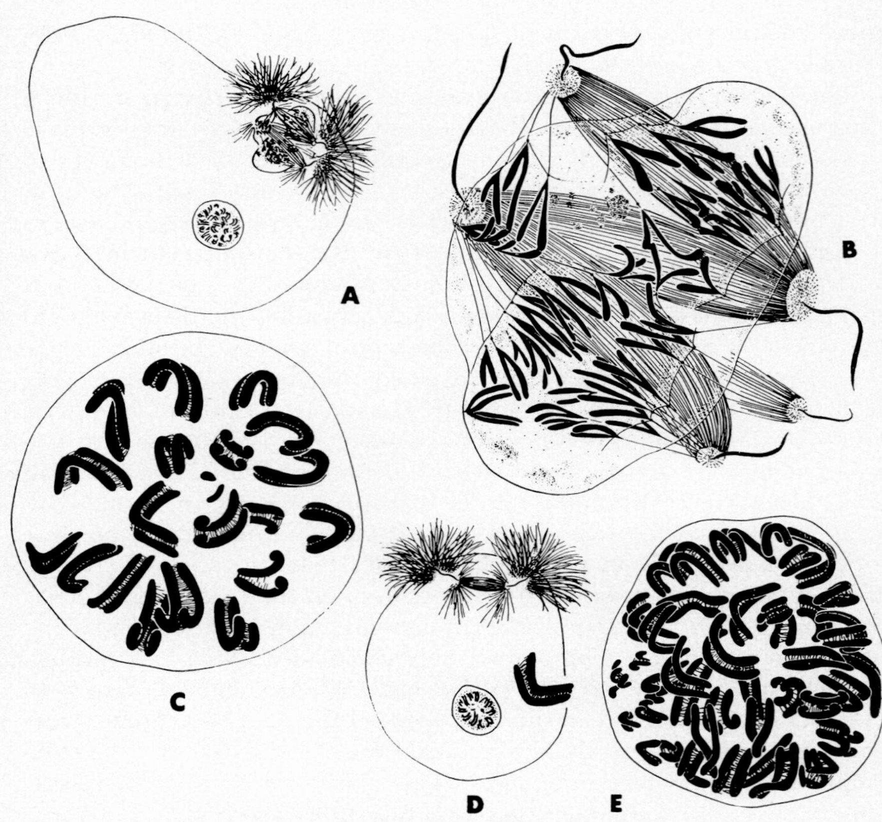

FIG. 12. *Barbulanympha*. The relationship of the central spindle and other components of the achromatic figure to the movement of chromosomes. A, low power of an entire cell. Five centrioles and two nuclei are present. All the centrioles lie near the upper nucleus; none near the lower one, which lies too far away to be influenced by the functions of centrioles in nuclear reproduction. B, detail of upper nucleus of A. Five centrioles and five central spindles are present. The chromosomes, except for one pair in the lower left-hand border of the nucleus, are in late anaphase and early telophase. They are moving to all five of the poles (distal ends of centrioles); some pairs in separating have moved along a central spindle, while others have not; some are moving at a 90° angle to a central spindle (upper left). C, detail of the lower nucleus of A. There is no tendency, in the absence of central spindles and of astral rays which may form chromosomal fibers, for the paired chromosomes to separate. Eventually the pairs will fall apart, but no nuclear reproduction will occur. D, low power of an entire cell. One nucleus and two centrioles are present, but this nucleus lies too far from the achromatic figure, which is in the anterior end of the cell, for astral rays to become chromosomal fibers and connect with the centromeres. Although the central spindle is the length of that of a late anaphase cell, the chromosomes, as a detailed inset of one pair shows, have not separated. This nucleus will not undergo reproduction; its chromosome number will be doubled. E, detail of nucleus of a cell similar to that of D except that the achromatic figure is in late telophase. The chromosomal pairs are falling apart without showing any tendency whatever to group themselves into two daughter nuclei.

out nuclear duplication in *Barbulanympha* enables one to see much more clearly than would otherwise be the case exactly what happens to the chromosomes of each nucleus. If the nuclear membranes of both nuclei were to break down, one would have considerable difficulty in tracing with certainty the behavior of the chromosomes of each nucleus.

Now, let's consider two examples where one nucleus and two centrioles are present. In both examples the nucleus lies much too far from the achromatic figure which the centrioles form for this structure to participate in chromosomal movement. The astral rays cannot reach half the distance to the nucleus, and hence no chromosomal fibers are developed (Fig. 12D). If the nucleus were near the achromatic figure, its chromosomes would be in anaphase. Instead, the one chromosome which is drawn in detail shows remnants of pairing between the daughters which still lie side by side for their entire length. If we take another similar example, except that the central spindle in this case is long enough for the chromosomes to be in telophase if the nucleus were close to it, we find the daughter chromosomes falling apart and there is no attempt whatever on the part of the chromosomes to group themselves in two nuclei (Fig. 12E).

We may now conclude that without centrioles no achromatic figure is formed and that without an achromatic figure there is no poleward movement of the chromosomes to form daughter nuclei. The chromosomes reproduce themselves but the nucleus does not. Hence, two or more centrioles must be present, and they must lie fairly close to the nucleus if the nucleus is to reproduce itself.

V. Birefringence of the Central Spindle

Polarized light has supplied conclusive evidence that the living central spindle is fundamentally fibrillar (Schmidt, 1939, 1941; Swann, 1951a,b; Inoué, 1953). In all of these studies except those of Inoué discrete spindle fibers, if observable at all, were exceedingly faint. And Inoué plainly states that the complete central spindle in his material, *Chaetopterus* eggs, could not be observed without greatly compressing the egg. Inoué's findings have been questioned by some — although I do not think they should be — because compression is known to produce birefringence as, for example, in a piece of rubber.

In *Barbulanympha* the birefringence of the central spindle may be seen plainly and easily and, without compressing the cell at all, in a rapidly swimming organism, although it is not a simple task to make a motion picture of such an organism owing to the problem of keeping it in the field. Considerable film is wasted in such an undertaking.

Since the central spindle and the other components of the achromatic figure are all so very plain in *Barbulanympha,* no matter what type of microscope is used, there is no difficulty, as the accompanying motion picture shows, in photographing the central spindle in color, employing a combination of polarized light and phase contrast (the microscope used was Zeiss Ultraphot II, 40-phase, oil-immersion objective, 6× Wild photographic ocular, Arriflex 16 camera at 24 frames per second, Zeiss carbon arc, and Ektachrome ER film).

ACKNOWLEDGMENTS

Acknowledgments for permission to republish previous figures of mine are due: *J. Protozool.* for Figs. 1-6, 8, 10, 11; *Biol. Bull.* for Figs. 7, 9; *J. Morphol.* for Fig. 12.

REFERENCES

Alexieff, A. (1913). *Arch. Protistenk.* **29**, 313.
Boveri, T. (1888). "Zellen-Studien," Heft 2. Fischer, Jena.
Boveri, T. (1895). *Verhandl. physik.-med. Ges. Würzburg,* [N.F.] **29**, 1.
Boveri, T. (1901). "Zellen-Studien," Heft 4. Fischer, Jena.
Cleveland, L. R. (1934). *Mem. Am. Acad. Arts Sci.,* **17**, 185.
Cleveland, L. R. (1935). *Biol. Bull.* **49**, 46.
Cleveland, L. R. (1938). *Biol. Bull.* **74**, 41.
Cleveland, L. R. (1949a). *Trans. Am. Phil. Soc.* **39**, 1.
Cleveland, L. R. (1949b). *J. Morphol.* **85**, 197.
Cleveland, L. R. (1951). *J. Morphol.* **88**, 199.
Cleveland, L. R. (1953). *Trans. Am. Phil. Soc.* **43**, 809.
Cleveland, L. R. (1954a). *J. Morphol.* **95**, 213.
Cleveland, L. R. (1954b). *J. Morphol.* **95**, 557.
Cleveland, L. R. (1955). *J. Morphol.* **97**, 511.
Cleveland, L. R. (1956). *Arch. Protistenk.* **101**, 99.
Cleveland, L. R. (1957a). *J. Protozool.* **4**, 230.
Cleveland, L. R. (1957b). *J. Protozool.* **4**, 241.
Cleveland, L. R. (1958a). *J. Protozool.* **5**, 47.
Cleveland, L. R. (1958b). *J. Protozool.* **5**, 63.
Cleveland, L. R. (1958c). *Arch. Protistenk.* **103**, 1.
Cleveland, L. R. (1960). *J. Protozool.* **7**, 326.
Cleveland, L. R. (1961). *Arch. Protistenk.* **105**, 149.
Cornman, I. (1944). *Am. Naturalist.* **78**, 410.
Costello, D. P. (1961). *Biol. Bull.* **120**, 285.
Heidenhain, M. (1894). *Arch. Mikroskop. Anat.* **43**, 423.
Hermann, F. (1891). *Arch. Mikroskop. Anat.* **37**, 569.
Hirasé, S. (1894). *Botan. Mag.* (*Tokyo*) **8**, 359.
Inoué, S. (1953). *Chromosoma* **5**, 487.
Janicki, C. (1915). *Z. wiss. Zool.* **112**, 573.
Kofoid, C. A., and Swezy, O. (1915). *Proc. Am. Acad. Arts Sci.* **51**, 289.
Meves, F. (1898). *Arch. Mikroskop. Anat.* **8**, 430.
Ris, H. (1955). "Cell Division, p. 91." Saunders, Philadelphia, Pennsylvania.
Schmidt, W. J. (1939). *Chromosoma* **1**, 253.
Schmidt, W. J. (1941). *Ergeb. Physiol. u. exptl. Pharmakol.* **44**, 27.
Schrader, F. (1951). *In* "Symposium on Cytology" (G. B. Wilson, ed.), p. 37. Michigan State Coll. Press, East Lansing, Michigan.
Sharp, L. W. (1912). *Botan. Gaz.* **54**, 89.

Sharp, L. W. (1914). *Botan. Gaz.* **58**, 419.
Specht, W. (1961). *Z. Anat. u. Entwicklungsgeschichte* **122**, 266.
Strasburger, E. (1895). *Jahrb. wiss. Botan.* **28**, 151.
Swann, M. M. (1951a). *J. Exptl. Biol.* **28**, 417.
Swann, M. M. (1951b). *J. Exptl. Biol.* **28**, 434.
Webber, H. J. (1897). *Botan. Gaz.* **24**, 225.
Wilson, E. B. (1931). *Science* **73**, 447.
Wolbach, S. B. (1928). *Anat. Record* **37**, 255.

Discussion by Arthur W. Burke, Jr.

Department of Oncology, Rhode Island Hospital, Providence, Rhode Island

I. Introduction

It is at once obvious, as well as meet and right, that a symposium discussant direct his remarks in channels of development of the theme of the contributor whose paper he has been invited to discuss, ramify upon it so as to interdigitate this with other contributions, and restrain himself from digressions beyond the scope of the topic at hand. Thusly delineated, my task is comparable to the role of a student disciple called upon to score the entre act for a complete and already satisfying opera; with this admission, I undertake a codetta on "Centrioles in Cell Reproduction."

The identity and role of the centriole in the whole life cycle of cell behavior was suspicioned in certain areas and documented in others by the time of the turn of the last century. Boveri (1895), who may be credited with the coinage "centriole," was a pioneer proponent of the concept that these structures retain their integrity without loss of identity as a permanent, autonomous cell organelle which always arises from a pre-existing body of the same kind. That this obtained for numerous examples of both plant and animal cells was indisputable, but the generality that perpetuity was universally the case met with great opposition. This was an era when that which could not be perceived on the basis of morphological criteria, was that deemed not to exist.

A broad concept of the multipurposefullness of the centriole was expressed some time ago by E. B. Wilson in his book, "The Cell in Development and Heredity" (1925, pp. 30–31): "The central bodies [meaning centriole surrounded by centrosome], in particular the centrioles, are undoubtedly organs of cell-division; but they have a broader significance than this. Even in the vegetative or non-mitotic condition of the cell the central body is sometimes surrounded by radiating fibrillae to form a more or less definite aster or astral sphere, as is shown conspicuously in leukocytes — and sometimes on a smaller scale in connective tissue cells; sometimes also in the early stages of the animal oocyte (surrounding the 'yolk nucleus'), and even in nerve cells. An interesting example of this is described by Del Rio Hortega (1915) in the cells of Purkinje, where a pair of centrioles, apparently always present, is surrounded by conspicuous, irregularly radiating wavy fibrillae to form an aster-like body. These cells, so far as known, are not capable of division. The function of the astral formations in these various cases is unknown; but it may possibly be connected with the fact that in the vegetative or non-mitotic phase of the cell the central apparatus often forms a focus about which are aggregated certain of the other formed elements, such as the Golgi-bodies and chondriosomes; and in the earlier stages of the animal egg it seems typically to form the original center of the yolk-formation. In the early stages of maturation of the germ cells the central bodies lie at or near that pole of the nucleus towards which the nuclear threads are polarized, and hence may play a part, if only indirectly, in the conjugation of the spirene threads during synapsis. Again, in the formation of the motile sperms of both plants and animals in many cases the centriole plays the part of a basal body or blepharoplast, from which grows forth the axial filament of a flagellum or cilium; and the same seems to be the case also in the flagellated cells of sponges and in many flagellated protista. In some cases, however, the blepharoplasts are quite separate from the centrioles; and it is probable that the centriole may appear as separate bodies. In any case, the facts enumerated above show clearly that the central bodies are concerned in many cell-activities that have no immediate connection with cell-division."

To understand better the role of the centriole in mitosis it would be well to examine this organelle's interplay in all cell functions, but I shall confine my remarks to certain activities of this structure in discrete circumstances. Further, I shall not embark upon a defense of the validity of the concept that the centriole is a morphological entity expressed as such in most living cells at some time or other in their life history, and in addition, is a functional unit which directs certain cellular activities in cyclic periodicity, not deriving proof from mere existence of a physically or physiochemically resolvable discrete organelle. The latter state may be called as Mazia suggested last year, "euphamism" (Mazia, 1960). Of the morphologically recognizable varieties of centrioles, we may dispatch their classification by saying that the centriole may exist as a relatively simple bar structure of the order of 0.2 micron length (Porter, 1957, 1961; Oberling and Bernhard, 1961), or as a complex organelle with a variety of attached and associated structures approaching 100 microns inclusive (Cleveland, 1957c), with a myriad of diversified forms and sizes between these extremes. That the centriole of any particular cell or cell type may fluctuate in appearance within certain limits during the history of the cell, is abundantly clear in many plant and animal forms; we shall consider such examples later in this discussion.

To further our conceptual understanding of the centriole and its function in cell division we could proceed one level higher in morphological organization and state that the varieties of division spindles, chromosome fibers, and astral rays produced by, or under the influence of the centriole are also numerous and should be collectively considered with the centriole as the achromatic apparatus as suggested by Wilson (1925, p. 117). The achromatic apparatus not including the centriole would be called the achromatic figure (Cleveland *et al.,* 1934). What behaviors of the centriole are concerned with cell division, whether it be mitotic, meiotic, or otherwise, as in certain unusual situations, could be related then to the function of the achromatic figure, while cell reproduction, in the broadest sense not requiring actual cytokinesis, or even karyokinesis, is related in part only to the function or lack of function of the achromatic apparatus, but is definitely related to the centriole. Finally, certain phenomena evolving out of the processes of cell reproduction must be examined with the situation of the centriole in mind.

II. The Centriole in Division Cycles

A. The Centriole in Mitosis

The most familiar role of the centriole is that expressed when, at mitosis, this organelle functions as the epicenter for the development of the central spindle. Whether the centriole "produces" (in the standard sense of the word) the spindle fibers, or serves to direct their formation, or on the other hand is entirely passive in the process, cannot be unquestionably deduced from all the information at hand. Certainly, the achromatic figure does not form in the absence of centrioles where the latter are normally distinct in the cell. There are so many variations in the substantiation of this fact that it could almost be accepted as a law (cf. Cleveland, 1934, 1955a, 1957d). That the spindle fibers and astral and chromosomal fibers are recognizable first in the immediate vicinity of the centrioles has been shown in a variety of cases, independent of whether or not the surrounding centrosome is present. The completely formed achromatic apparatus then functions to separate homologous chromosomes or daughter chromatids, and at the close of chromosomal movement "to the poles," ordinarily undergoes fracture, rupture, or other breakdown, such that the sister group of chromosomes and associated one-half achromatic figure no longer are bound to their counterpart. This obtains whether the spindle is intranuclear or extranuclear, and irrespective of the persistence, or lack thereof, of the nuclear membrane.

The question of the autonomy of this act is raised. It has been shown that the above described behavior of the achromatic apparatus takes place in entirety in the absence of chromosomes or of a definite nucleus. Dr. Cleveland has studied subsequent division of *Barbulanympha* and *Trichonympha* enucleated by manipulation of the gaseous environment of the symbiotic host and found that the achromatic apparatus functions in normal fashion (Cleveland, 1956b). Earlier, a similar type of situation had been demonstrated by Boveri (1895, 1903) and by numerous other investigators for sea-urchin eggs (cf. Wilson, 1925, pp. 176–177). The reverse proposition has also been described by Dr. Cleveland (1955a, 1957a) where duplication of chromosomes and centromeres takes place in the absence of the achromatic figure. Again, analogy is to be found in other circumstances, as in the so-called "monaster divisions" of sea urchin eggs. All of these unusual situations lead to unstable cell types. All these cell behaviors in the presence of separated chromatic and achromatic apparatus show two things: (1) relative temporary independence of the chromatic and achromatic apparati as regards apparent duplication and function; (2) final dependence of the cell upon the integrity of these two systems and their integrated function. Hence, the ultimate independence of the chromatic and achromatic apparati is restricted to the limits imposed by cell continuity.

B. The Centriole in Meiosis

A more precisely controlled asynchronous function of centriole and chromosome is to be found in meiosis. Where meiosis occurs in two steps, i.e., two nuclear divisions, the centrioles reproduce and function to form two achromatic figures. In the first meiotic division, the centromeres (kinetochores) and chromosomes do not become functionally duplicate, but rather homologs pair and advance to opposite poles (thus being segregated), while in the second meiotic division, corresponding to a mitotic division of haploids, the now functionally distinct daughter (sister) chromatids each with its independent centromere are separated to the poles of the second division figure. (A major difference between meiosis II and haploid mitosis is that chromosome duplication antedates centromere duplication in the former; chromosome and centromere duplication are approximately concomitant in the latter.) The net result of meiosis is a reduction in chromosome number per nucleus, and this is true when meiosis is concomitant with gametogenesis as is common in diploid organisms (so-called gametic meiosis,) as well as when meiosis occurs after fertilization as is seen in haploid organisms (zygotic meiosis). That meiosis does not precede fertilization in haploids is at once obvious — meiotic reduction of the haploid chromosome constitution would not lead to genetically stable cell lines. A schema for relating these key events would be briefly: (a) nonduplication of centromeres in meiosis I; (b) nonduplication of chromosomes in meiosis II; (c) duplication and function of centrioles in both divisions (cf. Cleveland

1947b, 1949a, 1953b). To understand what factors effect this shift in synthetic or functional activity would contribute to an understanding of the control of cell reproduction.

The whole process of reduction in numbers of chromosomes (and centromeres) may be accomplished simpler than this, however. By a complete repression of the duplication of both centromeres and chromatids, simultaneous function of the achromatic apparatus brings about reduction, now recognized as meiosis of the one-divisional type. This has been documented in the normal life cycle of several flagellates by Dr. Cleveland (Cleveland, 1950a,b,c, 1951a,b, 1956c). Here again, centriolar function is superimposed upon the duplication processes of the chromatic apparatus in a sequence which leads to the loss of a generation of chromatic apparatus. An alternate view would be to regard the net result as a gain of one generation of centrioles.

Where meiosis is gametic and fertilization ensues, the immediate zygote as a cell might contain the maternal as well as the paternal centrioles, being endowed as it were, with the extra centriole generation. In most instances the zygote is spared this burden by the previous loss, dissolution, or discard of one or the other centrioles during completion of fertilization. Examples of this are too numerous to cite. Less frequently this elimination is accomplished well after fertilization, but if not before the ensuing division, aberrations occur in the form of subsequent multipolar figures (cf. Cleveland, 1957d). The same obtains for dispermic sea urchin eggs where the first cleavage is most commonly tetrapolar and results in poorly viable monstrous larvae (Fol, 1879, quoted in Wilson, 1925). Summarily then, it would appear that homeostasis for the cell applies to numbers and kinds of centrioles as well as chromosomes.

So far in this discussion I have not presumed to assign the rank of autonomy to the centromere, or kinetochore, that we are bestowing upon centriole and chromatin or chromatid. Most certainly any full understanding of cell reproduction would involve an exhaustive examination of the properties of this organelle and its interrelationship to the other mentioned structures before, during, and after division. There are several indications that the centromere is indeed endowed with certain independence, e.g., its lag behind chromatid duplication at Meiosis I, the behavior of centric versus acentric fragments at anaphase, the supposed precocious development of centromeres induced by certain alkylating agents (Kato *et al.*, 1959), the development of the salivary gland chromosomes of insects, and the observation of early chromosome fiber organization in the field of the centromere with propagation of that fiber in the direction of the centriole (McMahon, 1956), to cite a few. The centromere may be regarded as a part of the chromosome; however, I have endorsed the terminological designation of chromatic apparatus to chromosome and centromere for the purposes of this discussion, and so I shall leave consideration of this organelle to other discussants. I cannot refrain from quoting Daniel Mazia on this point, however: "In fact, we may consider the kinetochore to be the only essential part of the chromosomes so far as mitosis is concerned . . . Indeed, the role in mitosis of the chromosome arms, which carry most of the genetic material, may be compared with that of a corpse at a funeral: they provide the reason for the proceedings, but do not take an active part in them" (Mazia, 1961, p. 212).

C. The Centriole in Endomitosis, Autogamy, Gametogenesis

As a process of cell reproduction without concurrent cell division, endomitosis occurs naturally in the life cycle of certain organisms, and in the protozoan material so carefully studied by Dr. Cleveland, it is the result of acceleration of the chromatic apparatus ahead of the function of the achromatic apparatus. In the organism *Barbulanympha* this phenomenon is beautifully cast in a setting where at the same time other individual cells of this same genus undergo cell reproduction leading to the formation of two complete gametic cells, and in still other individuals karyokinesis is completed but cytokinesis does not take place and autogamy results (Cleveland, 1954a). The role of the centriole is quite clear. When the gametocyte undergoes complete gametogenesis, and when fertilization ensues, all the male extranuclear organelles are discarded from the zygote including the centrioles. In the zygote thus formed the surviving centrioles are the old one passed from the female gamete and a new one generated by it. Net result: two centrioles in the resting zygote. Autogamy in contrast results, in part at least, from the precocious breakdown of the achromatic apparatus, and cytokinesis of the gametocyte does not occur. Also, the pronuclei thus formed are sufficiently advanced, i.e., differentiated, to be mutually attractive and so they fuse. Depending upon when one examines this cell, the two old elongate centrioles of the gametocyte, or one old elongate surviving centriole of the gametocyte, or the old centriole and a newly generated (usually smaller) centriole are present (Cleveland 1953a; 1954b). In contradistinction, when endomitosis occurs, the gametocyte centrioles do not form an achromatic figure, instead one of the two degenerates and the resultant cell (a sort of parthenogenetic zygote) only presents one elongate centriole. This is the extreme example of the acceleration of reproduction of chromosomes ahead of centrioles. It is worth noting that in the case of the endomitotic zygote the single centriole eventually produces a new centriole, as it does in the zygote resulting from autogamy, only the time sequences are slightly different (Cleveland, 1954b).

Here we encounter differentiation, or sexuality of cells and cell components. In order of familiarity they are: complete gametogenesis leading to the production of gametic cells; autogamy or the production of gametic nuclei within the same cell (conventionally known as "pro-nuclei"); and finally, endomitosis or the production of "gametic chromosomes" within the same cell and nucleus. Whether or not it is correct to view as a sexual phenomenon the act of endomitosis, is open to criticism. This is answered in part by several observed facts: *Barbulanympha* undergoes complete or incomplete (autogamy) gametogenesis, or endomitosis as a response to the actions of the insect molting hormone, ecdyson (Cleveland, 1947a; 1957b; 1959, Cleveland *et al.*, 1960). Whichever of these processes takes place is a matter of individual response to this potent growth and differentiation promoting substance, and where certain species of this genus are more prone to the simpler sexual acts, artificial increases in hormone titer induce more individual cells to speed up sex, as it were, and employ "short cuts to sexuality."

The role of ecdyson in inducing pronuclear differentiation (maturation to sexual attractiveness) is elucidated by the behavior of the genus, *Saccinobaculus*. After gametic fusion or autogamy in *Saccinobaculus*, the zygote persists naturally for about 30 days as a cell with partially fused pronuclei, until the titer of ecdyson is sufficiently high to induce pronuclear maturation (Cleveland, 1950b; 1956c; 1957b). This can be accomplished in a matter of hours experimentally, by elevation of the molting hormone titer (Cleveland *et al.*, 1960; Cleveland and Burke, 1960) and indeed *Saccinobaculus* can be rushed through the entire cycle from agamont through gametogenesis/autogamy to complete fertilization in a few days by the administration of suitable amounts of ecdyson (even in adult hosts where none of the protozoa ever start sexual cycles). A more exacting experimental situation is afforded by the genus, *Trichonympha*. This organism has almost never been observed to undergo a sexual cycle in any other way except by complete gametogenesis (Cleveland, 1949b). Here the two resulting gametes are morphologically dissimilar in many ways, so complete that this protozoan sports a male gamete which enters the female gamete only via a special structure called the fertilization cone which is surrounded by a ring of so-called fertilization granules. The enthusiasm for fertilization is so great that the whole male individual flings himself through the "vagina" of his mate. That slight confusion in the identity of male and female on morphological as well as fertilizable grounds also very uncommonly exists in *Trichonympha* (Cleveland, 1949b; 1957a) is ample subject for investigation in itself, especially by those interested in differentiative and sexual phenomena (cf. Cleveland, 1955a). One other intriguing fact in passing; the chromosomes which are destined to be incorporated in the male and female gametes respectively, are clearly discernible by staining as well as living-phase contrast methods at least as early as metaphase of gametogenesis (Cleveland, 1949b). Is this another example of sexual differentiation of chromosomes? The chromosomes are more advanced during prophase of gametogenesis than they are at a comparable stage of nongametic mitosis. This is especially well demonstrated by *Leptosprionympha wachula* where major coiling of chromosomes occurs quite early in prophase of gametogenesis (Cleveland, 1951a). Also, early duplication of chromosomes during gametogenesis is quite common in all these protozoa (Cleveland, 1950a). Most of the cellular differentiation of the gametes of *Trichonympha* can take place in the complete absence of a nucleus and chromosomes (Cleveland, 1956b; 1957a), so one is inclined to conclude that the molting hormone can act at three independent levels: on the cell, the nucleus, and on chromosomes. Finally, there are a very few individuals of *Trichonympha* which never undergo changes during molting of host, or as a result of the action of ecdyson, and these represent a race of the genus which is entirely refractory to the hormone (Cleveland and Nutting, 1955; Cleveland *et al.,* 1960).

Back to endomitosis — when an excess amount of ecdyson is presented to *Trichonympha* very early in its sexual differentiation, the organism begins to show types of differentiation never seen under natural conditions. These occur as autogamy and endomitosis, in order of increasing titer of the hormone (Cleveland *et al.,* 1960). This is sufficient evidence to consider endomitosis in this instance as an attempt at sexuality. Recall, if you will, that certain instances of parthenogenesis or reproduction without the benefit of completed fertilization, are based on suppression of the centriole of the egg and acceleration of the duplication of chromatic apparatus (cf. Wilson, 1925, pp. 467–486). I should like to call attention to the fact that this phenomenon, endomitosis — clearly the result of chromatic acceleration, stands in sharp contrast to endoreduplication induced by X-irradiation, where general growth and even deoxyribonucleic acid (DNA) synthesis is slowed (Whitmore *et al.,* 1958, quoted in Mazia, 1961). Could it be that ecdyson while apparently speeding up the morphological manifestations of chromatic behavior, actually impedes DNA synthesis? There is certainly no evidence to indicate this possibility, nor is there quantitation to refute it. The stimulative effect of ecdyson is more like the effects on chromosomes of plants cells produced by 2,4-D (cf. McMahon, 1956). Apart from any discussion of sexuality, a role of the centriole in the mechanics of cell reproduction — from complete cytokinesis to incomplete karyokinesis — is evident. Further investigation of the chemical and physiochemical forces operative in all these types of cell reproduction is urgently needed.

While we are on the subject of ecdyson effects on these cells, let us make small digression to entertain some interesting observations. Zygotic meiosis, occurring in certain species of these protozoa after ecdysis of the symbiont host, is inhibited by the administration of ecdyson and the duration of inhibition is proportional to the titer of the hormone imposed (Cleveland *et al.,* 1960). This is true in *Saccinobaculus,* its close relative, *Oxymonas,* and in *Notila.* All these organisms have a one-divisional zygotic meiosis, but in *Barbulanympha* and *Trichonympha* where meiosis is two-divisional, inhibition does not occur. On the other hand, when meiosis occurs, either one or two divisionally, as concomitant with gametogenesis and before ecdysis of the host, ecdyson exerts a powerful accelerating effect. Perhaps it is just as important to note that those genera which are inhibited have centrioles which conform to the same pattern of behavior, while those genera which are not inhibited actually have quite a different centriole cycle (cf. Cleveland, 1957c). As we see over and over again, not all divisional behavior is explicable on the basis of chromosome duplication alone. That ecdyson stimulates meiosis directly is improbable; what is more likely is that gametogenesis, autogamy, endomitosis — in a word, sexuality — is stimulated, and if meiosis is along for the ride, all so well. (Note that endomitosis precludes meiosis.) One thing though is abundantly clear, meiosis is in no way *a priori* a sexual process, and the forces which operate to cause dyssynchronization of function of chromatic and achromatic apparati are not identical with those which bring about cellular and nuclear differentiation or sexual phenomena. Ecdyson is certainly a growth and differentiation producing hormone, perhaps the most potent stimulator of sexuality yet discovered (as little as 60 Calliphora units[1] $= 6 \times 10^{-4}$ mg causing inception of sexuality in millions of protozoa in a medium-sized host of 300 mg, which corresponds to an efficacy of 2 mg/kg, host body weight). A careful examination of the biochemical events occurring in these situations would again go far in assisting our understanding of cell reproduction in the general sense.

D. The Centriole in Sexuality and Life Cycle Evolution

It would not be amiss to examine the roles of gametogenesis, or sexuality in the broadest sense, and meiosis in the evolution of cell behavior. This has been considered in detail elsewhere (Cleveland, 1947a,b; 1957b), and I shall only take inference from some of these observations. To begin with, if we were not aware of the occurrence of zygotic meiosis, we might overlook aspects of the independent, but integrated functions of the chromatic and achromatic apparati. The genus, *Barbulanympha*, provides an excellent example of such a life cycle. In addition it does so many interesting things, both naturally and in response to abnormal situations, that without fear of being repetitous, I will briefly restate its normal life cycle. Haploid agamonts undergo a series of differentiations under the influence of ecdyson which lead to complete gametogenesis or, to incomplete gametogenesis (autogamy) or, to endomitosis. Increases in ecdyson titer cause more the latter types of behavior, as mentioned before. The cell embarks on reorganization (which is so interesting in itself that it will be discussed separately) and following this renewal of its extranuclear organelles, remains unchanged for several days, until ecdysis of its host, after which it undergoes meiosis in two divisions producing four agamonts. Thus ends the cycle (Cleveland, 1953a; 1954a,b,c). That a haploid becomes by one of several ways a diploid and then undergoes meiosis to return to haploidy is easy enough to follow. What I want to dwell on here is the course of the centrioles, and this cannot be accomplished with the dispatch just given the chromosomes. For the purposes here, however, I can simply say that the agamont prior to becoming a gametocyte has two elongate centrioles which function to produce the achromatic figure of the gametogenesis division, which is mitotic of course, and during fertilization the male centriole disintegrates. Thus the zygote is "haploid" with respect to centriole, diploid with respect to chromosomes. Just as the agamont centrioles generate each a tiny new centriole which becomes visible

[1] Purified, crystalline α-ecdyson has a potency of 100,000 Calliphora units/mg (Cleveland *et al.*, 1960; Karlson, 1956).

about the time of prophase of mitosis, this female old centriole produces a tiny new one during gametogenesis which grows to a variable length during the completion of fertilization (pronuclear fusion). Very shortly after completion of fertilization, the old and the newly formed centrioles degenerate down to approximately the original tiny size of the newly formed centriole and much later both grow and attain full length. During this time the cell prepares for the two-divisional meiosis which follows, and during this process of reorganization there is no additional duplication of centrioles (Cleveland, 1954b). This zygote has the same number of centrioles as a haploid agamont approaching mitosis. After two-division meiosis is complete the four agamonts produced also have two elongate centrioles (in the resting, nondividing cell).

In summary: In gametogenesis (followed by fertilization) or autogamy, or endomitosis of haploid species there is a gain (doubling) in total chromatic material, since these types of cell behavior all lead to duplication and subsequent reamassing of chromosomes. The extra set of centrioles is rejected normally in fertilization and autogamy, and does not quite form in the case of endomitosis. Thus a cell results which is one generation ahead, so to speak, in chromatic apparatus. Nature, in her infinite wisdom, provides cells with a release from this situation. This is accomplished by meiosis: one-division meiosis gives the centrioles a chance to "catch up" in a single duplication, while chromosomes are suppressed; two-division meiosis allows the chromosomes and centromeres to duplicate only once, while the centrioles "catch up" by duplicating twice. By these interpretations, zygotic meiosis is the mechanism for restoration of cellular homeostasis.

High polyploidy seems to be difficult for cells to handle comfortably, especially during division. In some instances the spindle just does not seem to be adequate for the job. In fact it is much to the point that the achromatic figure (spindle) is not commensurate in size with the chromosome number (cf. Cleveland, 1957d; 1958a). One might foolishly exact a 1:1 ratio, if absolute dependence of spindle upon chromosome obtained. At any rate, the diploid cell which would undergo gametogenesis mitotically, would place itself, upon completion of fertilization, in the unhappy state of tetraploidy, and reduction back to the diploid constitution would pose problems (cf. Cleveland, 1958a). Perhaps nature has tried such experiments. Examples of union of gametes before completion of reductional division are numerous and widespread, and the life cycles of several of the roach poly- and hypermastigote flagellates bear upon this consideration. The polymastigote *Notila* exists as two forms on a chromosome basis, one a diploid and the other, a tetraploid ($N = 7$). Both of these forms commence their sexual cycle by gametogenesis and subsequent fusion of gametes, but neither pronucleus has yet undergone any sort of reductional division, nor do the pronuclei demonstrate any affinity for one another. In fact, the term pronucleus is not appropriate here. What happens next is a reductional division, a one-division meiosis, and the resultant pronuclei (four of them) pair and fuse to form two zygotic nuclei. This is a double zygote. Later cytokinesis brings a separation of these two nuclei and associated extranuclear structures from within a single cell to two cells, and the cycle is ended (Cleveland, 1950c). Aside from occasional interchange of pairing and fusion of the four daughter pronuclei (which may have some genetic distribution advantages), the cycle could be regarded as nuclear maturation of both sperm and egg after gamete fusion, but before completion of fertilization (nuclear fusion). The over-all economy of numbers of centrioles and chromosomes is preserved finally, by the degeneration of the male sets before pronuclear fusion and, by the cytokinesis not involving either chromatic or achromatic reproduction.

The most stratling and revealing life cycle of all the organisms described by Dr.

Cleveland is that of *Urinympha*. This diploid hypermastigote normally divides by mitosis, but in its usual response to the effects of ecdyson it undergoes meiosis in one division, the achromatic apparatus breaks down and one centriole degenerates, and the resultant haploid pronuclei differentiate and fuse without cytokinesis ever intervening (Cleveland, 1951b). This is autogamy concomitant with one-division meiosis. The economy is complete in one act: the chromosomes are not duplicated (nor are the centromeres); an old centriole is discarded and a new one produced; and the remainder of the extranuclear organelles are discarded and renewed. This organism forgoes many privileges for this economy in sexual cycle in that the zygote is genetically (i.e., chromosomally) identical to the precursor cell, and even crossing over is eliminated (cf. Cleveland, 1949a, 1960c). Only one very important event has occurred, that is reorganization with renewal of centrioles. Endomitosis occurs occasionally, and it may be that *Urinympha* is accelerated by artificially produced higher titers of ecdyson to undergo endomitosis instead of meiosis and autogamy. Also, autogamy is rarely observed to be mitotic rather than meiotic and results in a tetraploid zygote instead of a diploid one. These latter observed phenomena, mitotic gametogenesis and endomitosis, represent progressive accelerations of the chromatic ahead of the achromatic apparatus which result in unstable and poorly viable cell types. We are again confronted with some sort of cellular homeostasis. What teleological purposes I have injected into the foregoing accounts, you may judge for yourself, but it is still captivating that there should be such an economy of parts in nature.

III. The Centriole in Reproduction of Extranuclear Apparatus

A. Cycles of Centrioles and Reorganization

In the last sexual cycle accounted, that of *Urinympha*, I alluded to extranuclear organelle reorganization as the net result. Since the topic, cell reproduction, embraces all cell parts, and since in these organisms the centriole is so intimately concerned with most of the other extranuclear organelles, it seems appropriate to consider the processes by which these organelles are discarded and renewed. This cell behavior has been designated reorganization. The behavior of the centriole during reorganization may have nothing at all to do with the role of the centriole in division, but a careful examination of the observed facts is useful in our understanding of the role of the centriole in cell reproduction.

Since reorganization in the zygote of *Barbulanympha* is not confused by events of a concomitant division (as in the case in *Urinympha*), let us review the process in this genus. In the zygote, shortly after fertilization, the two centrioles degenerate to the point that only a tiny anteriormost portion remains of each. Later the other extranuclear organelles, flagella for example, begin to degenerate. The two tiny speck-sized centrioles migrate to the vicinity of the nuclear membrane and take up a position quite close to it. Here they produce an entirely new set of all that had degenerated, including elongate centrioles, which are of course, propagations from these generative portions. While the centrioles of *Barbulanympha* are in the abbreviated form of generative portion only, they do not and cannot produce an achromatic figure. Every attempt to make them do so has met with failure. They have to first elongate, that is elaborate the achromatic-figure producing portion (posterior ends) which bear centrosomes. Of course, the centrosome is not a *sine qua non* for spindle production in these protozoa, but some form of centriole, two or more of them in fact, has been shown to be (cf. Cleveland, 1958a).

The distinction between he generative portion of the centriole which is capable of producing all extranuclear organelles and that portion of the centriole which is only capable of producing the achromatic figure is at once clear then, in *Barbulanympha* (Cleveland, 1954b; Cleveland *et al.*, 1934). Also, in the genus, *Macrospironympha,* the achromatic-figure producing portion breaks away from the generative portion at each division and a pair of these migrate to the nucleus and produce the achromatic figure, quite disjunct from the generative portions (Cleveland, 1956c). This is perhaps an extreme distinction between the generative and the achromatic-figure producing portions of the centriole; a distinction, I might add, which is clearly to be made in almost all of these flagellates (cf. Cleveland *et al.*, 1934). Before continuing our discussion of reorganization, let us first reconsider the behavior of centrioles at times other than during reorganization.

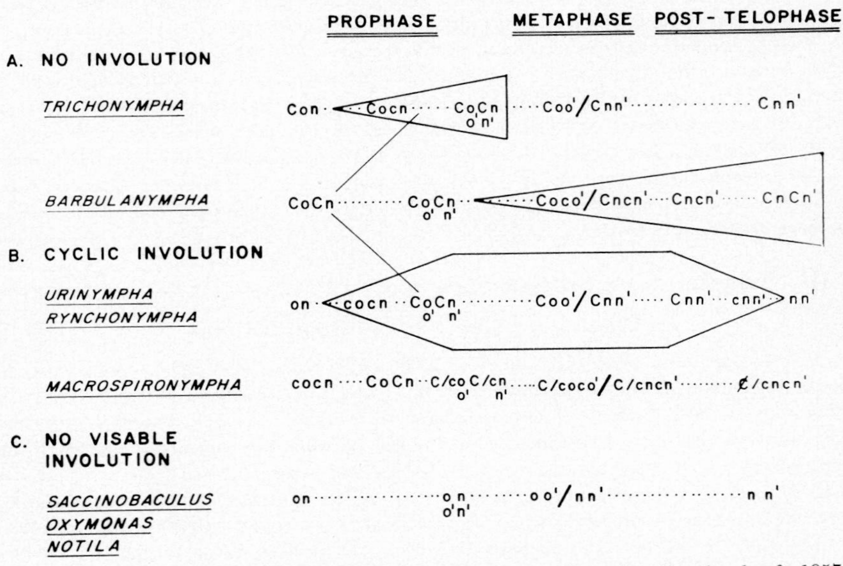

FIG. 1. Centriole life cycles. (Based on the studies of L. R. Cleveland, cf. 1957c.) Key: C, fully grown centriole, achromatic-figure producing; c, partial sized centriole; o, old centriole, generative portion; n, new centriole, generation portion (New centriole (n) becomes old centriole of next cycle); o′, first generation from old = new centriole in next; n′ first generation from new = new centriole in next cycle; ₵, discarded achromatic figure-producing portion; C/c, separation of achromatic-figure producing from anterior portion of centriole.

The poly- and hypermastigote flagellates studied by Dr. Cleveland have many variations in centriole behavior and he has classified the centrioles into five types on the basis of cylic behavior (Cleveland, 1957c). Consideration of these cycles reveals many interesting things and I shall attempt to point out a few. I have taken the liberty of representing the centrioles by a series of symbols and correlating the centriole cycles on the basis of these symbols, for the purposes of this discussion (Fig. 1).

In the genus, *Trichonympha,* it is easy to identify solely on centriole morphology which is old and which is new centriole from one division to the next at all times (cf. Cleveland, 1960b), except for a brief interval during prophase. It is to be noted that only the fully mature, elongate old centriole is capable of participating in achromatic-figure production, so that in order for a spindle to be produced, the new short centriole must first elongate and mature. This occurs during prophase. Also, the old elongate centriole never produces a second generation of the other extranuclear organelles, as these are carried over from the previous parent cell as attached to the old centriole. The new centriole on the other hand, regularly produces the new set of extranuclear organelles before maturing to participate in spindle formation. The same is true for *Barbulanympha* centrioles, only the new centriole matures and elongates during telophase and thus at the inception of prophase both the old and new centriole are elongate and therefore, indistinguishable. Accordingly, *Barbulanympha* has two elongate centrioles poised to form the achromatic figure earlier than those of *Trichonympha* (since the latter's new centriole must elongate first).

In two other types of centriole cycles the achromatic-figure producing portions of the centriole come and go, that is, are formed for each division and involute after functioning (cf. Cleveland, 1960d). As was just said a moment ago, in *Macrospironympha* the achromatic-figure producing portions actually fracture off from the generative portions during cell division. This behavior is a variant, perhaps, of the other type where continuity persists only during function, but not between divisions. Also, another interesting variation in centriole behavior is observed in relatives of *Macrospironympha*. The centriole loses all connection with other extranuclear organelles and functions in achromatic-figure production: only one centriole does this in *Spirotrichonympha bispira* and the daughter cells are decidedly unequal; both centrioles "free themselves" in *Rostronympha* (Cleveland and Day, 1958). In both, continuity persists at all times for the generative portions of the centrioles. In both of these types of cycles, just as in *Barbulanympha,* the new centriole is the only one that produces the other newly formed extranuclear organelles.

Finally, there are a number of polymastigotes and hypermastigotes which possess centrioles that function in a pattern very similar to those of *Barbulanympha* except there is not as clear-cut a morphological change when the new, generative centriole matures to become an achromatic figure producing one. In short, they do not elongate. Careful study of the connection of the centriole to the other extranuclear organelles reveals which is old and which is new, however. This is a type of centriole cycle which may well be representative of the behavior of centrioles in cells of higher organisms where no convenient tag is present to reveal the identity of centriole age. It is sheer speculation to conclude that this cycle is the most advanced in centriole behavior, for the original centriole may have been but a speck of cellular organization which did not undergo change in form, but in the processes of evolution convenient separation of generative and functional aspects became parts, and parts became outgrowths. As has been suggested by Dr. Cleveland, these protozoa may be surviving records of what has happened in the course of their evolution; records which are frozen, as it were, by the delicate interrelationships between host and parasite. The changing environment imposed by ecdyson has probably provided a powerful impetus to evolution, while successful adaptation to the stimulus has insured the survival. The involution of the centrioles in the sexual cycle just prior to reorganization in *Barbulanympha* may well represent just such a record. The facts have been presented already in this discussion. I now come to the questions, of how and why. It is quite clear in all other situations that the old centriole never produces new extranuclear organelles, except that the generative portion of the old centriole produces a new generative portion which in turn does produce the other extranuclear organelles. When both sets of the other extranuclear organelles are lost, as is the case during the reorganization in *Barbulanympha,* the cell is confronted with a different situation. It needs two new sets of extranuclear organelles and it would have only one new centriole to do the job, if the old centriole had not just been rejuvenated or reactivated. This is probably accomplished by involution of the centriole down to its generative portion (Cleveland, 1954b).

What makes the extranuclear organelles degenerate? Certainly there is no foolproof answer yet, but let us postulate that it is the result of the direct or indirect action of ecdyson. Do other genera lose their organelles, and if so when? The answer here is not complete (information on *Eucomonympha* is not yet available, cf. Cleveland, 1950c; 1960c), but *Rynchonympha,* a close relative of *Barbulanympha,* undergoes reorganization at much the same time, but as a zygote formed after two-division gametic meiosis (Cleveland, 1952). *Urinympha,* also a relative of *Barbulanympha,* undergoes reorganization during one-division meiosis and autogamy, as we have said before. The genera of another family embracing *Trichonympha, Macrospironympha,* and *Leptospironympha* undergo partial reorganization and it is significant here that these forms undergo reorganization during a period of encystment, the one exception being one species of *Leptospironympha,* namely, *wachula.* This species has no encystment in its sexual cycle, but it starts and completes the cycle much earlier than other forms of *Leptospironympha,* at a time when the titer of ecdyson is lower. In addition it demonstrates another thing, of which more will be made later, that is failure of the chromosomes to uncoil from late prophase of gametogenesis, through gamete formation, fertilization, to one-division zygotic meiosis (Cleveland, 1951a). Exception could also be made in that *Leptospironympha wachula* begins its sexual cycle by starting to reorganize before chromosome activity is apparent, although reorganization is not complete, in the sense that the extranuclear organelles do not completely degenerate. But in this case the final stage of reorganization does not take place until sometime after molting of its host, although the sexual cycle has long since ended.

All the foregoing are not incompatible with the suggestion that ecdyson incites reorganization, and that encysted forms would be spared, in part, this effect. The role of the centriole in replacing what becomes lost is a matter of anticipatory behavior. Those forms which in the course of evolution did not provide for renewal of extranuclear organelles by rejuvenation of old centrioles, did not survive, or evolved into another family, possibly like the Lophomonadidae which only possess one set of extranuclear organelles (of which there is one small-sized, living representative in the roach, *Cryptocercus*). By these interpretations, and others, the involution of centrioles which occurs only immediately prior to reorganization in *Barbulanympha* probably places this genus as more closely allied to the ancestral form of its family, Hoponymphidae. The other genera, *Rynchonympha* and *Urinympha,* in addition to incurring the same complete reorganization and antecedent centriolar involution under the influence of ecdyson, have a partial centriolar involution which occurs as a cyclic process and involves only the achromatic-figure producing portions during every division, and not induced by ecdyson.

One might ask, why does the second stage of reorganization, the regeneration of the extranuclear organelles, take so long in *Barbulanympha* and *Rynchonympha?* (thirty to 40 days.) I think that the answer lies in the fact that *Barbulanympha* can complete the second stage of reorganization in a matter of a few days, if the ecdyson titer is lowered in the host, or even sooner, if *Barbulanympha* in the first stage of reorganization is moved to a new host where there is no ecdyson being produced (Cleveland *et al.,* 1960). Meiosis ensues, by the way, immediately upon completion of reorganization in these cases (completion of regrowth of elongate centrioles), but not before (Cleveland and Burke, 1960). The suggestion is this: The inhibitory effects of ecdyson under circumstances of continued rising titer must be overcome before the cell can continue its cycle. Again, this should be easier for other forms which are encysted.

Recapitulating the centriole cycles, if in one stage or another the centrioles of a cell involute or mature out-of-phase, or the cell acquires a previously nonexistent enzyme for cyclic digestion of the achromatic-figure producing portions of the centriole, a series of interchanges in centriole cycles can evolve. If an accidental involution occurs

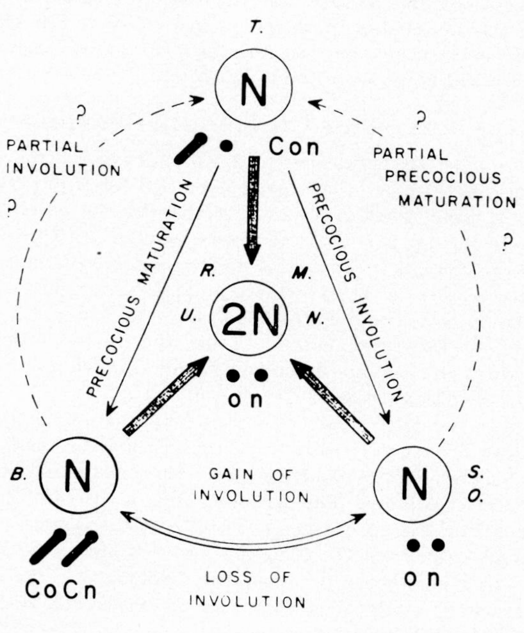

FIG. 2. Key: C, fully grown centriole, achromatic figure producing; o, n, old and new generative portions of centriole.

Abbreviations of generic names:

B. Barbulanympha	*R. Rynchonympha*
M. Macrospironympha	*S. Saccinobaculus*
N. Notila	*T. Trichonympha*
O. Oxymonas	*U. Urinympha*

at the right time with respect to chromosome duplication, irreversible diploidy could result from haploidy, a situation chromosomally identical with endomitosis, but different from a centriole cycle point of view. This is diagrammed in Fig. 2. By this schema, the predominant trend is towards diploid chromosome constitution, simply on the basis of involution of the centrioles at or just before metaphase, such that the centrioles must regrow or remature before another division can occur.

Finally, the observation that ecdyson inhibits meiosis in *Saccinobaculus, Oxymonas,* and *Notila,* becomes a relative one, according to the facts. The reason *Barbulanympha* and other genera do not undergo zygotic meiosis before ecdysis of the host may be a locking-inhibitory effect of ecdyson, which upon release of the locking mechanism, and adaptation to the inhibitory action, may afford the cell opportunity to continue its cycle. *Saccinobaculus, Oxymonas,* and *Notila* may have a more tenacious locking, or less cellular adaptability to the inhibitory effects of ecdyson. Remember, if you will, that the inhibitory effects need only be directed against the achromatic apparatus in these latter organisms, for meiosis here is one-divisional not requiring chromatic apparatus duplication. It will be a long time before correct answers can be given to these questions, but concealed in the confines of these problems lie many facts of cell reproduction not touched upon as yet; and I think that the centriole, as well as other cell parts, will be found to have roles which were previously unsuspect.

IV. The Role of the Centriole in Cytokinesis

I should like to turn for just a moment to the role of the centriole in cytokinesis. Nowhere can one imagine of more work being done and less really understood than in the endeavors of so many to show a casual relationship between the spindle and cytokinesis. I will refer to but a few of the observations of Dr. Cleveland that show in many cases, cytokinesis is either not related to the presence of the spindle, or even may be inhibited by it. Let us be brief and propose a classification. Cytoplasmic division may be: (1) "Cleavage and Furrowing" type; (2) "Simple Separation" form (pulling apart); or (3) "Rejection Forms." I use the pleural, "Rejection Forms" purposely, because there are several different examples which for simplicity's sake might be included under this category.

In the first category, we have many examples supported by that vast wealth of information concerning the type. I really think that we may be tempted to conclude that the achromatic figure, or some part thereof, participates in some way in the cleavage furrow type of cytokinesis. The centriole, then in its turn, is important. In the second type, where the one cytoplasm pulls into two, there are a number of cases at point which I could recall from Dr. Cleveland's work which show that the spindle plays no "lead role" in a mechanical or orientative fashion. One that comes to mind is the artificial termination of the sexual cycle of *Saccinobaculus* after formation of the zygote, but before final differentiation and fusion of the pronuclei, which results in cytoplasmic division of the zygote without any other event — a simple pulling apart of the two incompatible gametes (Cleveland and Burke, 1960). Another example, just as valid, is that normal cell separation which occurs in *Notila* after formation of the complete double zygote. This cell simply pulls in two (Cleveland, 1950c). I am even prompted to call attention to the persistence of the spindle after gametogenesis in *Leptospironympha wachula.* The cytokinesis which accompanies gametogenesis is so rapid that the spindle is still more or less intact after the two gametes are formed, and thus the central spindle binds the two daughter cells together (Cleveland, 1951a; cf. Heidenhain, 1907, pp. 310). Can we say the spindle has impeded cytokinesis? I think so. The same has been demonstrated for karyokinesis (Cleveland, 1960a).

In the third type of cytoplasmic division, many rejection forms occur. For example, in certain instances the male gamete or portion thereof is not acceptable to incorporation into the zygote, and a cytoplasmic division takes place (Cleveland, 1957a). In other forms, the male extranuclear organelles are rejected along with a bit of cytoplasm from the forming zygote instead of being digested as has been noted to occur in other genera (Cleveland, 1950d). Also, the division of giant multinucleate zygotes, so-called monster individuals, which occur spontaneously on rare occasion, takes place as a mutual rejection of individual components (Cleveland, 1955a).

As for the possible role of the breakdown products of the nuclear membrane in cytokinesis, I should like to call attention of proponents of this thought to the fact that persistence of the nuclear membrane with and without cytokinesis as seen in these organisms must be rationally explained. Besides, as was previously stated, cytokinesis can take place in the absence of any vestige of a nucleus – Boveri also showed that for *Arbacia* over 60 years ago. I will leave the subject with the thought that to implicate the achromatic figure in cytokinesis in general is premature, if not incorrect, despite some learned author's critique of the matter (cf. Mazia, 1961, p. 314).

V. The Behavior of the Centriole in Regulation of Cell Activity

A. Differentiation and Cancer

At this moment I find in myself a physician who is guilty of inflicting more sores than he is in healing. This is the plight of the discussant. I shall take the liberty to turn in defense and reflect some aspects of the behavior of the centriole in the regulation of cell activity which have been investigated by the medically minded.

Wolbach (1928) once described centriole behavior of the cells of two rhabdomyosarcomata, one a tumor of cardiac muscle, the other of skeletal muscle. He described the origin of the myofibrils of these myoblastic-primitive type cells as being from multiple centrioles and attributed the possible evolution of normal myofibrillar structure to proliferations from the centriole or centriole-like bodies of myoblasts. Very recently observations on another type of tumor cell with "pleuricorpuscular" centrioles were reported by Altmann (1961). He studied the fluctuations in numbers, forms, and behaviors of the centrioles of a "monster cell" sarcoma of human brain. These findings and their interpretations are so interesting that I will take them up in some detail.

The centrioles in these tumor cells are infrequently two in number, most often "pleuricorpuscular" in a variety of assemblages as bands, crescents, cluster, etc. Occasionally two cytocentra are present which function as two centrioles, but each center is seen to be composed of a number of smaller corpuscles. In the interphase of division cycle these corpuscles often tend to coalesce or group quite closely, with an over-all size range of 0.6 to 0.8 micron. When individual corpuscles are clearly distinguished, as in premitosis, they are about 0.3 micron diameter, and an interconnecting "centrodemose" is visualized. Throughout the mitotic cycle and between divisions, the pleuricorpuscular centrioles are seen to be surrounded by a very thin halo of PAS-positive material. In addition, during and just prior to mitosis, this entire body is further encased in a variable-sized sphere of PAS-positive material which has slightly different staining characteristics thought to reflect glycoprotein content. The accumulation of this so-called "spheroplasm" is premitotic, and Altmann feels that under the direction of the centriole at its center, the "spheroplasm" elaborates the astral rays, central spindle, and chromosome fibers, presumably via some enzyme-like substance, and thus he admits to support the "Theory of the Secreting Centriole." Aside from Altmann's deviations of terminology, he seems to be describing what is best called a two-part centrosome. He refers only to the inner and persistent halo adjacent to the centriole as the centrosome.

As to function [cf. Heidenhain (1907), p. 250, for normal ganglion cell behavior], when two or more achromatic-figure producing cytocentra are not present, he regularly sees endomitosis resulting, and even when two functional centriole centers do not have appropriate orientation with the nucleus, abortive types of mitosis occur. Further, chromosomes which are duplicating and dividing, do so independent of the position of any of these centrioles, and he correctly concludes that chromosome duplication and separation (daughters) have nothing whatever to do with the central spindle. What he reports in "unipolar mitosis" is a failure of chromosomes or daughters to progress to any pole; in fact just the reverse is said to occur. Where the cytocentrum is central in the cell, mitosis concludes with the chromosomes at the periphery of the cell and when the nuclear membrane reconstitutes (is reformed), variable numbers of chromosomes are incorporated into peripherally scattered nuclei, some of which may coalesce. What these chromosome movements suggest is, that when chromosomes are contacted by astral rays, if another ray from an "amphiastral pole" has not made contact with the sister chromatid, the whole chromosome (really a dyad) is simply pushed to the extreme limit of the cell which is the cell membrane. The scatter of chromosomes from original site of the nucleus to an eventual rotated position with respect to the centriole, is in accord with these mechanics and is reminiscent of motion of particles on the surface of sea urchin eggs during cleavage.

This, withal, suggests that we may have an explanation of how chromosome fibers may behave in typical mitosis. When an astral ray from a centriole contacts a chromosome at its centromere, it may push this centromere (and hence chromosome) along in the still intact nuclear membrane until the sister centromere is contacted by an astral ray from the other centriole. At this instant, for an unexplained reason, both astral rays cease to elongate and could now be called chromosome fibers. I would venture to say that the process is controlled by sister centromeres and may be mediated via change in astral ray characteristics by elaborations from the centromere itself. If, on the other hand, the sister centromere experiences no such contact (as is very unlikely with a bipolar spindle properly oriented with respect to the nucleus, and rather likely in abnormal positioning of nucleus and spindle, albeit certain with a unicentriolar, unipolar achromatic figure) the astral ray merely pushes the chromosome out of the field and as such should not be recognized as a "chromosome fiber." Since the centromere has been shown to consist in part, at least, of DNA, this seems hardly a supposition at all. DNA could accomplish almost anything these days (cf. Mazia, 1961, p. 213).

By this hypothesis then, the premetaphase movement of chromosomes to one pole or the other, as described in a variety of cells (cf. Mazia, 1961), would not be an unequal "pull" of a "pole," but a real push on the centromere by an astral ray of the far centriole until contact with the other (sister) centromere is made by a ray from the near centriole. Also, the occurrence of so-called nonconvergent, or divergent achromatic figures in other animal and certain plant cells may reflect a situation very much analagous to the "pleuricorpuscular cytocentra" here in the tumor cells (cf. Mazia, 1961). Further, the separation of chromosomes in the first spermatocyte division of *Sciara*, as described by Metz (cf. Mazia, 1961, p. 219), may be a unique example of this mechanism of astral ray "push."

I am supposed to be discussing Dr. Cleveland's paper and his contributions in this field. The description by Dr. Cleveland of abortive division forms with multiple centrioles, chromosome duplication and "splitting" into sister chromatids independent of the central spindle, and a failure of a unipolar achromatic figure to affect a nuclear bipartitioning is to be found in several of his papers published earlier (1934, 1954b, 1955a, 1956b, 1957d, 1958a), and in his presentation today. Thus the whole question of dependence of these results upon the material seems to vanish; the centrioles of *Barbulanympha* react in many ways like those of human cell origin, only the convenient size and exaggeration of separation into generative and achromatic-figure producing portions does not reflect this uniqueness in any significant way except to make them amenable to study. I should like to round off this comparison by pointing to the fact that Altmann observed pleuricorpuscular centrioles in a generative stage of development, while groups of generative centrioles were unable to "mature" to function in achromatic-figure production.

What suppresses the one function of the centriole while allowing the centriole to pursue its other generative course? Altmann (1961) found a gross proportional correspondence between ploidy of the chromosomes, size of these giant cells, and pleurality of the corpuscles comprising the centrioles. These tumor cells are thought to have arisen from interstitial cells of the brain. Is suppression of the achromatic-figure producing portion of centrioles the mechanism of tumor induction? Is this involution of centrioles?

Another paper was published from the same laboratory (Pathologisches Institut, Würzburg) which reported the abnormal mitoses resulting from "pleuricorpuscular cytocentrums" of glioma giant cells and the cyclic behavior of these centrioles in morphological fluctuation and in appearance of surrounding centrosomes and confirmed much of Altmann's findings (Palm, 1961). Elsewhere, the studies (Kinosita *et al.*, 1959) on differentiation of the thrombocytic cell series clearly point to a key role of the centriole. The results reported are as follow: a 2N megakaryoblast divides mitotically to produce two 2N daughter cells which fuse, and in most cases their nuclei also fuse. This 4N promegakaryocyte by a similar process produces an 8N megakaryocyte, in which cytoplasmic granules begin to develop. This 8N megakaryocyte twice repeats endoplasmic multipolar mitosis with considerable increase in cytoplasmic granules, eventually producing a mature 32N megakaryocyte. A plausible interpretation of these data which we would suggest is that one of the centrioles is suppressed after the first cell recombination which produced the 4N promegakaryocyte, and a similar suppression occurs for the 8N megakaryocyte, but these suppressed centrioles later function giving rise to the two multipolar divisions in the maturing megakaryocyte. It is very much to the point that centrioles of megakaryocytes were reported as "pleuricorpuscular" by Martin Heidenhain (1907, pp. 265-271) 50 years before the sequence of events leading to this condition was observed as reported above.

The list of studies concerning the morphology and behaviors of the centriole and its derivatives is certainly growing; it is not convenient in this discussion to even allude to most of them. However, inasmuch as we are discussing the centriole primarily based on the studies of centrioles of flagellate protozoa, it would be an oversight not to refer to the centriole and its derivatives in ciliate protozoa. A review of cilia, their morphology, related structures, and ultra structure has been recently made by Fawcett (1961). Also, the arguments for autonomy of the "kinetosomes" of ciliates have been taken up elsewhere (Nanney and Rudzinska, 1960), and so will not be undertaken here.

Jumping from the descriptive-morphological studies on centrioles to experimentation, we find that scattered observations already exist about chemical effects on centrioles. A host of studies have been undertaken in recent years to uncover and elaborate the toxic actions of so-called chemotherapeutic agents against cancer cells. The list of experimental drugs has become prodigious, but a small handful of these agents remain useful and have limited application. One group of chemicals, the so-called alkylating agents, seem at certain concentrations to specifically inhibit the achromatic apparatus without influencing function of the chromatic apparatus; examples of such agents being: triethylene melamine (TEM), triethylene phosphoramide (TEPA) (cf. Kato *et al.,* 1959), and actidine (cf. Karnofsky, 1960), to name a few. Nitrogen mustard (HN_2) is also reported to cause mitotic inhibition with a temporary delay, but no appreciable reduction in subsequent rate of cellular dry mass accumulation, (Lee *et al.,* 1961), and hence, query, protein synthesis? This result is similar to that obtained upon X-irradiation (cf. Caspersson *et al.,* 1958). A plant alkaloid, vincaleukoblastine (VLB), isolated from the periwinkle has been found of limited use in cancer chemotherapy, and this exerts an effect on cells similar to the action of colchicine, but probably more profoundly and less reversibly (cf. Cardinali *et al.,* 1961; Cutts, 1961; Palmer *et al.,* 1960; Vaitkevicius *et al.,* 1961).

Certain antimetabolite agents have the reverse effect, however, in that instead of inhibiting the achromatic apparatus, they specifically inhibit or interfere with deoxyribonucleic acid (DNA) synthesis and hence are thought to be specific for the chromatic apparatus. A well-understood example of this is 5-fluorouracil (5-FU) which specifically blocks DNA synthesis at a special step (Heidelberger *et al.,* 1957; Rutman *et al.,* 1954). It has been shown that protein and ribonucleic acid (RNA) synthesis continue, however, with the result that when a near tetraploid tumor cell strain is subjected to 5-FU, the amount of DNA per cell drops by one-half while RNA per cell doubles (Lindner, 1959). And, the listing could continue.

What I would suggest is that these and similar agents be tested on an organism which has unmistakable cellular organization, especially as regards the centriole, and which at the same time has morphologically discernible chromosomes. This is asking a lot of a biological assay system, but attempts directed along these lines using the protozoa of the wood-feeding roach might be rewarding.

Ecdyson prepared in this country by the methods of Karlson (Butenandt and Karlson, 1954) has been tested on several tumors and the results are conflicting (Burdette, 1959; Burdette *et al.,* 1960). At least the agent has been reported to be toxic for Sarcoma 180 cells in tissue culture at concentrations above 12 Caliphora units (Burdette *et al.,* 1960). Our experience with ecdyson supplied by Karlson has taught us that toxicity can easily accompany impure preparations of insect hormones and especially that ecdyson is quite subject to contamination and bacterial overgrowth in noncrystalline preparations, making them highly toxic, even for insects (unpublished, cf. Cleveland *et al.,* 1960). That tissue cultures of mammalian cells exhibit toxic changes from ecdyson is not surprising, in any event.

VI. Summary Hypothesis

A. Molecular Biology and the Centriole

It is extremely vogue to at least close any discussion about living matter with some thoughts and confusions of biochemical processes and in so doing, aspire to the heights of molecular biology. My closing summation is to attempt a hypothesis correlating some of what has gone before in this discussion of centrioles in cell reproduction. Figure 3 commits much of this hypothesis to diagram.

It has been amply demonstrated that a cell synthesizes DNA and RNA from a precursor pool, DNA being thought to be only synthesized in the nucleus, RNA predominantly there, and both syntheses taking place at certain rates and at certain times. Since more will follow in this symposium about these matters, I shall presume no review of this subject. There is a growing body of evidence that DNA, and more particularly RNA synthesis, is halted upon chromosome coil tightening and further, that the RNA synthesized in the nucleus rapidly diffuses into the cytoplasm (Taylor, 1960). From many points of view it would be attractive to think that the "information" of DNA is relayed to the cytoplasm via RNA (Crick, 1958; Goldstein and Plaut, 1955; Lockingren and DeBusk, 1955), and thus if ecdyson induces differentiation as we see visible evidence of in these flagellates, RNA would be a likely candidate to serve messenger. It is not too far afield to anticipate direct effects of ecdyson on RNA synthesis, in addition. Whichever way RNA has become "coded," it could direct syntheses, induce differentiations (cytoplasmic and pronuclear), direct dissociation of incompatible extranuclear organelles, etc. In addition, we must not rule out the possibility of a cytoplasmically synthesized RNA (cf. Brachet, 1961; Woodard, 1958), and if such be the case, ecdyson could affect its synthesis here in these protozoa. The almost complete maturation of anucleate gametes of *Trichonympha* (mentioned previously) would be comfortably explained thereby. On the other hand, "information" relayed to the cytoplasm by the nuclear RNA may precede enucleation and the expression of its direction could occur thereafter. Although, this would not obtain where the process of production of nonnucleate gametes is spontaneous (as rarely occurs).

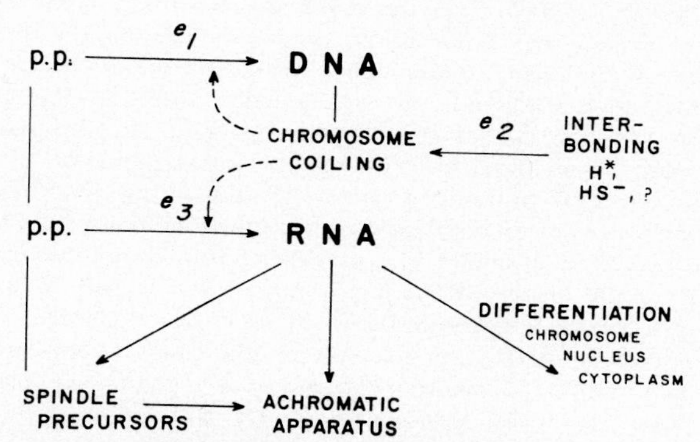

FIG. 3. Cellular effects of ecdyson. Key: $^e1$, $^e2$, $^e3$, ecdyson acceleration; p.p., cellular precusor pool substances; -----, inhibition.

There is no evidence to indicate that ecdyson accelerates spindle production beyond that expected of general cellular acceleration. Contrarily, ecdyson very definitely accelerates chromosome activity, probably via DNA synthesis, but even more notably via precocious coiling of chromosomes during gametogenesis (e.g., in *Leptospironympha wachula,* as mentioned previously). This in turn has been suggested to have an inhibitory influence upon DNA and RNA synthesis. If in a particular protozoan the predominant effect of ecdyson is to precipitate coiling of chromosomes, only that RNA which had been stored and previously synthesized would be present, barring cytoplasmic elaboration of RNA. If we go one step further, assuming that RNA participates in a spindle elaboration, and this is not a new idea (cf. Love and Suskind, 1960), we could deduce that this cell with the predominant ecdyson acceleration being directed at the chromosome coiling would be primed to undergo meiosis rather than mitosis and, of the one-divisional type, rather than the two. Such may be the adjustment of *Urinympha* to ecdyson (cited previously). Note in the foregoing argument we are endorsing the suggestion that the achromatic figure is an assembly of pre-existing units, or that the synthesis of spindle precursor moities occurs much ahead of actual appearance of that birefringent structure (cf. review by Mazia, 1960). RNA has been shown to promote mitotic division of an otherwise "amitotic clone" of cells, further implicating its role in spindle formation (Stone and Stevens 1960), while RNase has been shown to be of low activity in most cancer cells (Daoust and Amano 1961), and in cancer cells with a slowed mitotic rate, the RNA has been shown to be reduced in quantity (cf. Le Breton and Moule, 1961). I shall not attempt to procede through all the life cycles of these protozoa to validate this model, but let me say that the progressive acceleration of complete gametogenesis, through autogamy, to endomitosis produced by ecdyson is not incompatable with this hypothesis. Cytoplasmic RNA moities would be differentiated later where the titer and effects of ecdyson are gradual; acceleration would result in pronuclear differentiation by RNA which is associated with the nuclear membrane as has been shown to be present in cells not under the influence of ecdyson (cf. Brachet, 1961); and finally, "male" and "female" chromosome grouping in gametogenesis would be explained upon interbonding and differentiation effects of the chromosomes themselves, the latter possibly even induced by RNA (cf. Brachet, 1961; Hopkins, 1959).

There remains the unanswered question of why the achromatic figure is abortive in endomitosis. Simple acceleration of the chromosomes without some suppression of the centriole as an explanation is difficult to employ. What may be operative is the inhibitory effect of chromosome coiling upon RNA synthesis in the nucleus, hence faulty "information" to the centriole, or ecdyson may inhibit in a more direct way the assemblage by the centriole of the achromatic figure. In this connection, the possible ecdyson effects upon the achromatic-figure producing portion of the centriole prior to reorganization come to mind. This may be, as was suggested before, an effect mediated via an enzyme or enzyme system for controlled digestion of the centriole — an involution-producing effect. I should like to point out a major difference, however. The degeneration of the "male" centriole in fertilization as well as in endomitosis is complete and the process is the same in these analagous situations; the controlled digestion of the remaining female centriole and its new daughter does not involve the generative portions of either (as mentioned before in the section on Reorganization).

Two lines of information seem pertinent here: (1) the studies which indicate that the achromatic figure is an assemblage of pre-existing units (Went, 1960), part of which may be RNA (Zimmerman and Marsland, 1960); and (2) the intriguing suggestion that the kinetosome of *Trypanosoma* is composed of two distinct parts, a DNA (generative) portion and a RNA portion which at times becomes detached (cf. Mirsky and Osawa, 1961). Further the report that DNase induces post-telophase cytoplasmic recombination and other abnormalities of mitotic division in other cells (Montgomery and Bonner, 1959) suggests an effect of DNase on the achromatic apparatus as well as the chromatic apparatus. If we can carry the inference over, is it that the generative portion of the centriole is DNA, while the functional portion is RNA-protein? How eager everyone would be to accept the autonomy and perpetuity of the generative portion of the centriole if this were the case. It would have genetic rank with the "information content" of the chromosome. Could it be that the two distinct peaks in formation of deoxyribosidic components which occur prior to pollen microspore mitosis in *Lilium* anthers (cf. Stern, 1960) represent the accumulation of precursors for two different syntheses, i.e., chromosome DNA and centriole DNA-like substance?

The possibility that ecdyson may stimulate DNA synthesis raises the question of the stimulation of the generative portion of the centriole by ecdyson. Also, what about centriolar differentiation? Is this how even the generative portion of the centriole of the male gamete is recognized as incompatable? A continued study of the effects of ecdyson on the centriole, pursued to the macromolecular level might be rewarding.

From all the foregoing wherein I have placed the gospel before the epistle without benefit of an announced gradual, I derive the following collect: a particular act of nature, whether it be the function of an organelle, the sequence in development and generation of that organelle, or the forces which direct all these behaviors, need not be explicable on the basis of one catholic law. Since nature is only as logical and orderly as man conceives, let every bit of evidence be tracked down and all theories and combinations thereof pursued until as a first order approximation, we begin to see nature as she really is. In closing I ask that we regard for a moment the perverse pleasures of "StepMother" Nature as she reflects the efforts of man to fathom her secrets — the theories concocted upon flimsy evidence doled out to confuse him — the hasty complacency of science that theory is fact.

ACKNOWLEDGMENT

The author wishes to acknowledge the Rhode Island Hospital for support during the preparation of this manuscript, and particularly, the Department of Oncology and the staff of Pathology. Gratitude is expressed especially to Dr. L. R. Cleveland.

REFERENCES

Altmann, H. W. (1961). *Arch. pathol. Anat. u. Physiol. Virchow's* **334**, 132-159.
Boveri T. (1895). *Verhandl. physik.-med. Ges. Würzburg* **29**, (1).
Boveri, T. (1903). *Sitzber. physik.-med. Ges. Würzburg* [N. F.] **36**, 12-21.
Brachet, J. (1961). *In* "The Cell" (J. Brachet and A. E. Mirsky, eds.), Vol. II, pp. 771-841. Academic Press, New York.
Burdette, W. J. (1959). *Proc. Am. Assoc. Cancer Research* **3** (1), 11.
Burdette, W. J., Richards, R. C., and Moberly, M. L. (1960). *Proc. Am. Assoc. Cancer Research* **3** (2), 99.
Butenandt, A. von, and Karlson, P. (1954). *Z. Naturforsch.* **9b**, 389-391.
Cardinali, G., Cardinali, G., and Blair, J. (1961). *Proc. Am. Assoc. Cancer Research* **3** (3), 215.
Caspersson, T., Klein, E., and Ringertz, N. R. (1958). *Cancer Research* **18**, 857-865.
Cleveland, L. R. (1947a). *Science* **105**, 16.
Cleveland, L. R. (1947b). *Science* **105**, 287–289.
Cleveland, L. R. (1949a). *Trans. Am. Phil. Soc.* **39** (Pt. 1), 1-100.
Cleveland, L. R. (1949b). *J. Morphol.* **85**, 197-296.
Cleveland, L. R. (1950a). *J. Morphol.* **86**, 185-214.
Cleveland, L. R. (1950b). *J. Morphol.* **86**, 215-228.
Cleveland, L. R. (1950c). *J. Morphol.* **87**, 317-348.
Cleveland, L. R. (1950d). *J. Morphol.* **87**, 349-368.
Cleveland, L. R. (1951a). *J. Morphol.* **88**, 199-244.
Cleveland, L. R. (1951b). *J. Morphol.* **88**, 385-440.
Cleveland, L. R. (1952). *J. Morphol.* **91**, 269-324.
Cleveland, L. R. (1953a). *J. Morphol.* **93**, 371-404.
Cleveland, L. R. (1953b). *Trans. Am. Phil. Soc.* **43** (Pt. 3), 809-869.
Cleveland, L. R. (1954a). *J. Morphol.* **95**, 189-212.
Cleveland, L. R. (1954b). *J. Morphol.* **95**, 213-236.
Cleveland, L. R. (1954c). *J. Morphol.* **95**, 557-620.
Cleveland, L. R. (1955a). *J. Morphol.* **97**, 511-542.
Cleveland, L. R. (1956a). *Arch. Protistenk.* **101**, 99-168.
Cleveland, L. R. (1956b). *J. Protozool.* **3**, 78-83.
Cleveland, L. R. (1956c). *J. Protozool.* **3**, 161-180.
Cleveland, L. R. (1957a). *J. Protozool.* **4**, 164-168.

Cleveland, L. R. (1957b). *J. Protozool.* **4**, 168-175.
Cleveland, L. R. (1957c). *J. Protozool.* **4**, 230-241.
Cleveland, L. R. (1957d). *J. Protozool.* **4**, 241-248.
Cleveland, L. R. (1958a). *J. Protozool.* **5**, 47-62.
Cleveland, L. R. (1958b). *J. Protozool.* **5**, 63-68.
Cleveland, L. R. (1959). *Proc. Natl. Acad. Sci. U. S.* **45**, 747-753.
Cleveland, L. R. (1960a). *J. Protozool.* **7**, 326-341.
Cleveland, L. R. (1960b). *Arch. Protistenk.* **105**, 110-112.
Cleveland, L. R. (1960c). *Arch. Protistenk.* **105**, 137-148.
Cleveland, L. R. (1960d). *Arch. Protistenk.* **105**, 149-162.
Cleveland, L. R. (1960e). *Arch. Protistenk.* **105**, 163-172.
Cleveland, L. R., and Burke, A. W., Jr. (1960). *J. Protozool.* **7**, 240-245.
Cleveland, L. R., and Day, M. (1958). *Arch. Protistenk.* **103**, 1-53.
Cleveland, L. R., and Nutting, W. L. (1955). *J. Exptl. Zool.* **130**, 485-514.
Cleveland, L. R., Hall, S. R., Sanders, E. P., and Collier, J. (1934). *Mem. Am. Acad. Arts Sci.* **17**, 185-342.
Cleveland, L. R., Burke, A. W., Jr., and Karlson, P. (1960). *J. Protozool.* **7**, 229-239.
Crick, F. H. C. (1958). *Symposia Soc. Exptl. Biol.* **12**, 138-163.
Cutts, J. H. (1961). *Cancer Research* **21**, 168-172.
Daoust, R., and Amano, H. (1961). *Proc. Am. Assoc. Cancer Research* **3**, 218.
Fawcett, D. (1961). *In* "The Cell" (J. Brachet and A. E. Mirsky, eds.), Vol. II, pp. 217-292. Academic Press, New York.
Goldstein, L., and Plaut, W. (1955). *Proc. Natl. Acad. Sci. U. S.* **41**, 874-880.
Heidelberger, C., Leibman, K., Harbers, E., and Bhargava, P. M. (1957). *Cancer Research* **17**, 399-404.
Heidenhain, M. (1907). "Plasma und Zelle," Ab. I; (1). Fischer, Jena.
Hopkins, J. W. (1959). *Proc. Natl. Acad. Sci. U. S.* **45**, 1461-1470.
Karlson, P. (1956). *Vitamins and Hormones* **14**, 227.
Karnofshy, D. A. (1960). *Proc. Am. Assoc. Cancer Research* **3**, (2), 124.
Kato, R., Gagnon, H., and Yosida, T. (1959). *Proc. Am. Assoc. Cancer Research* **3** (1), 31-32.
Kinosita, R., Ohno, S., and Kakazawa, M. (1959). *Proc. Am. Assoc. Cancer Research* **3** (1), 33.
Le Breton, E., and Moule, Y. (1961). *In* "The Cell" (J. Brachet and A. E. Mirsky, eds.), Vol. V, pp. 497-544. Academic Press, New York.
Lee, H., Richards, V., and Furst, A. (1961). *Proc. Am. Assoc. Cancer Research* **3** (3), 244.
Lindner, A. (1959). *Cancer Research* **19**, 189-194.
Lockingren, L. S., and DeBusk, A. G. (1955). *Proc. Natl. Acad. Sci. U. S.* **41**, 924-934.
Love, R., and Suskind, R. G. (1960). *Proc. Am. Assoc. Cancer Research* **3** (2), 130.
McMahon, R. M. (1956). *Caryologia* **8**, 250-256.
Mazia, D. (1960). *Ann. N. Y. Acad. Sci.* **90**, 455-469.
Mazia, D. (1961). *In* "The Cell" (J. Brachet and A. E. Mirsky, eds.), Vol. III, pp. 77-394. Academic Press, New York.
Mirsky, A. E., and Osawa, S. (1961). *In* "The Cell" (J. Brachet and A. E. Mirsky, eds.), Vol. II, pp. 677-770. Academic Press, New York.
Montgomery, P. O'B., and Bonner, W. A. (1959). *Proc. Am. Assoc. Cancer Research* **3** (1), 44.
Nanney, D. L., and Rudzinska, M. A. (1960). *In* "The Cell" (J. Brachet and A. E. Mirsky, eds.), Vol. IV, pp. 109-150. Academic Press, New York.

Oberling. C., and Bernhard, W. (1961). *In* "The Cell" (J. Brachet and A. E. Mirsky, eds.), Vol. V, pp. 405-496. Academic Press, New York.

Palm, G. (1961). *Arch. pathol. Anat. u. Physiol. Virchow's* **334**, 160-172.

Palmer, C. G., Livengood, D., Warren, A. K., Simpson, P. J., and Johnson, I. S. (1960). *Exptl. Cell Research* **20**, 198-201.

Porter, K. R. (1957). *Harvey Lectures,* 1955-1956, **51**, 175-228.

Porter, K. R. (1961). *In* "The Cell" (J. Brachet and A. E. Mirsky, eds.), Vol. II, pp. 621-675. Academic Press, New York.

Rutman, R. J., Cantarow, A., and Paschkis, K. E. (1954). *Cancer Research* **14**, 119-123.

Stern, H. (1960). *Ann. N. Y. Acad. Sci.* **90**, 440-454.

Stone, D., and Stevens, D. (1960). *Proc. Am. Assoc. Cancer Research* **3**, (2), 153.

Taylor, J. H. (1960). *Ann. N. Y. Acad. Sci.* **90**, 409-421.

Vaitkevicius, V. K., Talley, R. W., and Brennan, M. J. (1961). *Proc. Am. Assoc. Cancer Research* **3**, 275.

Went, H. A. (1960). *Ann. N. Y. Acad. Sci.* **90**, 422-429.

Wilson, E. B. (1925). "The Cell in Development and Heredity," 3rd ed. Macmillan, New York.

Wolbach, S. B. (1928). *Anat. Record* **37**, 255-262.

Woodard, J. W. (1958). *J. Biophys. Biochem. Cytol.* **4**, 383-389.

Zimmerman, A. M., and Marsland, D. (1960). *Ann. N. Y. Acad. Sci.* **90**, 470-485.

The Central Spindle and the Cleavage Furrow[1]

Robert C. Buck

Department of Microscopic Anatomy, University of Western Ontario, London, Canada

I. Introduction

The process of cytokinesis has been carefully studied with the light microscope in many kinds of cells over a period of more than a century. Of the many theories which have developed in attempts to explain the process none, it seems, is entirely applicable to all types of cells or even to all animal cells. The existence of so many theories, two reviews of which have recently appeared (Swann and Mitchison, 1958; Wolpert, 1960), suggests that there are some important gaps in our knowledge of the facts about cytokinesis. In the light of the newer knowledge of cell structure which electron microscopy has provided it seems that an examination of mitotic cells by this method would provide some further answers by demonstrating the changes in membranes, fibrils, vesicles, and other submicroscopic structures during the cleavage period. And for the plant cell this has already been accomplished (Buvat and Puissant, 1958; Porter and Caulfield, 1958). A number of reports have appeared on the fine structure of animal cells in mitosis, but, except for a study by David (1959) of rat embryonic liver cells, they deal briefly, if at all, with cytokinesis.

The work which I am reporting was done in collaboration with Mr. James M. Tisdale, a senior medical student at our University. The normal cells examined are those of rat tissues, principally the erythroblasts of bone marrow but also the intestinal epithelial cells and the cells of regenerating liver. The tumor cells are those of Walker 256 carcinoma, mouse leukemia, and mouse sarcoma induced by the subcutaneous implantation of chemical carcinogens.

II. Development of the Midbody

The reality in mammalian cells of orientated material, commonly called the "spindle fibers," is now firmly established both by polarized

[1] The work was supported by grants from the National Cancer Institute of Canada, and the Medical Research Council of Canada.

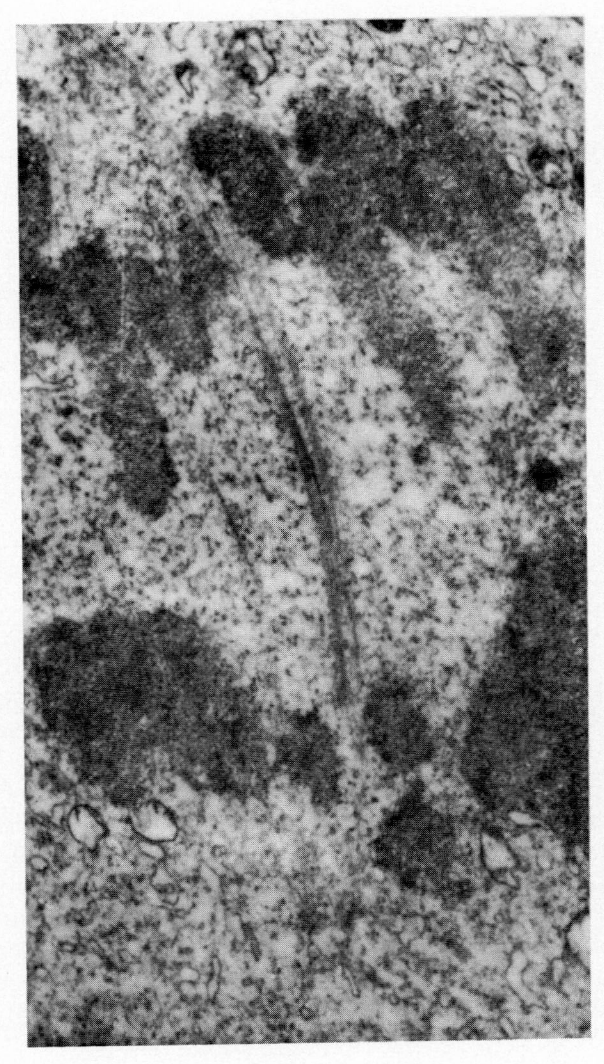

Fig. 1. Fibrillary material passes between the individual anaphase chromosomes of each daughter set, and extends towards the poles of the cell. This material is part of the system of continuous spindle fibers. At the plane of the equator a short length of fiber shows a pronounced increase in density. Mouse leukemia cell. Magnification: $\times$ 14,400.

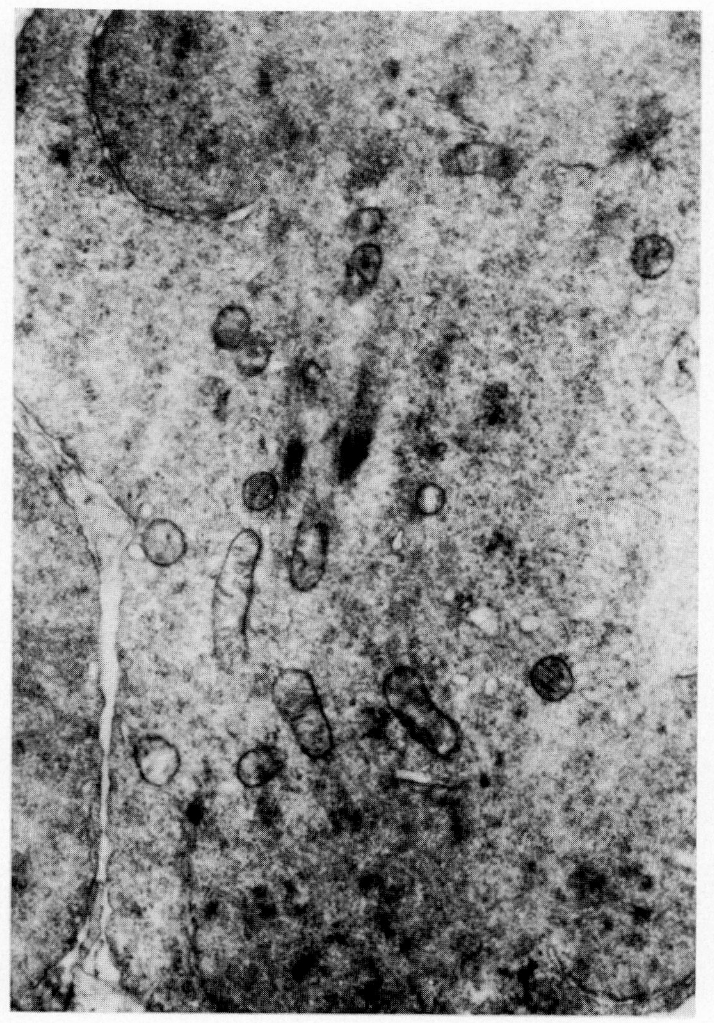

Fig. 2. This erythroblast, in early telophase, shows the nuclear membranes completely formed only on the polar surfaces. The cell is slightly narrower at the equator than at the level of the nuclei. In the equatorial plane there are two dense bodies from which fibrillary material radiates towards the daughter nuclei. These are interpreted as the aggregated continuous fibers, which now form the stem-body (Bělăr, 1927). Magnification: $\times$ 15,300.

light microscopy of the living cell (Hughes and Swann, 1948; Inoué and Dan, 1951) and by electron microscopy (Bernhard and DeHarven, 1958). According to Bernhard and de Harven (1958) the fibers of the aster, the continuous fibers and the chromosomal fibers have an identical fine structure.

In early anaphase the continuous fibers can be identified as they pass between the individual chromosomes of each daughter set and extend

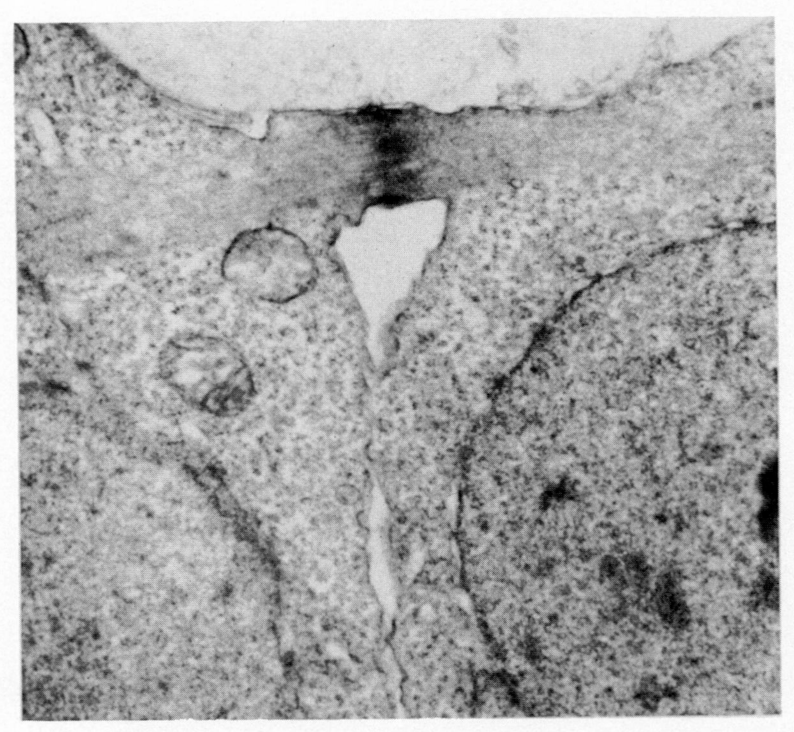

FIG. 3. Narrow intercellular bridge connecting daughter erythroblasts, showing a platelike midbody composed of dense fibrils. The fibrillary material of the continuous spindle fibers extends into the adjoining cytoplasm. Magnification: × 35,500.

across the equatorial plane (Fig. 1). This is the part of the spindle which is associated with cytokinesis. From the time of anaphase and later, short lengths of the continuous fibers show a distinctive character at the plane of the equator, which marks this part of the fiber as different from the rest of the continuous fiber. The distinctive feature at the equatorial plane is a localized increase in density. At anaphase (Fig. 1) the denser region is visible, although hardly remarkable, but with the onset of telophase (Fig. 2) an aggregation of the denser regions appears, and such a structure probably corresponds to the stem-body of Bělăr (1927).

Shortly after the cleavage furrow has developed a narrow bridge connects the daughter cells (Fig. 3). In images which show only a short bridge the denser parts of the continuous fibers coincide so as to form a plate at the midpoint of the bridge. This structure now corresponds to the midbody or Zwischenkörper or Flemming body (Flemming, 1891; Fry, 1937). At this stage it resembles the midbody which has been described by Fawcett and his co-workers in cnidoblasts of *Hydra* (Fawcett *et al.*, 1959). In many images of the midbody in our material quite a

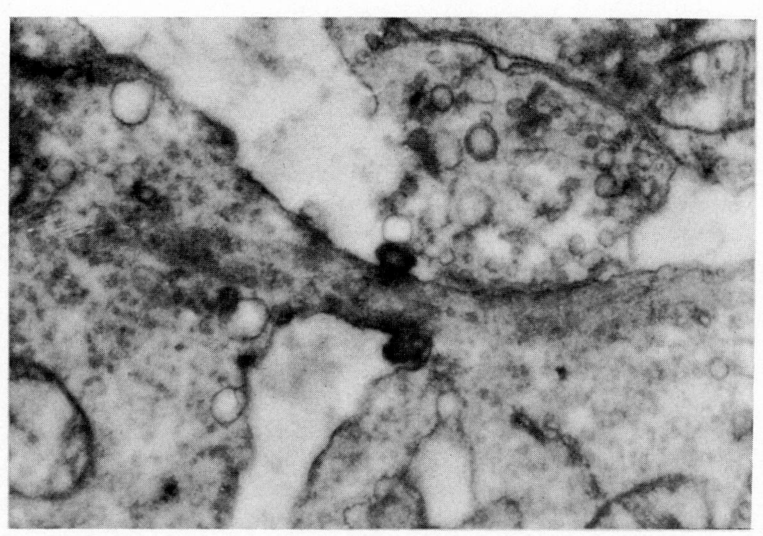

FIG. 4. Erythroblast midbody in which the dense material is amorphous and forms a bulging ring around the telophase bridge. Magnification: × 35,500.

different appearance is seen, and this is interpreted as representing later stages in the midbody development. It must be admitted, however, that we have no independent criterion, except the length and diameter of the telophase intercellular bridge, by which we can arrange the images of the midbody in a time sequence. Nevertheless, as it is probably valid to assume that long narrow bridges represent a later stage than short thick ones, a sequence of changes can be described on this basis. The changes consist of a loss of the fibrillary character of the midbody and a displacement of the dense material against the plasma membrane of the bridge, where it forms a ring. The plasma membrane bulges to accommodate the dense material, which is now amorphous (Fig. 4). While these pronounced changes are seen in the dense material of the midbody the remaining part of the bundle of continuous fibers may still be observed extending, as fibrillary material, into each daughter cell. A somewhat similar ring-shaped form of the midbody was also observed by Fawcett and his co-workers in secondary spermatocytes of mammalian testis (Burgos and Fawcett, 1955; Fawcett, 1959, 1961; Fawcett and Burgos, 1956).

Thus, although the midbody is formed in association with the continuous spindle fibers, apparently as a special part of them at the equatorial plane, it subsequently undergoes changes which are not shared by other parts of the spindle fiber, at least not at the same time in the mitotic cycle.

III. Development of the Cleavage Furrow

In early telophase some vesicular structures are observed in relation to the continuous spindle fibers, now massed together as the dense, fibrillary midbody or stembody. Usually a single membrane-bounded structure, the contents showing almost no density, lies against the midbody. In some sections the profile of a vesicle is observed at each side of it. Serial sections show that these profiles may represent a ring-shaped vesicle encircling or partly encircling the midbody. A structure of this shape cannot be illustrated adequately in a single electron micrograph and our interpretation is developed by examining several cells at this stage, some of which have been serially sectioned.

A slightly later stage is illustrated in Fig. 5. Here the midbody is recognized by the presence of dense, fibrillary material. Against it are the rather flat profiles of the ring-shaped vesicle. A row of smaller, spherical vesicles extends from the ring-shaped vesicle across the whole cleavage plane. These vesicles measure 40 to 70 millimicrons in diameter. Since about thirty of them are present in a single section there would likely be several hundred in the whole equatorial plane. Similar vesicles in a row on the equatorial plane have been seen in cells of the mouse leukemia.

It is hard to escape the conclusion that the new plasma membrane of the cleavage furrow is formed by the coalescence of these small vesicles. The ring-shaped vesicle appears to form first. It is also larger than the others. Probably the development of the smaller vesicles proceeds in a centrifugal direction, although we have scanty evidence on this point.

The formation of the cleavage membrane by the fusion of vesicles in mammalian cells resembles closely the process described by Porter and Caulfield (1958) in plant cells. A major point of difference seems to be the relationship to the midbody, which is not seen in *Allium* root tip cells.

The question of the origin of the ring-shaped and smaller vesicles cannot be fully answered from the evidence which we have at present. A striking feature of some cells at the time of cleavage is the appearance of an extensive system of double membranes or cysternae having a disposition which is obviously related to the cleavage plane (Fig. 5). I am indebted to Dr. Keith Porter for the suggestion that such membranes are

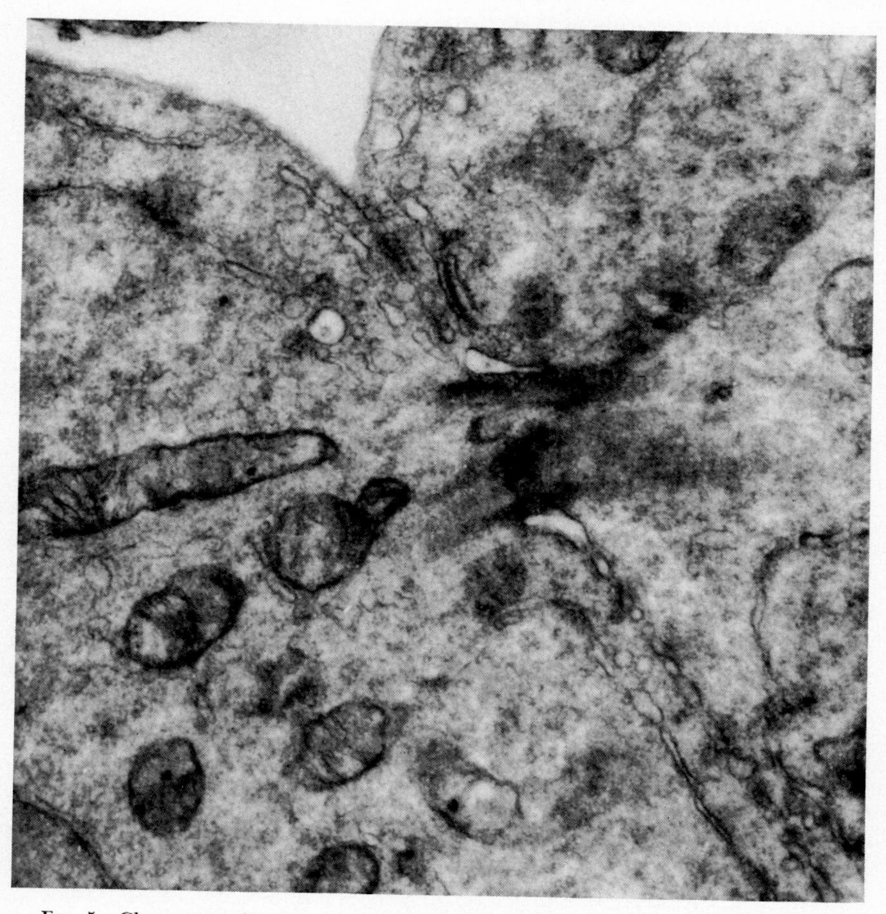

FIG. 5. Cleavage region of an erythroblast during the formation of multiple vesicles on the equatorial plane. The dense midbody lies in the bridge joining the daughter cells. On each side of the midbody is the profile of a rather large vesicle. It has an elongated shape, and the smaller vesicles show a circular profile. The latter extend as a straight row from the larger vesicle to the cell margin. On each side of the equatorial plane is seen part of a system of double-walled cysternae, the profiles of which are reflected away from the cleavage plane as they approach the spindle. Magnification: × 35,700.

probably a part of the endoplasmic reticulum (ER). The cysternae are particularly striking in cleaving erythroblasts because during interphase these cells show few cysternae of the ER, and they have none of comparable length to those present at the early cleavage stage. In each daughter cell the disposition of the cysternae is in relation to the cleavage plane, the plasma membrane, the bundle of continuous spindle fibers, and the nuclear membrane. This relationship is shown diagrammatically in Fig. 6, in which the cysternae are pictured as encircling the spindle and then

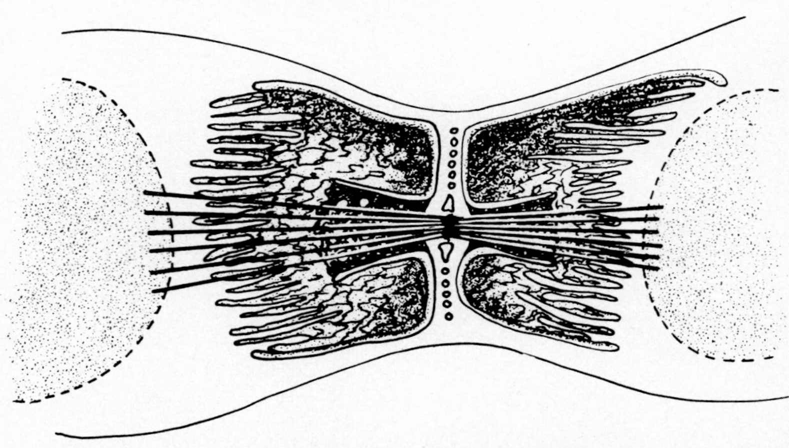

FIG. 6. Diagram of the cysternae seen at the time of cleavage. The cysterna of each daughter cell consists of a double membrane encircling the spindle, then extending centrifugally beside the cleavage plane almost to the cell margin, where it is reflected back towards the nuclei.

lying in close relation to the small vesicles of the equatorial plane. It seems reasonable to suggest that the small vesicles may arise from the cysternae, perhaps by a process of "pinching off," comparable to that seen in the formation of pinocytic vesicles from the plasma membrane of certain kinds of cells (Palade, 1961). At the present time this suggestion is a speculative one. Unfortunately, it is difficult to conceive of a method of establishing or refuting this suggestion with present-day techniques.

Figure 7 summarizes the cleavage process as we have observed it in these cells. In anaphase an increased density of the continuous fibers is seen at the equatorial plane. Some narrowing at the equator occurs, probably largely as a result of the plasma membrane following the contours determined by the nuclei at the poles of the cell. Vesicles begin to appear on the equatorial plane, particularly in association with the midbody, and they extend in this plane out to the partly indented margins of the cell. Judging from the relatively few cells which we have found in this state, the process of vesicle formation and fusion must be quickly over. When the vesicles fuse a cleavage furrow extends from the midbody to the cell margin. The bridge connecting the daughter cells lengthens. The material of the midbody becomes amorphous and distributes itself against the plasma membrane, which shows a local bulging. We do not have conclusive observations on the process of final rupture of the telophase bridge.

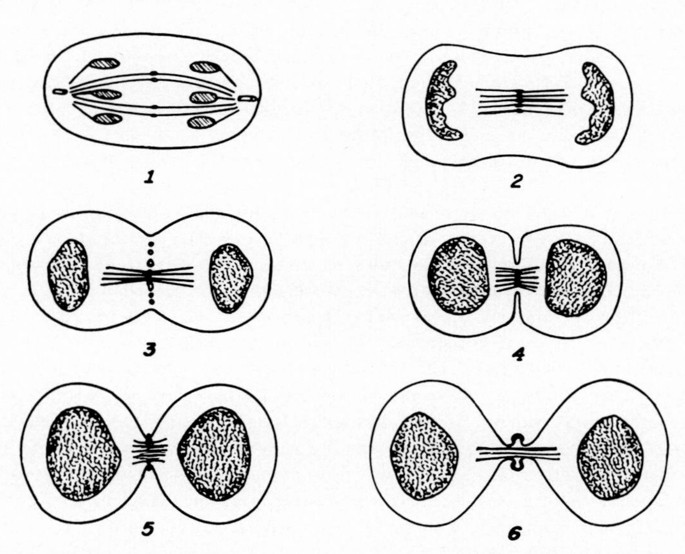

FIG. 7. Diagrammatic summary of the development of the midbody and the cleavage furrow. For explanation, see text.

REFERENCES

Bělăr, K. (1927). *Naturwissenschaften* **15**, 725.

Bernhard, W., and De Harven, E. (1958). *Proc. 4th Intern. Conf. on Electron Microscopy, Berlin* **2**, 217.

Burgos, M. H., and Fawcett, D. W. (1955). *J. Biophys. Biochem. Cytol.* **1**, 287.

Buvat, R., and Puissant, A. (1958). *Compt. rend. acad. sci.* **247**, 233.

David, H. (1959). *Acta. Biol. Med. Germ.* **3**, 330.

Fawcett, D. W. (1959). *In* "Developmental Cytology" (D. Rudnick, ed.), p. 161. Ronald Press, New York.

Fawcett, D. W. (1961). *Exptl. Cell Research Suppl.* **8**, 174.

Fawcett, D. W., and Burgos, M. H. (1956). *Ciba Foundation Colloq. on Ageing* **2**, 86.

Fawcett, D. W., Ito, S., and Slautterback, D. (1959). *J. Biophys. Biochem. Cytol.* **5**, 453.

Flemming, W. (1891). *Arch. mikroscop. Anat. u. Entwicklungsmech.* **37**, 685.

Fry, H. J. (1937). *Biol. Bull.* **73**, 565.

Hughes, A. F., and Swann, M. M. (1948). *J. Exptl. Biol.* **25**, 45.

Inoué, S., and Dan, K. (1951). *J. Morphol.* **89**, 423.

Palade, G. E. (1961). *Circulation* **24**, 368.
Porter, K. R., and Caulfield, J. B. (1958). *Proc. 4th Intern. Conf. on Electron Microscopy, Berlin* **2**, 503.
Swann, M. M., and Mitchison, J. M. (1958). *Biol. Revs. Cambridge Phil. Soc.* **33**, 103.
Wolpert, L. (1960). *Intern. Rev. Cytol.* **10**, 163.

General Discussion

DR. D. COSTELLO (University of North Carolina): Interzonal fibers are present in many materials including the Polychoerus that I am studying and the Coccid material of Sally Hughes-Schrader, and these produce density increases at the equator of the spindle during late anaphase. These interzonal fibers, which stretch between the separated chromosomes, produce what the Germans call "Zwischenkörper." Sally Hughes-Schrader (*Zeitschr. f. Zellforsch.* **13**, 742, 1931) has shown that the number of Zwischenkörper is exactly equivalent to the number of chromosome pairs which have separated. With the central spindle[1] arrangement found in Polychoerus, these Zwischenkörper are peripheral to the continuous fibers of the central spindle, rather than being derived from them.

A second point in relation to vesicles or vacuoles forming along the cleavage furrow is that this was the basis of a theory of cell division proposed by Henry van Peters Wilson in 1938. His theory relating these vacuoles to the cleavage furrow is contained in the abstracts of the National Academy of Sciences for that year (*Science* **88**, 435, 1938).

DR. R. C. BUCK (University of Western Ontario): In electron micrographs I have seen some fibers of the sort illustrated by Schrader, which are quite different from spindle fibers, and look exactly like Schrader's diagrams. I think they are interzonal fibers. In telophase, when you often see an irregular surface on the equatorial face of the chromosome mass, you may sometimes see a long spinning out of material which is rather amorphous, and not so fibrillary as the spindle fibers. I feel that this may represent the interzonal material, which has a different appearance from the spindle fibers in the electron microscope. At least, this is my interpretation. I haven't seen any relation of this material to the midbody. You don't see this material very often. The midbody is related to the continuous fibers.

DR. H. RIS (University of Wisconsin): The Schraders were working with Hemiptera where the interzonal fibers are especially pronounced. In some mitotic figures of milkweed bug testis, I found the meiotic chromosomes to be surrounded by several layers of membranes, apparently the endoplasmic reticulum, of the sort Dr. Porter described in the spindle. When the two homologs separate in anaphase these lamellae are drawn out between them and apparently form something like what is seen in the electron microscope. It is interesting that several workers in the past have found these to be basophilic, and one also finds RNP particles on them, as Dr. Porter has shown this morning. This would agree with the idea that these pieces of endoplasmic reticulum are forming interzonal fibers.

[1] E. B. Wilson (*in* "The Cell" published in 1925) defines the central spindle as "the primary spindle by which the central bodies are connected, as opposed to the 'contractile mantle-fibers' by which it is surrounded. (Hermann, 1891, *Arch. f. mikr. Anat.* **37**, 569–586)." Of course, many cells do not show a central spindle as such, but have continuous fibers and chromosomal fibers intermixed throughout. He also defines midbody (Zwischenkörper) as "a body or group of granules, probably comparable with the cell-plate in plants, formed in the equatorial region of the spindle during the anaphases of mitosis."

Dr. F. Porter (Harvard University): I thought that while we are awarding credits to people who have observed things before, we should mention that John Biesele, I think, described the Zwischenkörper in the electron microscope some 10 years ago. Is that right, John?

Dr. J. Bieselle (University of Texas): Thank you, but I didn't have the intelligence to interpret it.

Dr. J. G. Carlson (University of Tennessee): I wanted to comment also on the interzonal fibers in grasshopper neuroblast. I have been able to demonstrate with a microneedle that there are connecting fibers between the ends of all the separating anaphase chromosomes, so there definitely are interzonal fibers connecting each pair of separating chromosomes. But I have not as yet been able to identify anything that I can call continuous fibers. They may be there. My evidence is of a negative kind. I have not been able to demonstrate them but I can't say that I proved that they do not exist. I would like to ask Dr. Buck what evidence he has that the fibers he showed are continuous fibers rather than interzonal fibers which seem to be present in a great many forms.

Dr. R. Buck: The only evidence is from electron micrographs such as one I showed in early anaphase. As I can see spindle fibers running right from one pole to the other I assume that these must be continuous fibers. They do not connect with the chromosomes.

Dr. J. A. Carlson: Then these fibers do not appear until chromosomes are completely separated at the end of anaphase.

Dr. R. A. Buck: Yes, that's right. They are presumably present before that. I can't distinguish them in the metaphase plate. One would not expect to be able to visualize them. I haven't been able to, at any rate.

Dr. M. Rubinowitz (University of Wisconsin): What happened to those little vesicles that you showed between cells? Isn't there any connection with the formation of the plasma membrane?

Dr. R. C. Buck: I'm sorry I didn't make that clear. My interpretation is that the vesicles fuse together so as to form a continuous membrane, much as Dr. Porter has suggested for his plant cells.

Dr. M. Rubinowitz: But is this the plasma membrane?

Dr. R. C. Buck: Yes.

Dr. M. Rubinowitz: So the ergastoplasm in this case is giving origin to the plasma membrane.

Dr. R. C. Buck: It's a supposition that the ergastoplasm is concerned with the formation of the vesicles. In any case, it is not directly giving rise to the plasma membrane.

Dr. G. M. Mateyko (New York University): I was just wondering about the position of the interzonal fibers. Is there any regularity in their placement as in the case of the chromosomal fibers which are attached to the kinetochore at one end and the pole at the other? I wonder if anybody has studied that connection, that is, whether the kinetochore is directly opposite the interzonal fiber or whether it has another position on the chromosome.

Dr. D. Costello: This is the point I was trying to make. In my material there is a central spindle of continuous fibers. At metaphase the chromosomes are all hooked on the mantle fibers. Therefore, at anaphase, when the chromosomes separate, the interzonal fibers are peripheral to the continuous fibers. At telophase the Zwischenkörper are peripheral, until they are pressed together by the advancing division furrow.

Saltatory Particle Movements and Their Relation to the Mitotic Apparatus[1]

Lionel I. Rebhun

Department of Biology, Princeton University, Princeton, New Jersey and Marine Biological Laboratory, Woods Hole, Massachusetts

I. Introduction

THE PROPERTIES OF CYTOPLASM are often derived from the behavior of included particles, either natural or introduced. That is, conceptions as to whether cytoplasm possesses the properties of a Newtonian or nonNewtonian fluid are often derived from studies of the distribution of path lengths of particles moving under what is usually described as Brownian movement. As we shall review below, there may exist within the same cell at least two classes of particles. One exhibits only what is probably ordinary Brownian movement and the other can, on occasion, exhibit a kind of stop-go movement with long and rapid translatory motions which neither statistically nor casually can be conceived of as Brownian movement (saltations). This latter type of motion is often ignored in treatments of cytoplasmic particle movement because it usually occurs sporadically in only certain classes of particles and often enough, though present at other stages, is greatly accentuated in the presence of the mitotic apparatus. In addition, when they are not undergoing such saltatory movements the motion of particles can often be shown to fit a distribution corresponding to that expected for Brownian motion, and indeed viscosity values can be derived from such distributions (Taylor, 1957). In these cases the discontinuously moving particles yield maximum traverse distances far exceeding anything which can be expected in fluids with the viscosities derived from "Brownian" particles. The mechanism underlying these bizarre movements is obscure. However, as we shall outline below, we feel that with present evidence, these discontinuous movements can best be interpreted as arising from interaction of the given classes of particles with some physically existing substratum which transmits a motive force to them.

We have indicated (and will elaborate below) that saltatory particle

[1] Parts of this work were supported at different times by the National Science Foundation, American Cancer Society, and the National Institute of Health.

movements are influenced by the mitotic apparatus, primarily by the asters. Such influence generally, if not universally, manifests itself in increasing the number of particles moving at any one time and in directing the particle movements into the asters so that at mitosis the latter became surrounded by clouds of particles. It will become clear below, when the details of particle movements are discussed, that the asters are in some way involved in the transport of certain (but not all) classes of particles in some dividing cells, relative to other "stationary" particles and in directing such movements toward the centrosomes. It is thus germane in any study of the mechanism of saltatory particle movements to discuss our knowledge of the structure of the asters as seen in the electron microscope.

We shall present evidence in marine eggs that an important component of the asters is the endoplasmic reticulum (as it is in dividing plant cells as Dr. Porter has shown) and that, indeed, a portion of the endoplasmic reticulum of the cells is oriented radially in the asters. These observations, and others to be discussed below suggest that the endoplasmic reticulum may be the underlying force transmitting structure involved in the stop-go particle movements.

Thus, the general plan of this paper is to present three pieces of observational and experimental work which are reasonably well founded, namely: observations on discontinuous particle movements and their relation to and orientation toward the mitotic centers, observations on the distribution of endoplasmic reticulum during cleavage, and observations on the existence of certain linear moving fibrils in cytoplasmic droplets (Kamiya, 1959; Jarosch, 1956, 1957, 1958, 1960) and their possible relation to the endoplasmic reticulum and then to tie them together in a working hypothesis concerning saltatory particle movements.

Recently Ostergren *et al.* (1960) have presented a view of spindle function in chromosome movement, and in the movement of particles caught in or near the spindle, very closely parallel to that which shall be presented below for astral involvement in particle movements. They have also suggested that the structures involved in such "transport" mechanisms may be part of a general mechanism available to the cell for differential particle movements. Our results support such a hypothesis, although they cannot be said to add *direct* evidence for it.

II. Particle Movements

A. General Remarks

The general characteristics common to all the particle movements we shall discuss below can easily be listed here:

(*a*) The motion undergone by the particles is discontinuous (or saltatory) in the sense that a given particle may move little, if at all, for long periods of time, i.e., many minutes and then, suddenly undergo a 5 to 30 micron translatory and linear motion in a period of 2–5 seconds, terminating as abruptly as it started.

(*b*) Of two particles which can undergo this type of motion, separated by distances of less than a micron, one particle may move and the other remain stationary. On the other hand, groups of 2–5 particles may move as a whole, with total motion having property (a).

(*c*) Only restricted classes of particles in a given cell type may undergo these movements, e.g., "lipid" droplets but not mitochondria in newt heart fibroblasts in mitosis (Bloom *et al.,* 1955; Taylor, 1957), or the particles described in detail below in a variety of cells.

(*d*) In cells showing saltatory particle movements, the number of particles moving varies with the stage of the mitotic cycle but always shows a considerable increase in the presence of the mitotic apparatus which, when present, orients the movements (statistically) toward the centers.

(*e*) Participation in this movement or maximum velocities attained during movement, are not dependent on particle size. That is, where a distribution of particle sizes exists (beta granules of *Spisula,* melanin granules of *Fundulus* melanocytes), no differences in saltatory behavior exist.

(*f*) The particles when moving appear to do so with approximately uniform velocity so that the force applied in overcoming viscous forces, is applied constantly and uniformly along the whole path of movement, i.e., the motion is not one of "jet" propulsion.

B. Specific Observations

We shall now describe the behavior of particles in three separate cases we have directly studied and then briefly discuss similar phenomena reported in other cells by other investigators.

1. *Metachromatic Beta Granules in Eggs of Spisula solidissima (Surf Clam)*

The basic facts concerning vital staining procedures for visualizing the beta particles in *Spisula,* general descriptive results, electron microscope identification of the particles as multivesicular bodies, work of other groups, such as that of Dalcq, Pasteels, and Mulnard, speculations, etc., are covered by Rebhun (1959, 1960). Briefly, eggs treated with dilute methylene blue or toluidine blue in sea water, take up dye in a particulate form. There are some differences in staining results depending upon how the dyes are used (Rebhun, 1959). In the present studies methylene blue was used as in Rebhun (1959), i.e., unfertilized eggs were treated in solutions of 1 part methylene blue to 250,000 parts of sea water for several hours, after which they were washed free of dye. This procedure was checked for effects on cleavage rate, percentage of eggs cleaving, length of development time, etc., and no significant differences with controls were seen as long as the stained eggs were kept in somewhat dim light. In bright light a photodynamic effect apparently causes injury to the eggs.

After this treatment particles can be seen in the unfertilized egg, ranging in size from those appearing as bluish points (i.e., about the size of the Airy disc) to those 1–2 microns in diameter, relatively uniformly distributed in the cytoplasm.

a. Behavior in Unfertilized Eggs. Beta particles in unfertilized eggs show the characteristics listed in (*a*), (*b*), (*c*), (*e*), and (*f*) in part IIA above. With respect to (*c*) observations can be made on other types of particles in the cytoplasm, especially if the eggs are somewhat flattened. Extensive observations have been made on "refractile granules" (probably mostly yolk and lipid) which can easily be seen in the egg and in no case have particles moving with the listed properties been seen other than those marked by the stain.

Interesting phenomena occur in *Spisula* eggs which are allowed to "age," i.e., to remain in sea water for 12–24 hours after staining (at 18–20°C). Such eggs are 95 to 100% fertilizable after this period. Observations of these aged eggs reveal that in most of them the particles are no longer uniformly distributed but are now mostly gathered in one or two (rarely three) large masses pressed against the nuclear envelope, and resembling, closely, asters of the mitotic apparatus in stained cells (Fig. 1) (the germinal vesicle normally breaks down only after fertilization in this species). That is, the particles tend to aggregate in spherical clusters, excluding a more or less clear center, and the particles, although retaining movement characteristics (*a*), (*b*), (*c*), (*e*), and (*f*) of part IIA, now have their movements directed more or less radially with respect to these masses. However, particles some distance from the masses, e.g., on the opposite side of the nucleus (see arrows, Fig. 1), do not appear to be influenced by the centers in short term movements, with respect to the direction of this movement. Entire regions have been seen to fuse or single regions to split into two (the latter especially upon fertilization and prior to germinal vesicle breakdown).

b. Behavior in Fertilized Eggs. Detailed descriptions of the behavior of the particles after fertilization have been given in Rebhun (1959). Briefly, the randomly directed saltatory movements begin to become oriented relative to the asters of the first polar body spindle, which forms in the center of the egg after the germinal vesicle breaks down (starting about 10 minutes after fertilization) and migrates to a position at and perpendicular to the egg surface at about 20 minutes after fertilization. This orientation of particles is, at first, variable in extent in different eggs. As the events of meiosis (formation of polar bodies) proceed, more and more particles enter the astral regions, until by the end of second polar body formation, few if any of the particles may be seen in regions other than those surrounding the central aster.

As the female pronucleus forms and migrates centrally, it carries the massed particles with it on its centrally directed pole (Fig. 2). As the male and female pronuclei come into contact, the particles form a (usually) complete ring in the plane tangent to this contact point. This ring soon divides into two arcs which then reform to two partial spheres surrounding the centrosomes of the first cleavage spindle (Fig. 3).

After first cleavage, the particles do not disperse, but remain aggregated about the peripheral poles of the blastomere nuclei. Just prior to nuclear breakdown and formation of the second cleavage spindle in both the AB and CD cells, the mass of particles divides into two masses which migrate in opposite directions over the nuclear surface, each mass now outlining a centrosome of the second cleavage spindle. The mass outlining the peripheral pole of the excentrically located spindle in the CD cell is smaller than that outlining the central pole. The masses about the ends of the centrally located spindle of the AB cell are of approximately the same size (Fig. 4). These events are repeated in essentially the same form at least through fifth cleavage and most likely beyond.

We have spoken above of an aggregation of particles as a whole. We wish to emphasize again that the individual particles appear to move independently and in saltatory manner and that particles, once in the astral regions, need not necessarily remain there. Indeed, they continually move out radially from the centers, although, fewer do so and less frequently than those moving in. Also, movements away from the astral centers are rarely for distances of more than about 5 microns. However, during a part of the interphase, e.g., between first and second cleavage and between second and third cleavage, the particles may show great activity (i.e., many more than usual participate in the movements) which manifests itself as shuttling of particles radially between the blastomere nucleus and the internal surface of the egg, again with movement characteristics listed above as (*a*), (*b*), (*c*), (*e*), and (*f*) in part IIA.

The number of particles participating in the movements at any one moment appears to vary with the stage of the mitotic cycle the cell is in, as indicated above with respect to an increase in the middle part of interphase. Another period of considerable activity in all cleavages observed, is that after spindle translation to the periphery and just prior to and during cell elongation, corresponding to late metaphase or early anaphase.

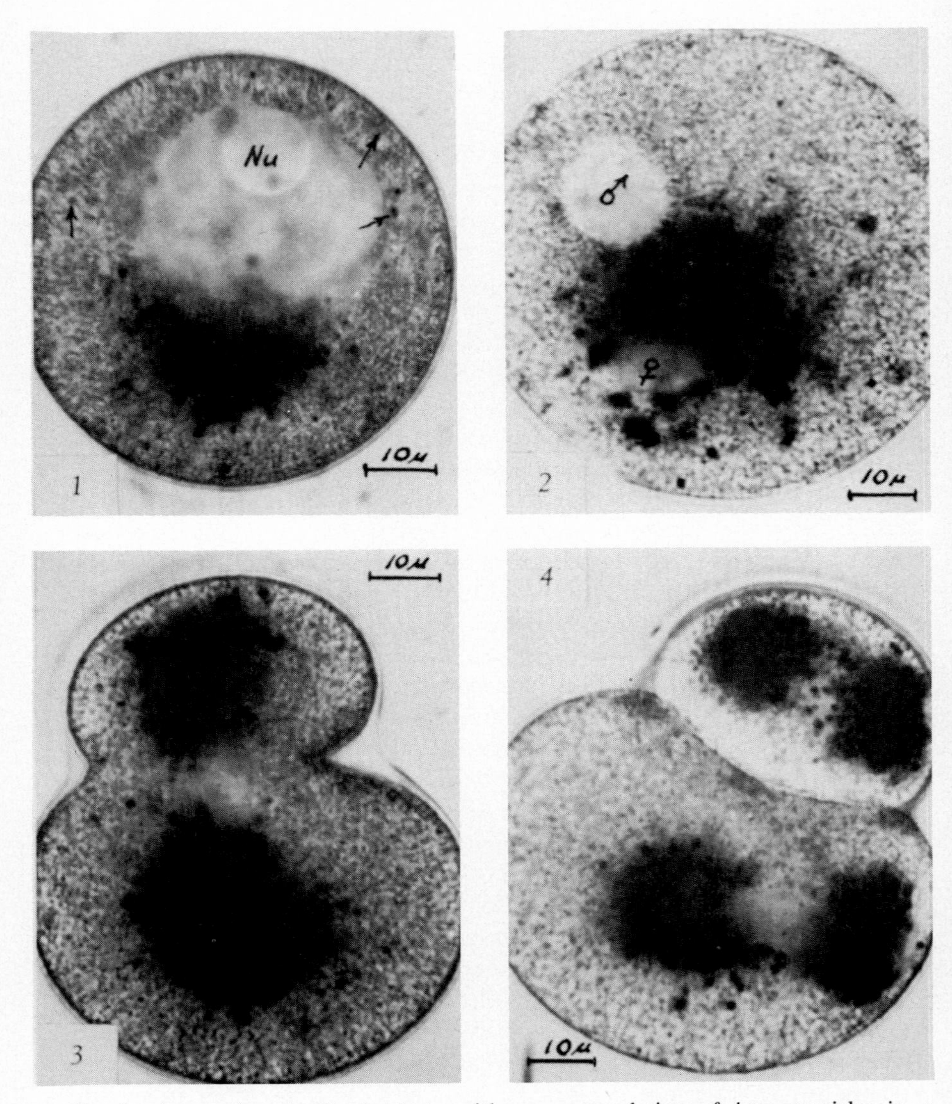

FIG. 1. An aged unfertilized oöcyte with an accumulation of beta particles in a single juxtanuclear mass. Arrows point to stained particles not in the mass. Note the nucleolus, Nu. *Spisula.* Bright field microscopy.

FIG. 2. A fertilized egg at about 50 minutes post-fertilization age. The mass of beta particles is gathered at the central pole of the female pronucleus. Syngamy takes place about 2–3 minutes after this stage. *Spisula.* Bright field microscopy.

FIG. 3. A fertilized egg at midcleavage. *Spisula.* Bright field microscopy.

FIG. 4. A fertilized egg at beginning of second cleavage. Note eccentrically located spindle in CD cell and centrally placed spindle in AB cell. *Spisula.* Bright field microscopy.

The particles which are not close to the asters at this time rapidly move in, the impression, in time lapse films being that a suction has suddenly been turned on with the centrosome as center.

It should also be pointed out that particles, in much smaller and variable numbers, can be strung in lines parallel to the spindle axis and on the spindle surface. Such particles participate in the motions we have been discussing and eventually move closer to the asters, with saltatory and individual movements, especially during anaphase. In many cases, linear strings of particles may move from one pole to the next across the spindle to be followed seconds later by a string moving in the opposite direction (see also Ostergren *et al.*, 1960).

c. Stratified Eggs. *Spisula* eggs both fertilized and unfertilized may be stratified by application of centrifugal forces of the order of 10,000 g for approximately 5 minutes. If stained unfertilized eggs are subjected to such a regime, the particles may be seen to gather in a layer just centripetal to that occupied by the mitochondria and centrifugal to the germinal vesicle. Observations of these particles show that they undergo the same characteristic, saltatory motions after centrifugation as before.

Light microscope observations on eggs stratified after germinal vesicle breakdown reveal the egg contents to be separated into six layers; from centripetal to centrifugal poles these are lipid cap, upper hyaline zone (to which pronuclei stratify if they are present), lower hyaline zone distinguished from the previous zone by a difference in optical refraction, mitochondrial layer, yolk layer, and cortical granule layer (usually, not all the cortical granules stratify under our conditions). In stained eggs the beta granules gather primarily in a narrow layer separating the upper and lower hyaline zones. If stratification has been accomplished at times when a mitotic apparatus is present in the cell, the particles may be seen to move from this layer into the asters by saltatory motion within 10–50 minutes after centrifugation and often form very dense aggregates about the centers. The occurrence of this redistribution is dependent on the stage at which stratification is performed and although the detailed work remains to be done, it appears that redistribution is most likely to occur in eggs stratified at about metaphase of first polar body, second polar body, or first cleavage divisions.

2. *Yolk Granules in Egg of the Annelid, Cistenides*

The particles discussed in the previous section could be seen to be involved in unorientated saltatory movements in the absence of a mitotic apparatus, the latter organizing and orienting, but not altering the character of these movements when it appeared. The yolk granules in the egg of *Cistenides gouldi* show the saltatory movements sporadically in the unfertilized egg, but such movements become extensive in the presence of a mitotic apparatus. Indeed, many individual particles begin the saltatory movement at the time of first polar body spindle formation, about 10 to 15 minutes subsequent to shedding when the germinal vesicle breaks down (this is highly variable and some batches of eggs will retain the germinal vesicle until fertilization). At the time of first polar body formation the number of yolk particles involved in these movements is substantial and the distances traversed can be up to 20 microns, but it is only after formation of the asters of first cleavage spindle, concomitant with pronuclear breakdown and fusion that the yolk becomes extensively involved in the saltatory movements. The characteristics of these movements are those listed as (*a*) to (*f*) in part IIA above. In addition, in the early stages of spindle formation just subsequent to pronuclear breakdown, yolk particles can often be seen moving at angles to astral radii though the most prevalent direction of motion is radial. Such transverse movements do not occur after a recognizable spindle is seen.

Movement velocities appear to be in the same range as those of beta particles in *Spisula,* i.e., up to 5–10 microns per second. However, in *Cistenides* much of the yolk is still not astrally located at the time at which the maximal number of particles move (i.e., at about late metaphase to early anaphase) so that many more particles than in *Spisula* undergo long translations (20 microns or so) at the critical periods and the effect is quite dramatic.

After cleavage, the blastomere nuclei reform by fusion of karyomeres. The yolk particles in *Cistenides,* unlike the beta particles in *Spisula* do not remain aggregated at the peripheral pole of the blastomere nucleus, but scatter in the cytoplasm and usually, concentrate to some extent in the periphery. Here, the movements partially subside in that fewer particles move. The movements of those that do move are again saltatory and of about the same velocities as before but are now undirected. From prophase of second cleavage on to early metaphase the presence of asters again appears to influence the number of moving particles and their direction of motion and groups of them may move into the asters relative to others which are stationary at the movement. By late metaphase most of the yolk particles have left the periphery and are in or near the asters. From late metaphase to early anaphase the particle movements are extensive and dramatic and can easily be followed by direct visual observation with the light microscope. They result in a tight ingathering of yolk granules in the asters (excluding the centrosomal areas) (Fig. 5).

Close observation of the cytoplasm in these very transparent and small eggs (they are prolate spheroids 55 by 25 microns before polar body formation and round up to about 40 microns in diameter afterwards although they are often compressed to 70–80 micron diameters for observation and movie making) reveal a number of other inclusions. The most easily identifiable of these are minute particles appearing as dark points in the cytoplasm (with dark contrast phase microscopy) undergoing rapid motion which appears to be "Brownian," although we have not checked this statistically. However, in no case have any of these particles ever been seen to move in saltatory fashion and they show a uniform distribution in the cytoplasm not occupied by the mitotic apparatus. Close observation shows the yolk particles to move through areas occupied by such dustlike particles with no coupling of motion to them, just as yolk particles themselves may move relative to one another when less than a micron apart, with no apparent influence upon each other.

The yolk particles continue the sequence of motions described above until at least fourth cleavage beyond which we have not specifically studied them (third cleavage spindles and yolk orientations may be seen in Figs. 6, 7, 8). In all cases the characteristics (*a*) to (*f*) above are observed. Interphase of later cleavages (which, at least until the fourth is at most 15 minutes long) is characterized by saltatory particle movements, extensive, though reduced in the number occurring per unit time as compared to that of mitotic stages, and unorganized in direction.

3. *Melanin Granules in Developing Melanocytes of the Top Minnow, Fundulus*

Our observations have been primarily on embryos of 3–4 day post-fertilization age. In such embryos the melanocytes have spread over the yolk sac and are still generally flat although some have begun to associate with blood vessels and to develop projections in several planes. The melanin granules are primarily about 1 micron in diameter, although it is not uncommon to see granules very much smaller than this.

Melanin granule movements in these nondividing eggs are easily seen and quite striking and again show properties (*a*), (*b*), (*e*), and (*f*) of part IIA similar to the movements shown by beta granules in unfertilized *Spisula* oöcytes [we could not check (*c*) since no other granules were visible in the cytoplasm with our techniques]. The melanocytes at the age studied are quite densely packed with melanin granules. In addition to the general characteristics (*a*), (*b*), (*e*), and (*f*), there are some distinctive ones which we shall now discuss.

While it is true that the saltatory movements are extensive, the number of particles involved varies greatly from cell to cell, and, indeed, even within different regions of one cell. Thus, in a given cell one may find areas many microns in diameter in which very few particles move and adjacent to them, areas with considerable activity occurring. In the latter

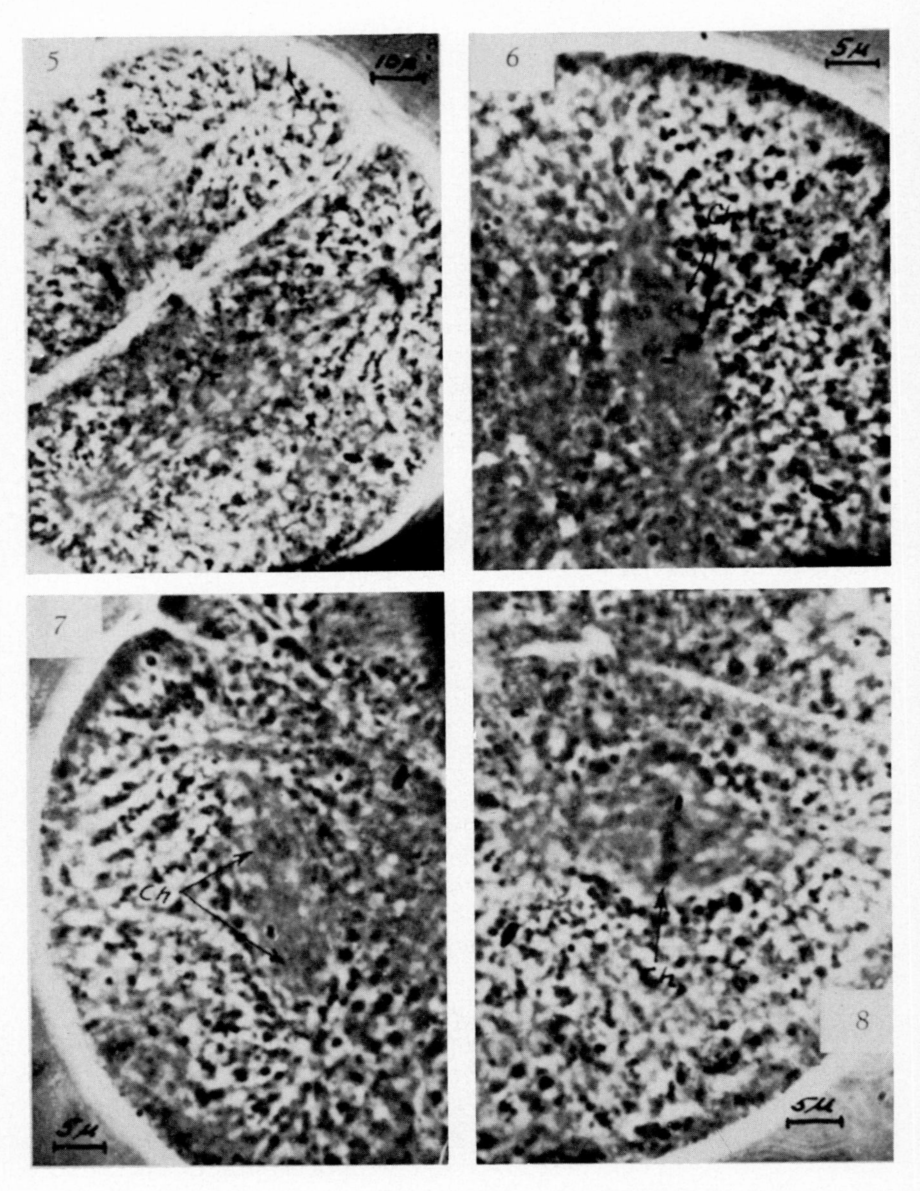

Fig. 5. Beginning second cleavage in a *Cistenides* egg. Note radial arrangement in the dark granules (yolk) in the asters. Dark contrast phase microscopy.

Fig. 6. Spindle in blastomere at beginning of third cleavage. Note, again, the radial arrangement of yolk. *Cistenides* egg in early anaphase. Chromosomes marked, Ch. Dark contrast-phase microscopy.

Fig. 7. A stage similar to that in Fig. 6 *Cistenides* egg with blastomere in mid-anaphase. Dark contrast-phase microscopy.

Fig. 8. Another *Cistenides* egg in similar stages to those of Figs. 6 and 7. A blastomere at metaphase. Dark contrast-phase microscopy.

regions, a melanin particle may be seen to readily move distances of 10–15 microns passing within less than a particle diameter of many melanin particles along its path without influencing their movement in any discernible fashion. In addition, long blocks of melanin granules may begin to move in unison, only to have individual granules often in the center of the moving group, move suddenly in a direction 180° opposite to that of the group as a whole.

Older melanocytes begin to develop thin cytoplasmic extensions, often 20–25 microns long and 2–3 microns wide. Granule movements in such pseudopods often result in particles rapidly moving past each other in opposite directions. Such opposed saltatory motions also appear in the main body of the cells, occasionally giving rise to 4–5 parallel, counter flowing streams of total breadth, less than 10 microns.

The direction of movements is probably not completely random since they appear to occur more often in a direction radial to the surface than in any other specific direction, but this has not been checked statistically (a formidable task since each cell contains thousands of moving melanin granules). Observations of nuclei in these cells did not consistently reveal orientation of movements toward them but on occasion, in some cells, such direction clearly existed.

C. Discussion of Particle Movements

We have discussed three different types of cells which we have directly observed. In these certain classes of particles were seen to undergo motions always having the properties of sudden onset and termination of periods when velocities increased manyfold and in which traversed distances could be up to 30 microns for a single, continuous movement. In unfertilized *Spisula* eggs, beta granules, in *Cistenides,* yolk granules in interphase of later cleavages (and to a small extent in the unfertilized egg), and in *Fundulus* melanocytes, these movements occurred in the absence of a mitotic apparatus. In both *Spisula* and *Cistenides* these movements were highly influenced by the presence of asters both as to number of particles moving and direction of movement. In *Fundulus,* no visible organization was correlated with areas that did and those that did not show extensive granule motion. We wish now to show that some of these considerations extend to a rather wide class of subcellular particles and of cell types other than eggs, although we shall start with some further phenomena in eggs.

The first we wish to consider are echinochrome granule movements in unfertilized *Arbacia* eggs. The basic properties of these movements are again, those listed in (*a*), (*b*), (*c*), (*e*), and (*f*) of part IIA. As for (*c*), the remaining particles, yolk, etc., can be clearly seen, especially with the television microscope (Parpart, 1953; Parpart, personal communication and direct observation). No other visible particle participates in the saltatory movements in the unfertilized egg and, indeed, the movement of echinochrome granules, appearing to push their way past relatively stationary yolk, is startling.

The events on fertilization again indicate the ability of processes in the cell to organize the saltatory movements of granules. In this case, the granules begin to move individually and sporadically to the cortical regions (i.e., toward the surface) and after about 10 minutes up to 80% of them have so migrated. After migration their motions subside. The remaining, interior particles become astrally located at cleavage spindle stages.

There are three other important observations concerning these granules which should be mentioned. One is that the motions are strongly temperature dependent and the echinochrome granules reversibly all but cease movement at about 10°C (Parpart, 1953). A second and very important observation has to do with the centrifuged eggs. In stratified *Arbacia* eggs the echinochrome granules go to the most centrifugal pole of the egg and the saltatory motions cease (Parpart, 1953). Further observations on the stratified echinochrome granules reveal that after a period of 60 to 80 minutes, the granules begin saltatory movements again (Parpart, 1961). The third observation has to do with partially fertilized eggs. *Arbacia* eggs may be sucked into capillaries and drawn out to lengths of several hundred microns (normally they are spheres approximately 75 microns in diameter). Such eggs may be activated from one end in such a way that the cortical granule reaction travels over only a part of the surface pressed against the capillary wall. In these eggs echinochrome granule movement to the cortex takes place only within the end of the cell which has been activated and for at most 25 microns distal to the last portion of "activated" surface (Allen and Rowe, 1958). That is, the cellular event which appears to be mostly directly associated with organizing the individual and saltatory movements of the echinochrome granules is the modification of the cortex involved in preparing the fertilization membrane.

A last case in eggs which is important though, unfortunately, not reported in detail is that of *Urechis* eggs as studied by Taylor, 1931. Taylor reported on the movement of certain bright red granules in the cytoplasm, identified by Horowitz and Baumberger (1941) as heme containing granules, relative to the asters. He describes these bodies as "sliding down the astral rays." The totality of his description makes us feel that the movements he describes are undoubtedly of the character we have been describing.

In cells other than eggs, the most elaborately studied case appears to be that of the "lipid" granules in newt heart fibroblasts in tissue culture, as studied by Bloom *et al.* (1955), Bloom and Zirkle (film, undated: see reference list), and especially by E. W. Taylor (1957). Taylor was interested in studying viscosity changes in the fibroblasts during the mitotic cycle. Since the cells are quite flat, he was able to make good statistical studies of particle movements in the cells. He found that the majority of the lipid particles had movements which statistically fit that expected from two dimensional Brownian processes. Occasionally, however, the lipid granules underwent saltatory movements with traverse distances so large that they were essentially impossible if the assumptions were made that the cytoplasm is Newtonian and the viscosity value is that derivable from the motions of the other lipid granules. Furthermore, the lipid granules in many cases show increased numbers in saltatory motion in the presence of asters; they move into the asters with the same individual and saltatory motions as described for *Spisula* and *Cistenides,* but also with the exceptions that occasionally groups of particles move as a whole.

There can be little doubt that the same over-all verbal description can be used for lipid granules in newt fibroblasts as for yolk granules in *Cistenides,* or beta granules in *Spisula.* Namely that a class of particles, showing saltatory movements in interphase experiences an increase in their numbers undergoing saltations in the presence of asters, and that such movement takes place radially and predominantly but not exclusively, toward the astral centers, resulting in an accumulation of the particles in these regions.

A probable second case in tissue culture is that of the "microkinetospheres" reported by Rose (1957) in Hela cells. The description of the motion of these particles is not complete enough to judge with certainty, but since they are described as moving with sudden motions either with or against the stream of material entering the cell by pinocytosis and destined for the region of the cell center, it would seem reasonable to assume that a fuller analysis of the particle movements would reveal the characteristics of saltatory motion we have been discussing. This seems especially so to us since films of fibroblasts which we have seen, have occasionally shown particles with saltatory behavior. It would not be amiss to predict that were such motions looked for in cells with time lapse speed-ups of 10–16 times rather than 100–500 times, the phenomenon would be more widely observed.

In protozoa, autonomous movements have occasionally been reported, e.g., in *Amoebae* (Allen, 1961), *Stentor* (Andrews, 1955; Andrews quoted by Seifritz, 1952), and, (if one may call it a protozoan) the slime mold, *Physarum cephalum* (Stewart and Stewart, 1959) where the description of particle movements is extensive and similar to that given above. Indeed, in *Physarum,* such motions were observed in both "stationary" ectoplasm and moving endoplasm. No difference in viscosity could be found between the two regions of the organism using particle movements as indicators of viscosity. However, it isn't clear as to whether saltatory particles were used in these calculations.

Finally, we shall mention another "probable" case, that reported by Holt (1957) (see also Fig. 4 in Holt's article in Danielli, 1958) and partially repeated by Dr. A. Novikoff (1961) and in our laboratory by Mr. William Dougherty. In this work, Holt showed with acid phosphatase and esterase techniques, that particles positive for these enzymes and which ordinarily line the peribiliary regions of the liver parenchyma cell, leave these regions and accumulate at the ends of the spindle (and to a smaller degree at the equator) during mitosis induced by partial hepatectomy. Novikoff (1961) and Dougherty have found this to be a somewhat variable phenomenon in adult liver but the former has shown it to be consistent in young postnatal rats. The striking similarity to the phenomenon in eggs of the orientation of these particles relative to the spindle ends and, indeed, of only these particles (e.g., mitochondria do not show this orientation) is very strongly suggestive of the identity of this process with those in which observations of the living cell have been made as discussed above.

The totality of evidence we have reviewed (for similar phenomena in plant cells, see review in Kamiya, 1959; Seifritz, 1952) suggests that the phenomenon of saltatory particle movement is of widespread occurrence in cells and has at least the following two important properties: (1) in a given cell type it may occur in only restricted classes of particles; (2) the motion may become oriented relative to the mitotic apparatus and in the presence of the latter results in the nonuniform distribution of the affected particles into the asters.

It should be pointed out that more than one class of particles may show these properties in a given cell. For example, in *Spisula,* there are a set of cortical granules which do not break down upon fertilization and which do not normally undergo saltatory movement. However, when the peripheral aster of the first (or later) cleavage spindle comes near the surface, the cortical granules near it leave the surface and "slide down the astral rays" and behave like the beta granules with which they comingle. Despite the fact that more than one class of particles may participate in the movement, (1) above still has considerable force since observations reveal that other classes definitely do not so participate.

We will not review the hypotheses which have been suggested to "explain" these phenomena but will refer the reader to the short summaries in Seifritz (1952), Kamiya (1959), and Andrews (1955). Suffice it to say, that the fact that a relatively large number of particles, many of which can be described as relatively inert, e.g., pigment granules, yolk granules, heme granules (with the reservations for the latter mentioned above), lipid droplets, etc., undergo similar kinds of motion relative to other particles in the same cells (e.g., mitochondria in newt fibroblasts) which do not, plus the specific characteristics of the movement, (*a*) to (*f*) of part IIA, suggests strongly to us that the term "autonomous" movement for this phenomenon is a misnomer, and the neutral, descriptive term, saltatory, should be used. Indeed, we subscribe to the general body of opinion (see, e.g., Seifritz, 1952; Andrews, 1955) which suggests that the particles are moved by some agent external to the particles themselves and which is able to transmit some motive force to them. We further suggest that the agent is probably some organized, partly connected, physically existing component of the cell which can be strongly oriented relative to the mitotic centers. The next section will tentatively suggest that this system is, indeed, associated with the endoplasmic reticulum in eggs and evidence will be presented there for this contention.

We cannot, however, leave this section without some description of the observations of saltatory movement in extruded droplets of *Chara* and *Nitella* cytoplasm by Jarosch (1956, 1957, 1958, 1960) and Kamiya (1959), since these movements have been related to another component of the cytoplasm which appears to have surprising properties. Basically, the observations are the following: particles in the cytoplasmic drops occasionally undergo saltatory movements such as described above. When these drops are observed by dark-field microscopy, thin, filamentous rods may be seen which appear to be in rapid translatory movement parallel to their long axes. Such filamentous rods may aggregate temporarily and disperse, the aggregates moving with lower velocity. Such aggregates may take the form of polygons which rotate, the whole figure being built up and gradually broken down by filaments adhering to and then dispersing from the figure. In addition, currents may be seen in the immediately adjacent cytoplasm, which go in a direction opposite to the direction of movement of the filaments (or rotation of the polygons). On occasion, cytoplasmic particles may be seen to intercept the filaments, in which case they are stabilized relative to the slight Brownian movement they normally undergo and are rapidly translated to the end of the filament in a direction opposite to the direction of filament movement, after which they resume their normal, slight Brownian movement. That is, their saltatory movements are related to very close proximity to the moving filaments (Kamiya, 1959).

We have been assuming that saltatory movement with the characteristics listed in a previous section is basically the same phenomena wherever it is found in cells; that is, we assume that the mechanism of movement is the same for all cells. If this hypothesis is correct, it is clear that the observations described in the previous paragraph take on considerable meaning in suggesting that filamentous structures capable of moving the medium surrounding them (and particles in it) may be of widespread occurrence in cells in general. Jarosch (1956, 1957, 1958, 1960) and Kamiya (1959) have used such ideas to explain diverse phenomena such as cyclosis (filaments built into the stationary ectoplasm), nuclear roto-oscillation (filaments built into the nuclear envelope), chloroplast revolution (filaments built into the chloroplast surface), etc., and, indeed, have amassed observational evidence for these contentions. However, direct evidence for the existence of such structures in other cells is not at hand which considerably weakens the hypothesis. We feel, however, that the totality of evidence presented above (and to some extent later) makes this hypothesis of the propulsion of particles by filaments capable of exerting shearing forces at their surfaces, the most attractive one at present. We shall return to this discussion after the next section.

III. Electron Microscope Studies of Dividing Eggs

The previous sections contain the evidence which makes us suggest that saltatory particle movements in cells are probably caused by forces exerted on particles by some underlying physical structure. The minimum requirements that such a structure should possess are that it show distributional changes during the middle cycle similar to those shown by the particles. That is, (1) it should be pervasive in penetrating to all parts of the cytoplasm, (2) it should show continuity over distances of up to 20–30 microns, at least in some parts of its distribution, (3) it should be relatively unoriented in interphase cells and finally, (4) it should show an orientation radial to the centrosomes in cells with a mitotic apparatus. The burden of the next section is to show that the endoplasmic reticulum (ER) in *Spisula* and *Cistenides* possesses the requisite properties. We shall then attempt to bring other more indirect evidence to bear on the hypothesis that the ER is, indeed, the force-transmitting structure involved in saltatory particle movements.

A. Techniques

The fixation techniques used with marine eggs have been discussed elsewhere and consisted primarily of the use of either Dalton's or Palade's fixatives (Rebhun 1960, 1961b), a 5% potassium permanganate solution buffered to pH 7.4 with 1% potassium dichromate (Rebhun 1961c), or, freeze substitution, Fig. 9 (Rebhun 1961a), and embedding in Araldite 502 (except Figs. 13 and 16, where the material was embedded in metha-

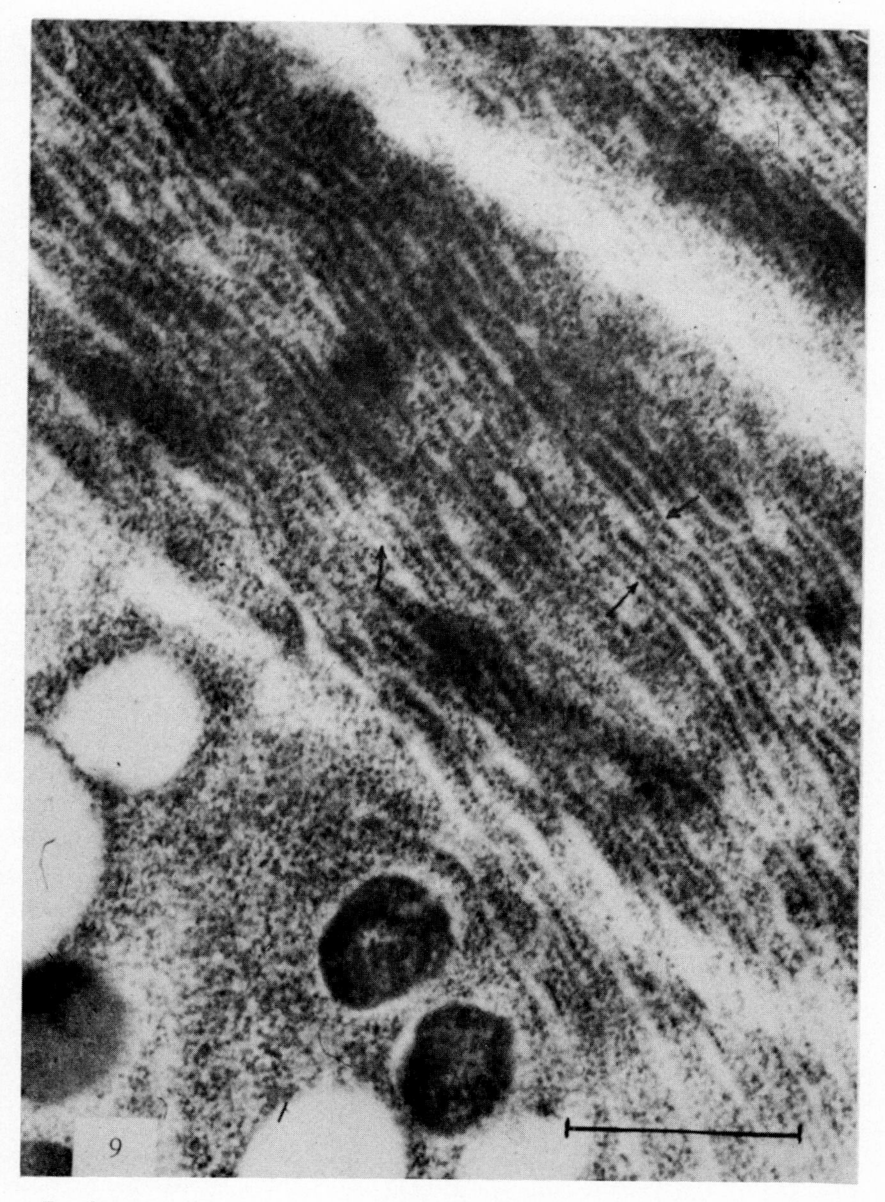

FIG. 9. A parallel array of cisternae of the endoplasmic reticulum in a *Spisula* oöcyte. Note the double membrane sandwich (arrows) limiting the cisternae and the 150 Å particles throughout the cytoplasm (presumably all ribosomes). Prepared by freeze-substitution, stained with potassium permanganate after sectioning (Rebhun, 1961a).

crylate). Most of the material was viewed with a Hitachi HS-6 microscope although some of it was viewed with an RCA, EMU 2d microscope.

B. Observations

1. *Endoplasmic Reticulum in the Unfertilized Egg*

The distribution of ER in *Spisula* eggs has been dealt with at some length in Rebhun (1956a,b, 1961b). Briefly, the ER, elements of which penetrate to almost every cubic micron of the egg, appears to consist of three morphological forms; annulate lamellae, individual cisternae whose profiles may be up to 15–20 microns in length, and smaller flat or vesicular elements. In addition, the cisternae can exist stacked as parallel sheets or arranged in whorls. The three types of elements are often morphologically interconnected and a model for the details of such connections have been presented (Rebhun, 1956b). With permanganate fixation similar elements can be seen. The most evident components of the ER in the cytoplasm in these preparations are the cisternae (as can be seen in Fig. 10) of a partially centrifugally stratified egg. Permanganate fixation does not preserve nucleic acids of the cytoplasm (Jansen and Molemaar, 1961, Luft, 1956) and thus the ribosomes are missing. Other techniques must be used to visualize these. Figure 9 is a micrograph of a frozen-substituted *Spisula* egg and shows a stack of parallel cisternae. It can be seen that each membrane of the ER is a double-layered sandwich about 75 to 100 Å thick and that ribosomes coat the outside layer. The elements of the ER in nonstratified eggs are seen to be randomly oriented except that larger cisternae or stacks of them appear to show a preferred orientation parallel to the surface layer (or nucleus). This is undoubtedly a mechanical orientation deriving from the fact that the cisternae are often longer than the 15 micron wide space between nucleus and egg surface. Figure 11 is an electron micrograph of a stratified fertilized egg at about first cleavage, showing the upper hyaline zone which again illustrates the general form of the cisternae present in these eggs and may be compared with Fig. 10 to illustrate the general similarity of these structures in fertilized and unfertilized eggs.

Figure 20 is an electron micrograph of an unfertilized egg of *Cistenides* and shows some elements of the ER. They are long cisternae (C), parallel arrays of long cisternae (C), and short pieces of cisternae (SC). The nature of the small vesicles (V) we do not know. Again as in *Spisula*, there is a preferential orientation of long, but not short, cisternae parallel to the nucleus and the surface of the eggs and as in *Spisula* almost every cubic micron of the egg has some representative of the ER in it.

In summary, unfertilized eggs of both *Spisula* and *Cistenides* have extensively deployed elements of the ER in the cytoplasm. The shorter elements of the ER show no preferential orientation in the cytoplasm, but the longer element, some of which may reach 20 microns in length, are generally parallel to the nuclear and egg surfaces, probably for geometrical reasons, although possibly because of their mode of origin from the nuclear envelope.

2. *Changes on Fertilization*

The eggs of *Spisula* are obtained in the germinal vesicle stage and do not undergo further developmental processes unless activated by either

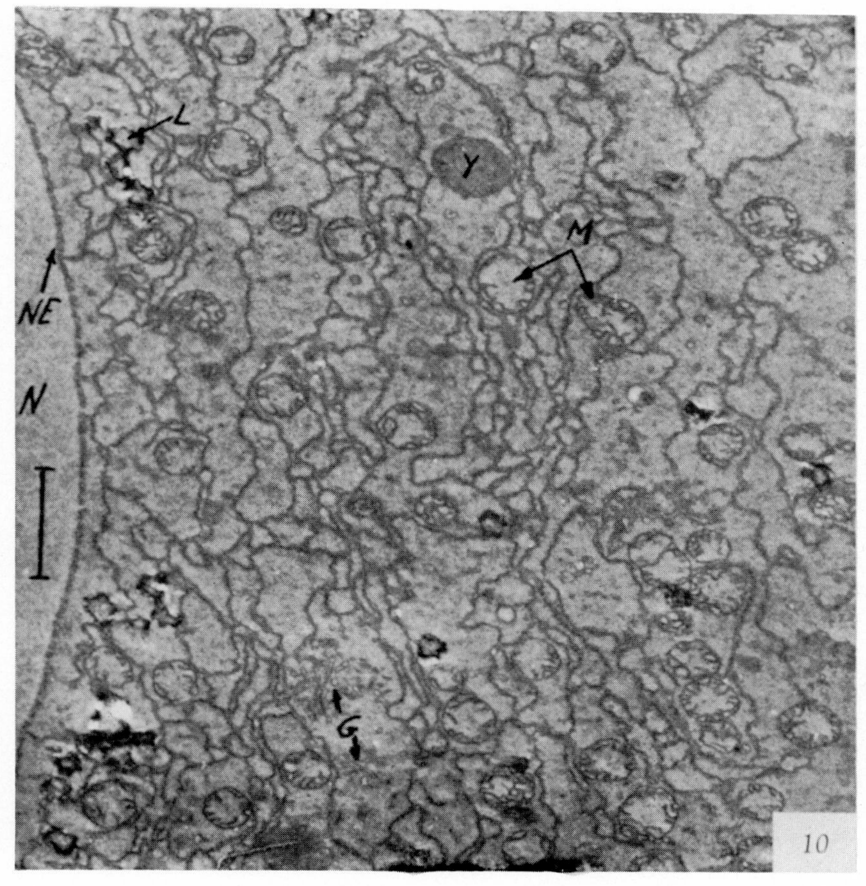

FIG. 10. A mass of cisternae of the endoplasmic reticulum in a partially centrifugally stratified, unfertilized *Spisula* oöcyte. The apparent orientation of the endoplasmic reticulum is due to the application of centrifugal force. Mitochondria, M; Golgi bodies, G (partially leached) ; lipid droplets, L; and yolk particles, Y, are marked. Permanganate fixation.

natural or artificial means. *Cistenides* eggs are obtained in the germinal vesicle stage and depending on the female they are taken from, may undergo no further change until stimulated or may progress to metaphase of the first maturation division. For both animals, however, the events up to the formation of the zygote nucleus are similar. Thus, in both, the first polar body spindle forms in the center of the egg and then migrates to a position perpendicular to the surface. The polar bodies form and the female pronucleus migrates to the center of the egg. No true female

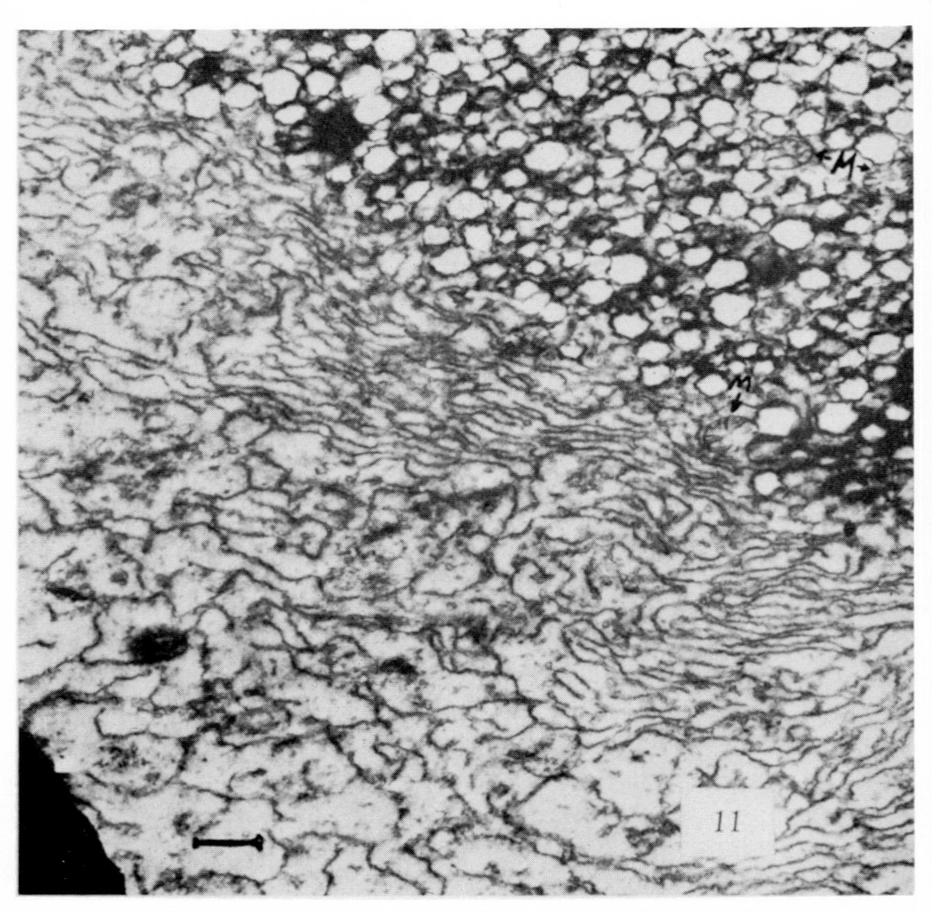

FIG. 11. The upper hyaline zone and lipid layer of a fertilized *Spisula* egg at about first cleavage. The lipid droplets are almost completely leached out. They show great variability in different blocks of permanganate fixed cells. Note that the cisternae are parallel to the lower border of the lipid layer with their surfaces perpendicular to the direction of applied centrifugal force. An occasional mitochondrion can be seen in the lipid layer.

pronucleus is formed in *Cistenides,* however, and in this organism, the chromosomes form karyomeres which begin to fuse with each other and with the male pronucleus to form the zygote nucleus. In *Spisula,* a conventional female pronucleus is formed. At the next stage in both eggs the zygote nucleus breaks down as the first cleavage spindle forms in the center of the egg. The spindle undergoes a sudden translatory motion to the periphery and cleavage then ensues giving rise to a larger CD and smaller AB cell (see Fig. 17). Further cleavage events are similar to those of most spirally cleaving eggs (see, e.g., Raven, 1958).

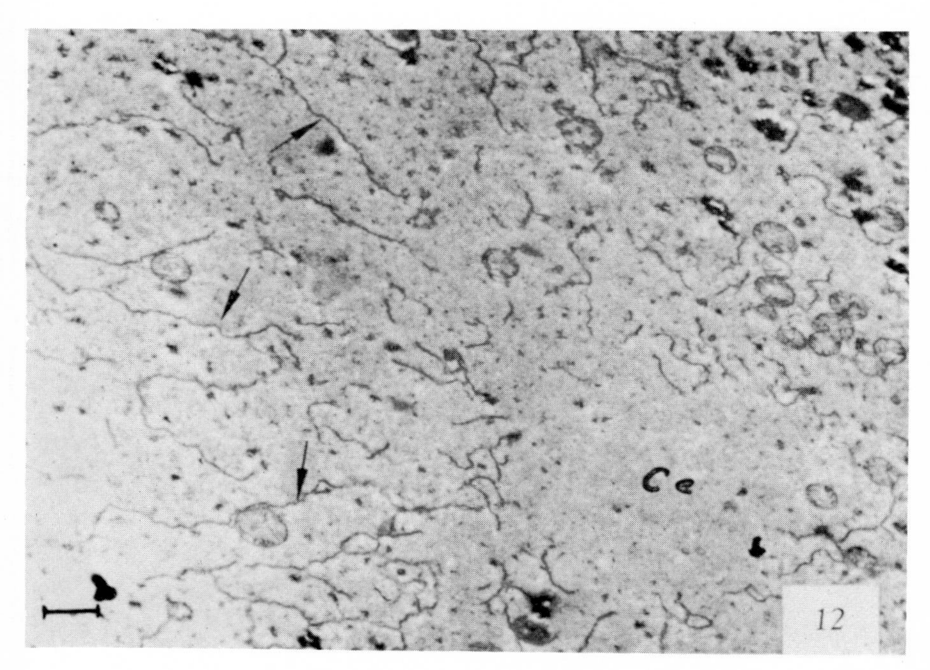

FIG. 12. Electron micrograph through the centrosome (Ce) of *Spisula* egg just subsequent to the formation of the first polar spindle. Note the approximately radial orientation of cisternae. Fixation with permanganate.

The periods when we would expect to find strong orientation in any system responsible for saltatory particle movements are those during which the mitotic apparatus is well on its way to being formed or in the case of *Spisula* during interphase of later divisions, since, as we have seen in the latter, the beta particles remain at the peripheral pole of the blastomere nucleus and show radial saltatory movements between nucleus and cell surface. We have, therefore, examined eggs fixed at periods just after first polar body spindle formation, just after zygote nucleus formation and breakdown, at first and second cleavage and in *Spisula* during the interphase between first and second cleavage.

Figure 12 is a micrograph of a permanganate-fixed *Spisula* egg at about 15 minutes after fertilization when the spindle and asters are already well formed. The clear area is the centrosomal region (Ce) and

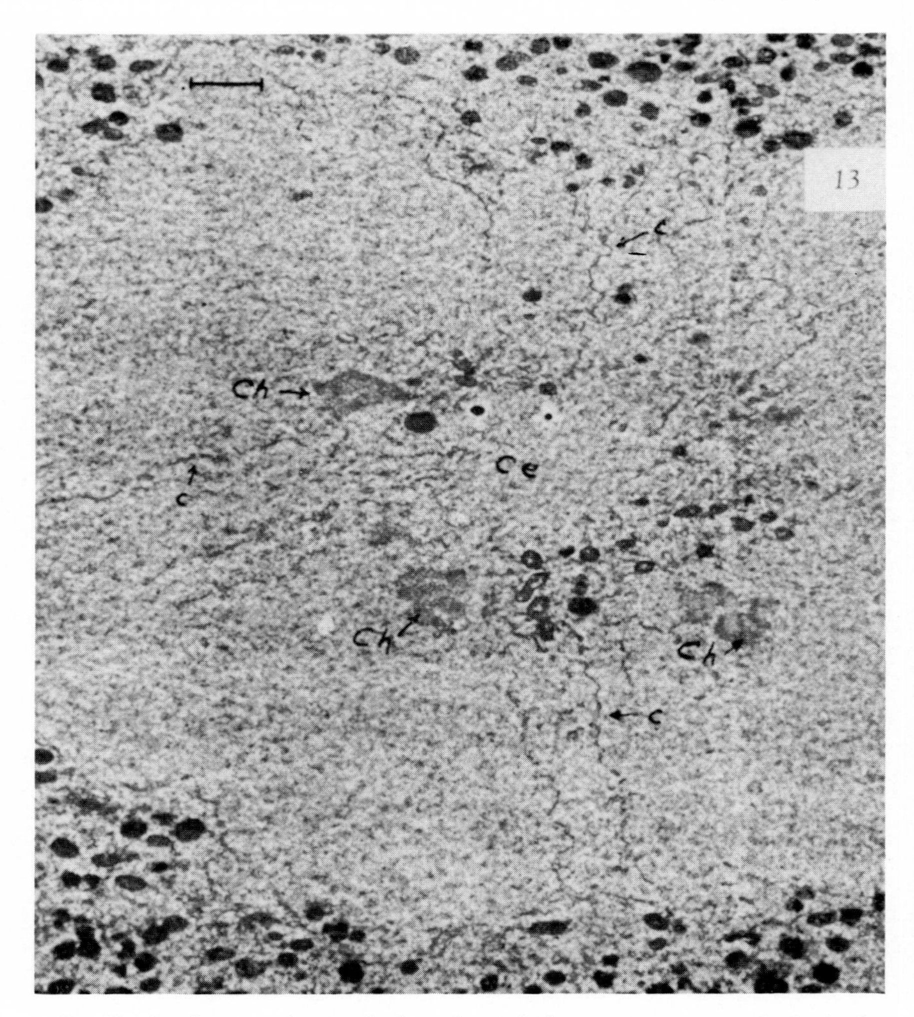

Fig. 13. An electron micrograph through a *Spisula* egg centrosome at the beginning of first polar body spindle formation. The chromosomes are near the spindle poles prior to the establishment of metaphase. The nucleoplasm of the germinal vesicle has not yet dispersed in the cytoplasm. Note the cisternae (some are marked C) and their arrangement. Such cisternae are indistinguishable from remnants of the nuclear envelope. Fixed in Dalton's fluid and embedded in methacrylate.

the arrows point to some of the elements of the ER which are radially oriented to the centrosome. Figure 13 represents a *Spisula* egg fixed with Palade's fluid and shows a similar orientation of ER (arrows) radial to

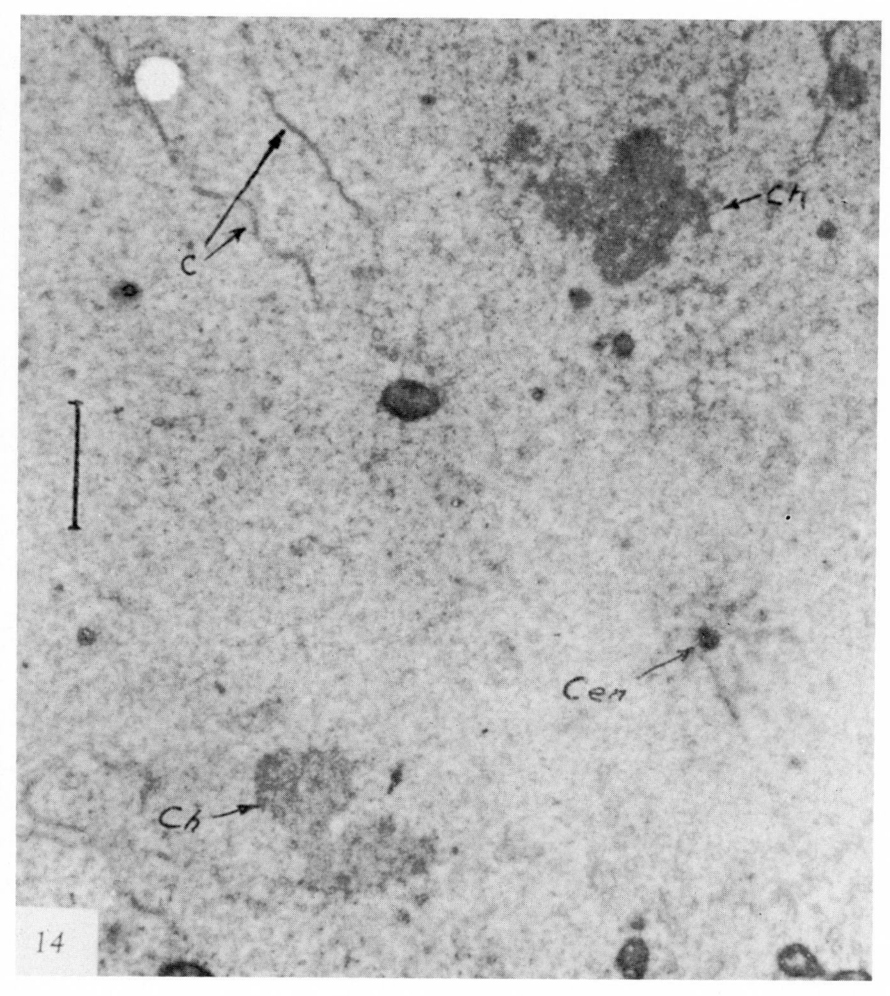

FIG. 14. A *Spisula* egg at the same stage as that in Fig. 13 and prepared in the same way. Note the centriole, chromosomes, and cisternae.

a clear centrosomal area surrounded by bodies, including chromosomes (Ch). The large, clear region surrounding the centrosome is the remnant of the nucleoplasm of the germinal vesicle, which has not yet mixed with the cytoplasm.

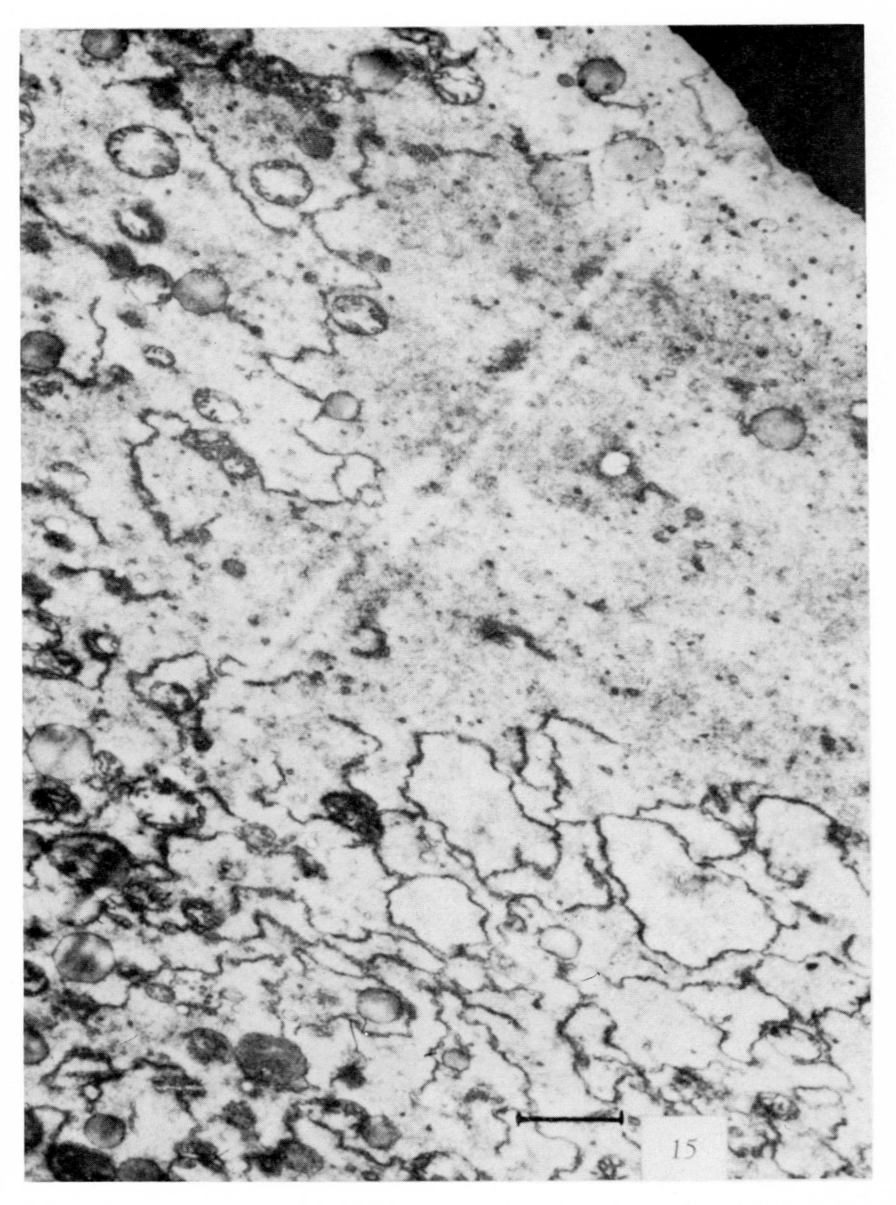

FIG. 15. *Spisula* egg at about first cleavage. The clear area is the spindle area and the section probably does not intersect the centrosome. Fixed in permanganate.

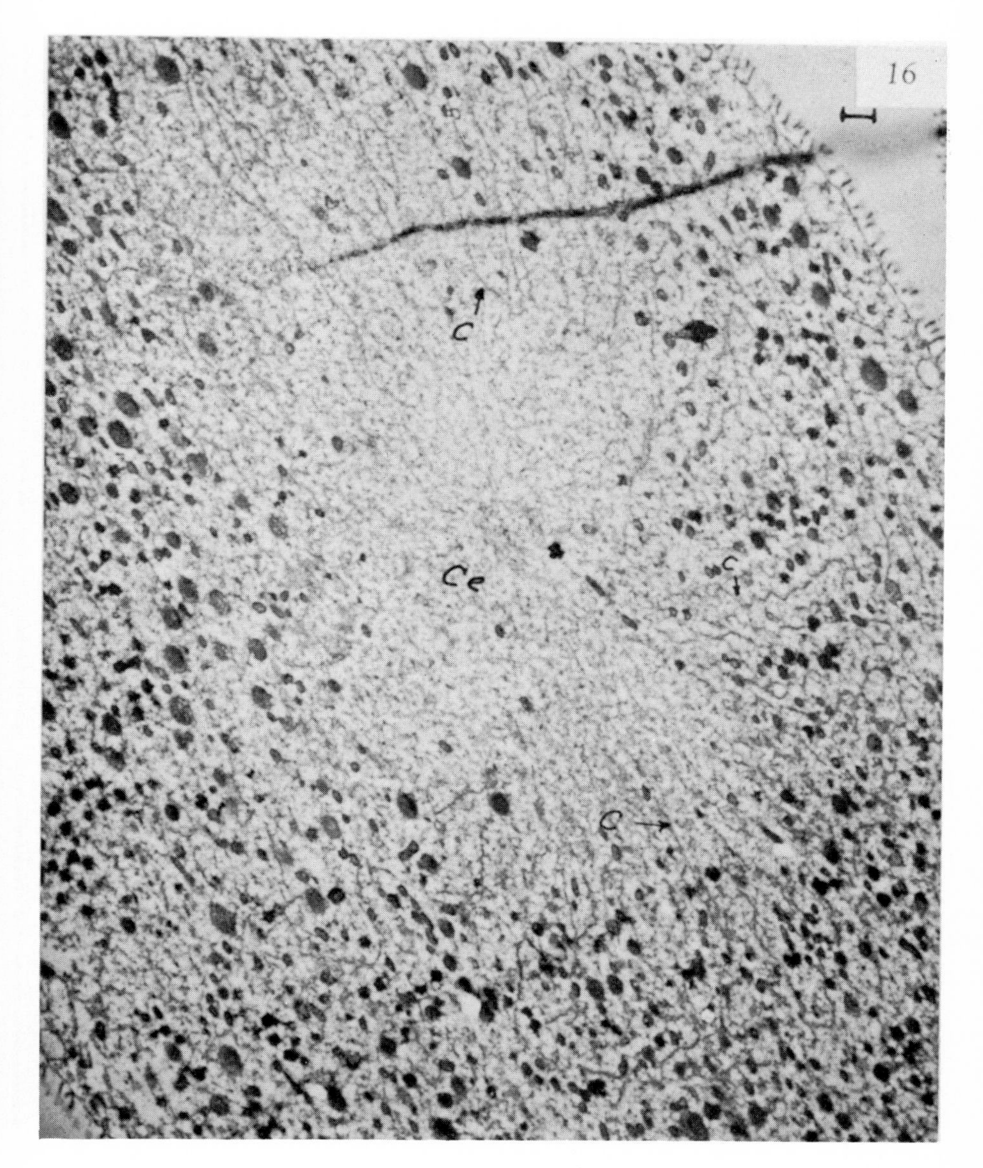

FIG. 16. A *Spisula* egg at first cleavage. Note the radial arrangement of cisternae. Fixed in Palade's fluid and embedded in methacrylate.

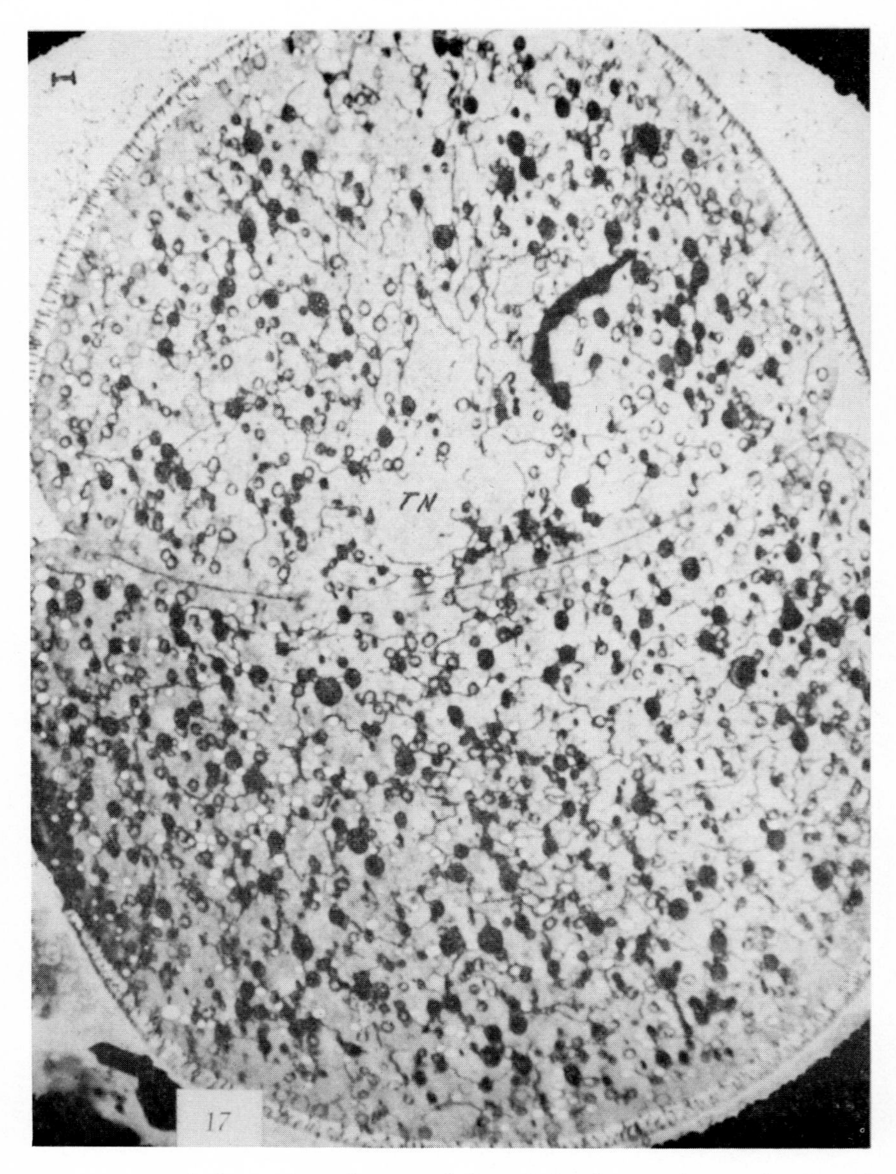

FIG. 17. A *Spisula* egg just after completion of the cleavage furrow. Smaller cell is AB cell and the clear area, TN, is probably just outside of the forming telophase nucleus (not in the section). Note the arrangement of cisternae pointing toward the periphery. The microvillated vitelline membrane surrounds the egg. Fixed in permanganate.

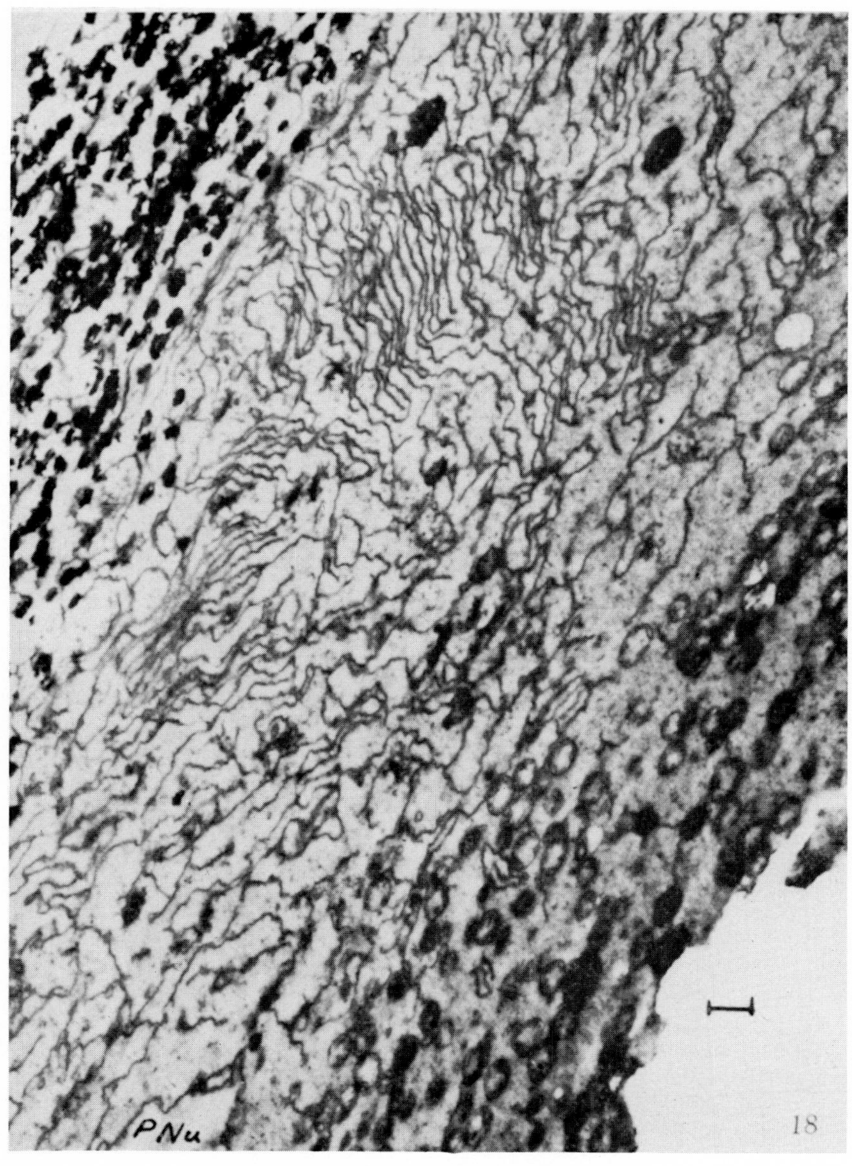

FIG. 18. An electron micrograph of a stratified *Spisula* egg after redistribution has begun to take place. The egg is at a stage just prior to pronuclear breakdown (i.e., at about the stage of the egg in Fig. 2) when the asters of the first cleavage spindle are just forming. Note that the cisternae are beginning to orient radially to a region just outside a pronucleus (PNu). The mitochondrial layer has begun to disperse as has the lipid layer (upper left of the micrograph). Fixed in permanganate.

The identification of the clear region toward which the elements of the ER point as a centrosome is based on micrographs such as that in Fig. 14. A centriole (Cen) is seen surrounded by a small dense area about 1 micron in diameter embedded in a clear area about 4–5 microns across. We consider the whole region as the centrosome. Some radial elements of the ER are pointed out by arrows.

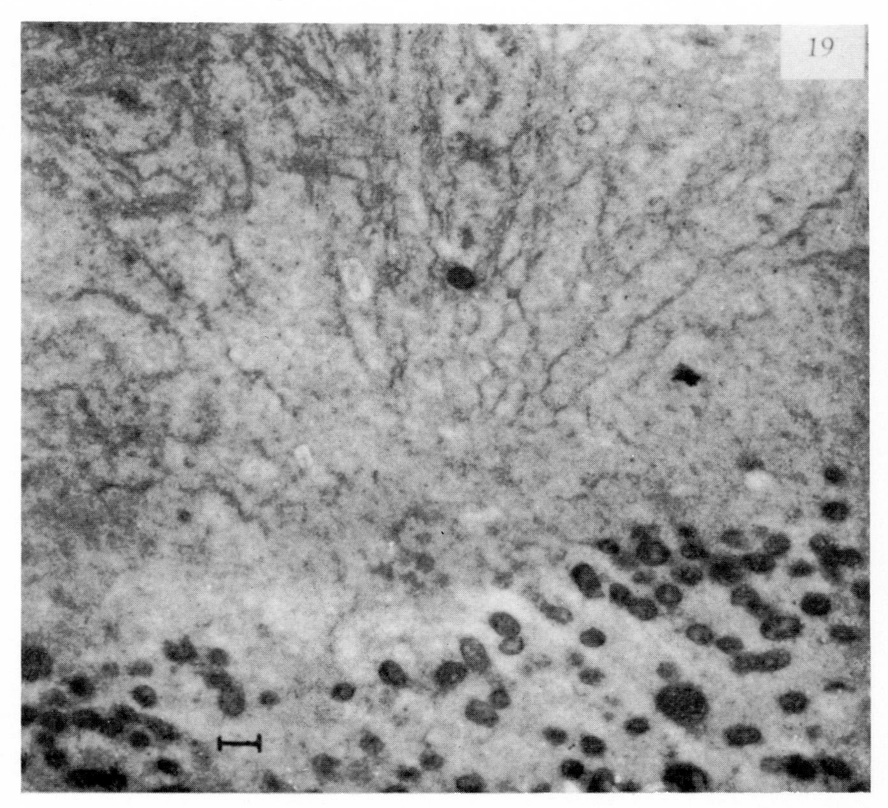

FIG. 19. A stage similar to that in Fig. 18 but fixed with Dalton's fixative and embedded in Araldite 502. The redistribution of mitochondria has not progressed as far in this egg as in that of Fig. 18.

Figures 20–22 show elements of the ER in the egg of *Cistenides*. The radial orientation is established at about the time of zygote nucleus formation, a period of the onset of extensive saltatory yolk particle movement. Elements of the ER point toward regions just outside the zygote nucleus and such elements may extend for distances of 10–15 microns.

Figures 15 and 16 show *Spisula* eggs at about the time of first cleavage.

Figure 16 is an electron micrograph of an egg fixed in Palade's fluid (the egg represented in Fig. 15 was fixed in permanganate) and the cisternae (arrows) radiate from a clear centrosomal region, the central part of

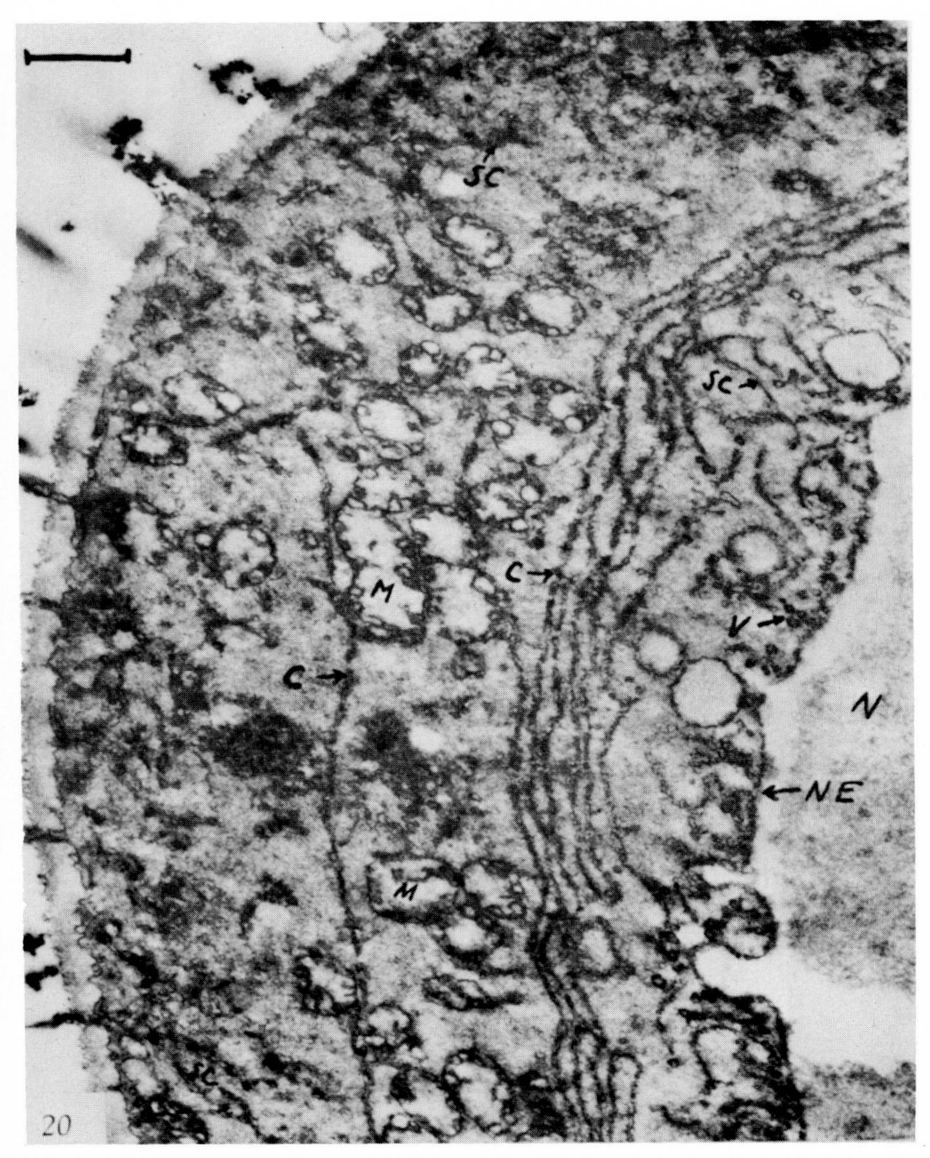

FIG. 20. An unfertilized oöcyte of *Cistenides*. Note the long cisternae (C), short cisternae (SC), mitochondria vesicles (V), etc.

which is occupied by the denser region which immediately surrounds the centriole (the latter was not included in this section).

Figure 17 is a low-power micrograph of a *Spisula* egg immediately after first cleavage. The clearer area in the small (AB) blastomere is probably immediately adjacent to the blastomere nucleus. An approximately radial

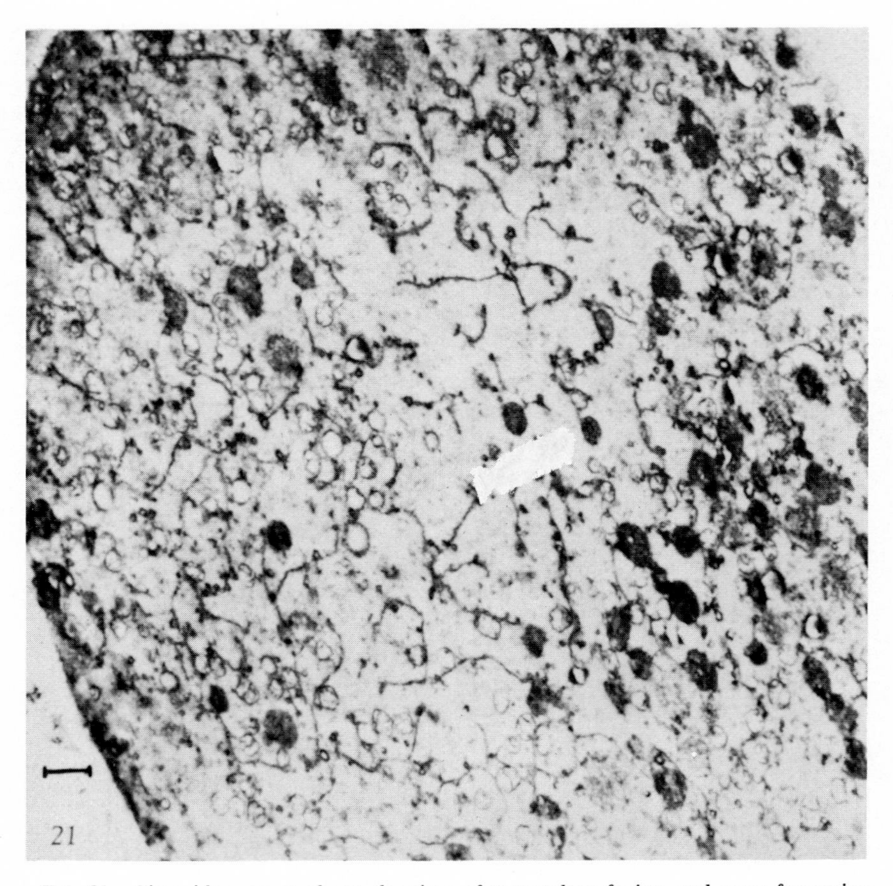

FIG. 21. *Cistenides* egg at about the time of pronuclear fusion and aster formation for the first cleavage spindle. Note the radial arrangement of cisternae pointed toward the clearer central area. Fixed in permanganate.

orientation can be seen in the ER, especially in the more peripheral parts of the cell where we would expect radial orientation on the part of a system for moving beta particles.

To complete this section we wish to discuss some events which occur in centrifuged *Spisula* eggs. If eggs are stratified when a mitotic apparatus is present and are then examined at various times afterwards, a succession of stages in the reformation of the asters can be seen. That is, at first either no asters can be seen or those present appear to be "one-sided" with radial striations descending from the upper hyaline zone where the ER accumulates (Fig. 11). Gradually, this one-sided aster

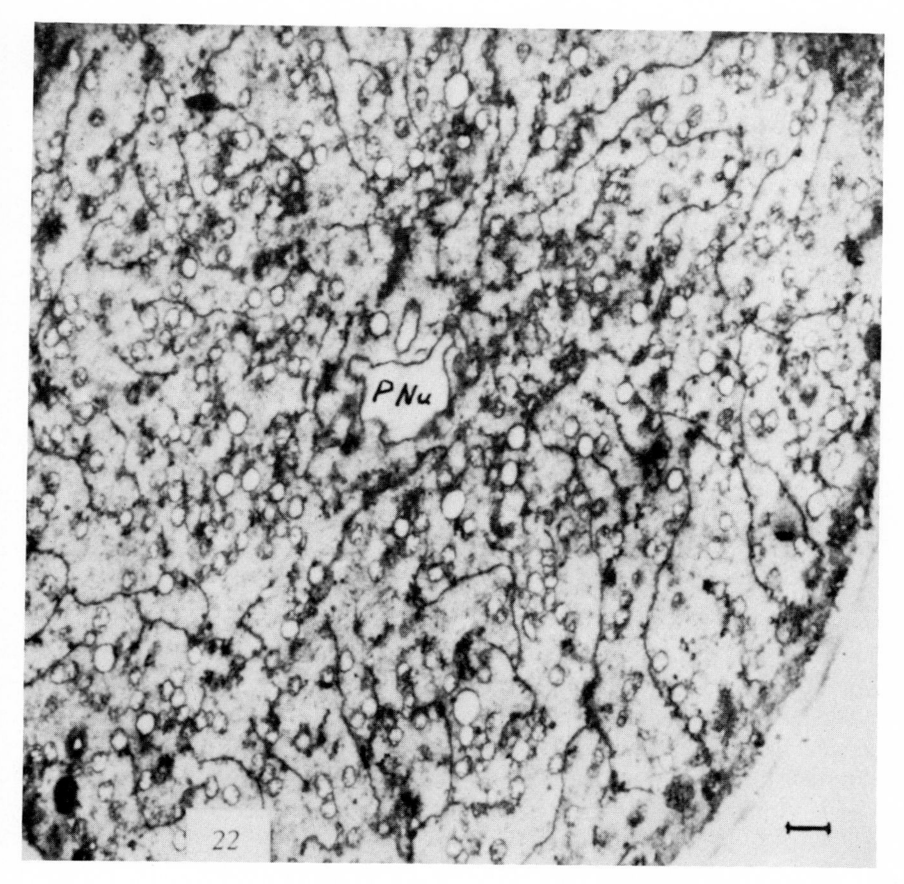

FIG. 22. A *Cistenides* egg at about the same time as that in Fig. 21. Note radial orientation of cisternae pointed toward the nuclear area. Fixed in permanganate.

becomes clearer and in 10–50 minutes (highly variable with egg and stage at which centrifugation is performed) a normal, completely radial aster is formed.

In electron micrographs of eggs fixed when "one-sided" asters are present, the ER can also be seen to be oriented in a radial pattern extending toward the centers from the upper hyaline zone [see Figs. 18 (permanganate fixation) and 19 (Dalton's fixative)]. The distribution of the ER thus follows the distribution of the asters visible with the light microscope and, indeed, is again in the proper position to account for the redistribution of beta particles in stratified eggs, described earlier.

3. *Discussion of Orientation of ER in relation to Particle Movements*

The totality of the evidence presented above indicates that during critical periods in the development of the egg an orientation of the ER occurs similar to that which we would require of any force-transmitting structure which might be involved in saltatory particle movements. In addition, in stratified eggs saltatory particle movements become directed toward the centers of the reforming asters and electron micrographs indicate that the ER is again in position to be a candidate for the force-transmitting structure. It is clear, however, that as almost circumstantial evidence has been adduced for the hypothesis and it will behoove us to see if a more substantial scaffolding can be reasonably built. We will return to this task after some remarks on the nature of the process of orientation of the ER.

That orienting forces can arise from mitotic centers has been suggested for many years. A particularly clear case relevant to our discussion is the orientation of elongated mitochondria (chondrioconts) in the permanent asters of amphibian leucocytes (for review, see Pollister, 1941). The mitochondria accurately mirror the orientation of the astral rays and in living cells can be seen to be relatively quiescent on direct visual observation. Pollister (1941) has suggested on the basis of fixed cells that diffusion streams of material may be running into the centers orienting the nonmoving mitochondria caught in gelled regions between the streams. Our own observations of these cells in the living state have revealed particles other than the elongated mitochondria which are related to the asters — namely small (less than a micron) spherical bodies, dark in dark contrast phase. These particles show constant saltatory movements directed both toward and away from the center of the aster but always radial to it when the particles are in the asters. The particles may move in and then suddenly out again past the oriented, but relatively quiescent mitochondria. It is difficult to interpret the shuttling of particles in and out of the center as due to diffusion streams and it seems more reasonable to interpret this case within the general framework we are considering — that is, that the particles are moved by a stationary and oriented force-transmitting structure.

The observations just discussed as well as the details of orientation of the ER described above, and many other observations in classical cytology (Wilson, 1925) suggest to us that at least three major considerations are involved in the accumulation of particles in the asters of dividing cells: namely, an orienting force emanating from the mitotic centers (probably from the centriole), which can cause radial patterns in elongate objects such as threadlike mitochondria or cisternae of the ER; a mechanism for moving particles in a saltatory manner, such mechanism being associated with at least one class of the center-oriented structures and finally, some mechanism, again probably emanating from the center, for making the direction toward the center the preferred (though not exclusive) one for saltatory movement once the force-transmitting structure is radially oriented.

We feel that the occurrence of saltatory particle movements in interphase, i.e., in the absence of a mitotic apparatus, allows us to eliminate the mitotic center as the direct cause of saltation and thus to identify structure-orienting and particle-moving forces as originating from separate mechanisms.

We can now fruitfully return to a discussion of other evidence suggesting the possibility that the ER is involved in saltatory particle movements.

IV. General Discussion

We have earlier discussed some of the work on *Chara* and *Nitella* by Jarosch and Kamiya which indicates that saltatory particle movements in protoplasmic droplets from this material are correlated with the linear translatory movements of filaments visible in the dark-field microscope. In the same preparations both the chloroplasts and nuclei show a rotating motion and many workers have observed streamlets moving in directions opposite to that of the rotating bodies in the immediately adjacent cytoplasm (for review, see Kamiya, 1959). Kamiya and Jarosch also report that the moving fibrils, which can aggregate laterally and disperse in the cytoplasm, show similar behavior with respect to chloroplasts and more germane for our discussion, nuclei. That is, such filaments can be seen to attach to and detach from the surfaces of rotating nuclei. The hypothesis suggested by the above workers for nuclear rotation is thus, most reasonably, one in which the moving fibrils are built into the nuclear surface in such a way that they all exert forces on the cytoplasm in the same direction, resulting in a rotating nucleus and counter rotating stream of adjacent cytoplasm.

We are fortunate to be able to discuss this material in the present volume since Dr. Porter has presented (1961) some of the evidence indicating that the endoplasmic reticulum is very closely related to the nuclear envelope. In many cells, there is a definite morphological connection between them and in egg material the interrelations of annulate lamellae, nuclear envelope, and ER suggest that the ER may interchange (reciprocally) with the annulate lamellae, the latter being formed under the influence of the nuclear envelope (Rebhun, 1961b). At any rate, the evidence that the ER may be ultimately derived from the nuclear envelope is compelling and suggests to us that if the "moving" fibrils are a component of the nuclear envelope, they may, indeed, also be built into or attached to the membranes of the ER. We, therefore, suggest, as a working hypothesis, that saltatory particle movements universally occur by the transmission of a force from a particular kind of fibril in the cytoplasm to particles and that these fibrils are usually morphologically associated with the membranes of the ER. Some direct visual evidence for the participation of sheetlike structures in cell cytoplasm in particle movement appears to be contained in moving pictures of plant cytoplasm obtained by Honda *et al.* (1961) and Hongladarom *et al.* (1961). Such structures apparently have rapidly moving cyclotic currents on their surfaces, which can, indeed, move particles in a saltatory manner. The identification of these structures as ER, however, is not absolutely certain, though highly suggestive.

Finally, we wish to discuss the work of Parpart (1953) mentioned earlier. Parpart showed that echinochrome granules, when centrifuged to the centrifugal pole in *Arbacia* eggs did not resume saltatory movement until periods of 60–80 minutes after stratification. Since stratified eggs are capable of cleaving quite regularly, centrifugation is clearly not an irreversibly damaging process to the cell and it is of interest to note that the ER of the *Arbacia* egg (which exists in considerable, though lesser, amounts than in the *Spisula* egg) stratifies to the hyaline layer (unpublished observations and Pasteels *et al.*, 1958). Thus, the echinochrome granules are separated by the mitochondrial and yolk layers (a distance greater than 30 microns) from the ER which, if it is the force-transmitting structure, again tends to support the general hypothesis we are discussing. In *Spisula* the ER ultimately redistributes after stratification (in from ½ to 1 hour) and if similar redistribution occurs in *Arbacia,* the resumption of saltatory particle movement may be linked to the renewed contiguity of ER and echinochrome granules.

It may be of use to summarize the observations and hypothesis we have gathered together in this paper to see how much of a house of cards we may have constructed. The first section of the paper reviewed work on saltatory particle movements in a wide variety of cells and led to the formulation of the following hypotheses:

(1) The descriptions of saltatory movement in these cells possess so many similarities that as a working hypothesis we assume that there is

a common underlying mechanism to all saltatory particle movements no matter where they occur in cells.

(2) The kinds of particles which move in saltatory fashion include many which may be described as relatively inert metabolically. These include pigment granules, lipid droplets, yolk, and heme granules. This plus the details of movement suggests the working hypothesis that the particles are moved by application of forces outside themselves, probably transmitted to them by an organized, physically existing structure in the cell.

(3) The first working hypothesis (1), above, of a uniform mechanism for saltatory movement leads us to suggest that the moving fibrils of Jarosch and Kamiya known to be involved in saltatory movements in *Chara* and *Nitella* cytoplasmic droplets probably have a wide distribution in cells and are probably the force-transmitting structures postulated in (2) above.

The second part of this paper dealt with the distribution of the ER in egg cells and pointed out in detail that this distribution fits the characteristics necessary for a structure producing saltatory particle movements, that is:

(4) The hypothesis is that the ER in cells can produce the forces involved in moving particles in a saltatory fashion.

As other evidence possibly supporting this hypothesis we discussed Jarosch's and Kamiya's work on the rotation of nuclei and using a current conception that the ER is a differentiation of or, at least, morphologically very similar to the nuclear envelope we suggest that:

(5) The ER is the force-transmitting structure in cells involved in saltatory particle movements by virtue of structures similar to the moving fibrils, which are built into or otherwise associated with the membranes of the ER.

Two auxiliary hypotheses we have used are:

(6) During mitosis the ER is oriented by forces (or other processes) emanating from the mitotic centers, and

(7) Once orientation is established, some process induces the prevailing (though not unique) direction of force transmission to particles to be toward the centrosomes.

It is clear that the set of hypotheses above are to be considered as much opinions as to where to look experimentally for the structural basis of saltation as a statement of what we feel contains the essence of truth concerning this phenomena. Also, it must be remembered that the cell has considerable facility for organizing fibrillar systems, such as the mitotic spindle, which may show a liability just beyond the capabilities for preservation shown by most of our present fixation techniques for electron microscopy. Thus, it is not impossible that systems other than the ER have the distributional properties required of the movement-inducing structures but are not visualized in our preparations. For example, the 150 Å tubules seen in spindles of protozoa by Roth and Daniels (1962) and in sea urchin eggs by Harris (1961) and especially Kane (1961), have been shown by Kane to be present in asters of the isolated mitotic apparatus. However, since they have not yet been identified in the cytoplasm, it is by no means clear as to where (if anywhere) these structures are in interphase. It is necessary that they be present in the nonmitotic cell if they are to be involved in the particle movements because of the existence of interphase saltatory movements.

Despite the reservations we (and others) may express, we feel the totality of evidence presented in this paper supports the working hypothesis that the ER is the underlying force-transmitting structure involved in causing saltatory particle movements in cells. It is possible to speculate extensively on the uses in normal physiological and developmental processes to which the cell may put a system for differentially distributing one particulate component of the cell relative to others. We would guess, however, that our quota for speculation has been temporarily filled by the contents of the previous pages.

Acknowledgments

The author wishes to thank Mrs. Harriet T. Gagné for the patience and skill which she brought to the tedious job of sectioning rather difficult material and for her excellent job of editing this manuscript.

References

Allen, R. D. (1961). Personal communication.

Allen, R. D., and Rowe, E. C. (1958). *Biol. Bull.* **114**, 113-117.

Andrews, E. A. (1955). *Biol. Bull.* **108**, 121-124.

Bloom, W., Zirkle, R. (not dated) "Mitosis of Newt Cells in Tissue Culture" (Film, Print No. 64). Univ. of Chicago Bookstore, Chicago, Illinois.

Bloom, W., Zirkle, R., and Uretz, R. (1955). *Ann. N. Y. Acad. Sci.* **59**, 503-513.

Harris, P. (1961). *J. Biophys. Biochem. Cytol.* **11**, 419-431.

Holt, S. J. (1957). Presented at *9th Intern. Cong. for Cell Biol., St. Andrews.*

Holt, S. J. (1958). *In* "General Cytochemical Methods" (J. F. Danielli, ed.), Vol. I, 388. Academic Press, New York.

Honda, S., Hongladarom, T., and Wildman, S. (1961). Presented at *1st Meeting Am. Soc. Cell Biol.*

Hongladarom, T., Honda, S., and Wildman, S. (1961). Presented at *1st Meeting Am. Soc. Cell Biol.*

Horowitz, N. H., and Baumberger, J. P. (1941). *J. Biol. Chem.* **141**, 407-415.

Jansen, M. T., and Molemaar, I. (1961). *J. Biophys. Biochem. Cytol.* **9**, 716.719.

Jarosch, R. (1956). *Phyton, Ann. rei botan,* **6**, 87-107.

Jarosch, R. (1957). *Biochim. et Biophys. Acta.* **25**, 204-205.

Jarosch, R. (1958). *Protoplasma* **50**, 93-108.

Jarosch, R. (1960). *Phyton, Ann. rei botan,* **15**, 43-66.
Kamiya, N. (1959). *In* "Protoplasmatologia (L. V. Heilbrunn and F. Weber, eds.), Vol. 8/3a, p. 1. Springer, Vienna.
Kane, R. E. (1961). Personal communication.
Luft, J. H. (1956). *J. Biophys. Biochem. Cytol.* **2**, 799-802.
Novikoff, A. (1961). Personal communication.
Ostergren, G., Molè-Bajer, J., and Bajer, A. (1960). *Ann. N. Y. Acad. Sci.* **90**, 391-408.
Parpart, A. K. (1953). *Biol. Bull.* **105**, 368.
Parpart, A. K. (1961). Personal communication.
Pasteels, J. J., Castiaux, P., and Vandermeersche, G. (1958). *Arch. biol.* (*Liege*) **69**, 627-643.
Pollister, A. W. (1941). *Physiol. Zoöl.* **14**, 268-280.
Porter, K. R. (1961). *In "The Cell"* (J. Brachet and A. E. Mirsky, eds.), Vol. II, p. 621. Academic Press, New York.
Raven, C. P. (1958). "Morphogenesis: The Analysis of Molluscan Development." Pergamon, New York.
Rebhun, L. I. (1956a). *J. Biophys. Biochem. Cytol.* **2**, 93-104.
Rebhun, L. I. (1956b). *J. Biophys. Biochem. Cytol.* **2**, 159-170.
Rebhun, L. I. (1959). *Biol. Bull.* **117**, 518-545.
Rebhun, L. I. (1960). *Ann. N. Y. Acad. Sci.* **90**, 357-380.
Rebhun, L. I. (1961a). *J. Biophys. Biochem. Cytol.* **9**, 785-798.
Rebhun, L. I. (1961b). *J. Ultrastruct. Research* **5**, 208-225.
Rebhun, L. I. (1961c). *Biol. Bull.* **121**, 404-405.
Rose, G. G. (1957). *J. Biophys. Biochem. Cytol.* **3**, 697-704.
Roth, L. E., and Daniels, E. W. (1962). *J. Biophys. Biochem. Cytol.* **12**, 57-78.
Seifritz, W. (1952). *In* "Deformation and Flow in Biological Systems" (A. Frey-Wyssling, ed.). North-Holland, Amsterdam, p. 107.
Stewart, P. A., and Stewart, B. T. (1959). *Exptl. Cell Research* **17**, 44-58.
Taylor, C. (1931). *Physiol. Zoöl.* **4**, 423-460.
Taylor, E. W. (1957). Ph.D. dissertation from Committee on Biophysics, University of Chicago.
Wilson, E. B. (1925). "The Cell in Development and Heredity," 3rd ed. Macmillan, New York.

Discussion by Arthur Zimmerman

State University of New York, Brooklyn, New York

I should like to congratulate Dr. Rebhun on his lucid discussion of the saltatory movements of cytoplasmic particles, their relationship to the endoplasmic reticulum and their role in mitosis.

The presence of the endoplasmic reticulum in cells shown by Dr. Rebhun suggests that an internal cytoskeletal structure may exist. The evidence for such a network may be supported by the non-Newtonian flow characteristics found in some of the low-speed centrifugal studies on eggs.

The spatial proximity of the endoplasmic reticulum to the asters during the mitotic events and its radial orientation strongly suggest that this is not a coincidental arrangement but probably involves an active system.

The elegant observations of Dr. Robert Chambers some years ago suggested that astral fibers may orient solational flow along the fibers themselves and thus transport material to and from the mitotic cell centers. This relates to the general pattern proposed by Dr. Rebhun.

The saltatory movements of cytoplasmic particles is a most intriguing phenomenon. Dr. Rebhun has discussed the cytoplasmic particle movements in several different cellular species. Besides pointing out the similarities of movement in the various forms, he attempted to find the mechanism concerning these movements.

Certainly the extensive studies of Kamiya (1960, *Ann. Rev. Plant Physiol.* **2**, 323) in which it is postulated that there is a parallel displacement force in the cytoplasm which is responsible for cytoplasmic movement are of interest. Kamiya has applied this theory in studying cytoplasmic movement in plant cells. Let us consider these studies and those of Jarosch (1958, *Protoplasma* **50**, 93) who has reported the formation, movement and disruption of fibril components within cytoplasmic droplets, from which we gain more insight into possible movement of the particles. It is possible that these cell particles which are frequently observed in the region of the mitotic apparatus are working in a similar nature.

The endoplasmic reticulum as shown by Dr. Rebhun and other speakers this morning, suggests very strongly an orientation in the astral fibers, and perhaps this endoplasmic reticulum is responsible for the translation of movement to the particles.

Some recent work conducted by other investigators has elucidated the possible role that these particles play in cell division. I have a few slides with me that I will keep in my pocket concerning this. However, I would like to state that there has been some study by Dr. Marsland and myself showing that metachromatic granules may be responsible in initiating cytokinesis. These granules accumulate around the astral areas during the formation of the mitotic apparatus and are similar to those shown by Dr. Rebhun. Just to briefly go on, we have given cells high speed centrifugal force under high pressures, and have been able to initiate cytokinesis (*Ann. N. Y. Acad. Sci.* **90**, 470, 1960). We believe that it is these metachromatic granules that undergo this dancing motion, and when these granules are disrupted or broken, they contribute to the initiation of cytokinesis. Dr. Rebhun has proposed that the granules may move about normally by a reaction with endoplasmic reticulum. Furthermore, they may be involved directly in cytokinesis, as well as the preceding mitotic events.

I thank you for letting me give my remarks.

General Discussion

Dr. A. W. Burke (Rhode Island Hospital): For just a moment, if I may, I should like to recall some points from elementary physics, which may bear analogy to the collection of cytoplasmic particles in the aster bodies of these dividing sea urchin cells. I make reference to the very neat little experiment used in demontrations of the behavior of the so-called perfect black body with respect to visible light. The model black body may be constructed by stacking double-edged razor blades so that their sharpened edges are aligned, and thus the space between any two blades, at the edged surface, is a very narrow wedge of space — in cross section, a very acute angle "V." Light which is incident to almost any angle of the open end of these wedges will be trapped by the wedges, and does not reflect outwards. Now, if the astral body is viewed as a similar arrangement of "wedges" (note that the astral rays form wedges in 3-dimensional, rather than just 2-dimensional planes), it would be quite natural that cytoplasmic bodies which are in motion could be trapped by the astral body. This trapping would be independent of either motion stemming from within the particle, or motion caused by forces external to the particle. Further, if the endoplasmic reticulum is propagated or sequestered off from the rest of the cytoplasm in some way, the pattern of its alignment could be explained in exactly the same way.

I do not mean to negate the hypothesis of Dr. Rebhun, but it is possible by this sort of explanation, to show that the motion of the particles may be quite independent of the reticulum, per se. If one wanted to be erudite about this analysis, one could resort to an analysis of the situation of degrees of freedom of these particles. A particle relatively free in the cytoplasm would have a certain number of degrees of freedom, but as the particle approached the center of the astral body, the degrees of freedom would diminish, so that finally, at the center of the astral body, the particle would have at most one axis of translation, and the degrees of freedom would diminish. It follows then, that a particle which is free to communicate between these two points would tend to flow to the position of least degrees of freedom, hence these cytoplasmic particles would aggregate in the aster bodies in time.

Another similar situation presents where the particles are trapped within the continuous fibers of the achromatic figure (so-called, central spindle), as in these beautiful movies which we have just seen. They behave according to the degrees of translational freedom which seem to be dictated by the central spindle, in part at least.

DR. L. I. REBHUN (Princeton University): Your discussion seems to imply that the number of degrees of freedom of a particle in the periphery should be greater than the number for a particle in the central aster in order for trapping to occur. In the early stages of spindle assembly there are particles which undergo movements which are not radial to the forming asters; but after the spindle is fully assembled, such movements do not occur, as I have pointed out (at least, I have not seen them). Thus a particle near the periphery moves in a radial direction as does one at the center. The number of degrees of freedom in the periphery would appear to be almost as restricted as in the astral center.

DR. A. W. BURKE: Yes, that is correct.

DR. L. I. REBHUN: Would your analysis predict that particles in the periphery would show other than radial movements?

DR. A. W. BURKE: In our mechanical concept of the configuration of the central spindle and astral rays (collectively the achromatic figure), we have said nothing about the total zone of physical forces about that birefringent structure as we view it. It would not be surprising that the astral rays, for example, exert forces of repulsion and attraction beyond the bounds of our microscopically resolved "fiber." In any event, as the astral rays grow out to the cell surface, the lateral motion of the particles now trapped would be expected to diminish in favor of almost exclusive radial motion with respect to the astral body. Another way of looking at this new situation would be that radial motion is nonturbulent in character while lateral motions of the particles would be of a turbulent nature. If there is any point in bringing this argument to bear on the situation, nature seems to prefer non-turbulent flow patterns in cytoplasmic streaming of normal cells.

DR. L. I. REBHUN: With respect to the particle movements associated with the spindle itself, in the original eggs, it can be seen that the particles move on the spindle surface not within the spindle. I don't know what argument involving degrees of freedom would move the particles back and forth over the spindle surface.

DR. K. R. PORTER (Harvard University): I wondered if actually these particles may not be moving on the surface of the membranes? Did you see Honda's film (S. Honda *et al., 1st Meeting Am. Soc. Cell Biol.*) that showed protoplasmic streaming of plant cells?

DR. L. I. REBHUN: No, I did not see it.

Dr. K. R. Porter: They show that the particles come down to the surface of the ER and then they go shooting right along fast. They then leave it and slow down and this streaming is not inside of the ER in plant cells but most directed along the surface of the lamellae.

Dr. L. I. Rebhun: I'm sorry if I gave you the impression that the particles move within the cavities of the ER. I've been thinking of them as moving along the surface.

Dr. K. R. Porter: So actually the particle may come up to the membrane and slide along quite rapidly to be trapped in the middle. What particles are these incidentally that you're watching?

Dr. L. I. Rebhun: In the case of *Spisula,* I believe the metachromatic particles to be multivesiculate bodies as seen in the electron microscope. This is based on electron micrographs of unstained eggs fixed at stages when the metachromatic granules were known to be aggregated. Thus, the particles gather at the spindle ends in first and second cleavage and stay on the peripheral pole of the blastomere nucleus during interphase. Electron micrographs of the eggs at these stages show multivesiculate bodies in the same locations as metachromatic particles in living cells whereas other particles (yolk mitochondria, etc.) are more uniformly distributed. Also, in centrifugally stratified fertilized eggs the metachromatic granules layer at the junction of the upper and lower hyaline zone as do the multivesiculate bodies in electron micrographs. The total evidence, therefore, leads us to hypothesize the identity of multivesiculate bodies with those stainable with methylene (or toluidine) blue.

Dr. S. Gelfant (Syracuse University): Could these particles be related to the sulfhydryl staining material that Kawamura and Dan (1958, *J. Biophys. Biochem. Cytol.* **4**, 615) have shown to accumulate in the astral regions.

Dr. L. I. Rebhun: As I recall that work, the sulfhydryl material was distributed throughout the centrosomal region. The particles we have been studying do not enter the centrosomes and, indeed, the boundary of the latter acts as a sharply defined barrier to the particles. If they are watched (in either *Spisula* or *Cistenides*), particles can be seen to bump into an invisible barrier and actually appear to "bounce" on it before they come to rest. Thus, the distribution of these particles does not overlap that of the — SH materials.

A group of investigators at the University of Brussels [Dalcq, Pasteels, and Mulnard; e.g., J. J. Pasteels *et al.,* 1958, *Arch. biol.* (*Liege*) **69,** 627] have produced some evidence that the metachromatic particles are acid phosphatase positive. Unfortunately, despite great effort on the part of myself and three other people in our lab, we have not been able to repeat this work even though at least one organism (*Arbacia*) was used by Mulnard and by our group. We occasionally find acid phosphatase (using both Gomori and azo-dye techniques) which shows a distribution similar to that of metachromatic granules but this is only in one out of every four to five batches of eggs and only in one out of five or ten eggs in each batch. What this means we do not know since there is evidence from other cells which make us sympathetic to the idea of acid phosphatase in these particles. However, our results in eggs are not encouraging.

The Fine Structure of *Tetrahymena pyriformis* during Mitosis[1]

Alfred M. Elliott

Department of Zoology, University of Michigan, Ann Arbor, Michigan

I. Introduction

Only a few protozoa have become popular research tools in cellular biology. Among these, the holotrichous ciliate, *Tetrahymena pyriformis,* has found a secure place, owing primarily to its capacity to grow axenically in a chemically defined medium. This characteristic has made it especially useful in physiological and biochemical studies. Little attention has been given to its morphology, probably because other, much larger ciliates have been more useful in studies of morphogenesis (Tartar, 1961). With the advent of the heat-shock treatment devised by Scherbaum and Zeuthen (1953) which induces over 80% of the cells to divide synchronously, *T. pyriformis* immediately emerged as a valuable organism for studying intracellular differentiation. This treatment apparently induces no cytological abnormalities (Williams and Scherbaum, 1959), hence is a convenient method for obtaining vast numbers of cells for investigating specific stages of the division cycle.

Associated with division in *T. pyriformis,* as in all ciliates, there is a precise duplication of the cytoarchitecture which can be followed in synchronized mass cultures. Several studies have contributed to our understanding of these events (Elliott, unpublished data; Elliott *et al.,* 1962; Furgason, 1940; Holz *et al.,* 1957; Williams and Scherbaum, 1959). Before reviewing our present knowledge of mitosis in *T. pyriformis,* it is useful to examine the morphology of a typical interphase cell.

II. General Morphology

T. pyriformis is of average size as protozoa go, approximately 50 × 30 microns, and is pear-shaped (Elliott and Hayes, 1953). Its cilia are arranged in eighteen meridianal rows, called kineties, each consisting of numerous units. Electron microscope studies of these units demonstrate that they consist of a kinetosome, to which the cilium is attached, and a tapered kinetodesmal segment (Metz and Westfall, 1954). The kine-

[1] This investigation was supported in part by a research grant E-1416 (C8) from the National Institute of Allergy and Infectious Diseases, N.I.H., U.S.P.H.S.

todesmal segments overlap to form a common filament, the kinetodesma, which lies to the right of the evenly spaced kinetosomes. The fine structure of these as well as the cortical architecture as a whole of several small ciliates has been thoroughly examined recently by Pitelka (1961).

Bacteria-feeding ciliates usually possess a complex oral apparatus. In *T. pyriformis* it is composed of four membranes, each a compound ciliary structure, which arise from the wall of the buccal cavity. Three of the membranes known as the "adoral zone of membranelles" (Corliss, 1953; Furgason, 1940), consisting of three rows of cilia, lie on the left wall of the buccal cavity whereas the undulating membrane, composed of a single row of cilia, is located along the right side of the cavity. The buccal cavity terminates posteriorly in the cytostome which opens into the cytopharynx where food vacuoles are formed. The structural and functional details of the oral apparatus are not, as yet clearly defined.

Additional cortical structures are the two contractile vacuole pores located in the posterior region on the right side, the ventral cytoproct located posteriorly in the post-oral kinety, and the protrichocysts which open at the surface by means of a permanent pore. The protrichocysts have been described by Pitelka for other closely related ciliates. Whereas the pore seems to be a permanent cortical structure, the pocket, usually identified as the protrichoyst, consists of a membraneless sac and appears to be transitory, depending on the tonicity of the environment. They are conspicuous in cells maintained in distilled water for several hours, as well as in stationary phase cells (Fig. 1). They are few in number or totally absent during log growth (Fig. 2).

Membranes of the endoplasmic reticulum are inconspicuous and the ribosomes are evenly dispersed in strain WH_6 (Fig. 16). This may be a strain characteristic since some derived strains of variety 9 possess well-formed membranes with adhering granules (Fig. 4). Moreover, Williams *et al.* (1960) report well-developed endoplasmic reticulum localized around the stomatogenic kinetosomes during cilia formation in strain GL. When membranes are present in strain WH_6 they appear in the vicinity of the macronucleus, kinetosomes, and peripherial mitochondria (Fig. 3).

Interphase cells taken during the stationary growth phase cultured in axenic media, show scattered oval-shaped mitochondria and lipid granules (Fig. 1). During log growth, however, the mitochondria become elongated and migrate to the vicinity of the cell membrane (Fig. 2, 3). This movement of the mitochondria is probably correlated with the sites of metabolic activity. The mitochondrion resembles that of other protozoa, its internal structure consisting of microvilli (Fig. 3). Other oval bodies are seen in stationary phase cells which may be lysosomes. Bodies that might be interpreted as centrioles have not been seen.

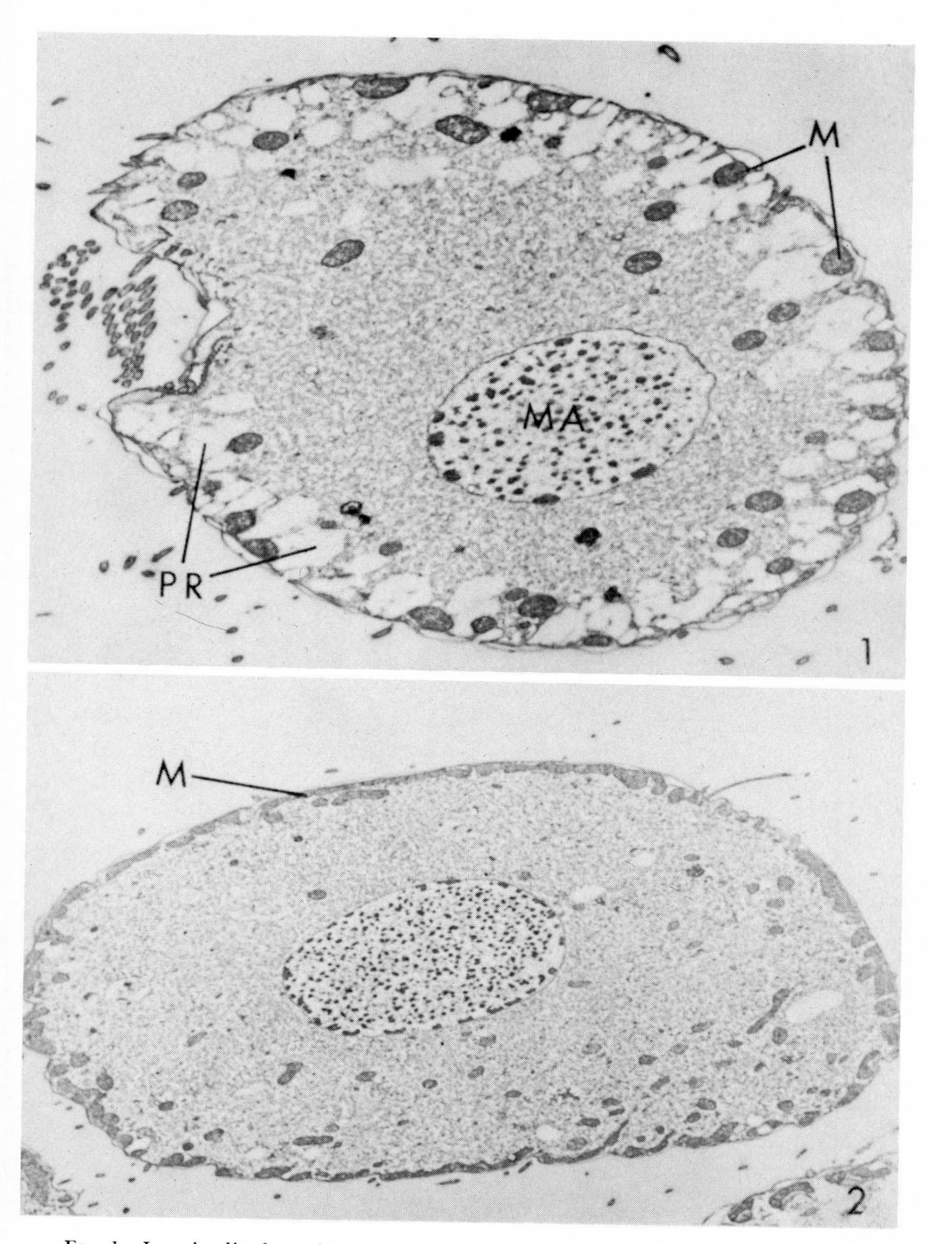

FIG. 1. Longitudinal section of whole cell during stationary growth phase. Note the numerous, peripherally located protrichocysts (PR) and the oval mitochondria (M), few in number and distributed throughout the organism. Macronucleus (MA). $\times$ 4000.

FIG. 2. Longitudinal section of whole cell during logarithmic growth phase. Note the absence of protrichocysts and the peripherally located mitochondria (M) which are greatly elongated. $\times$ 4000.

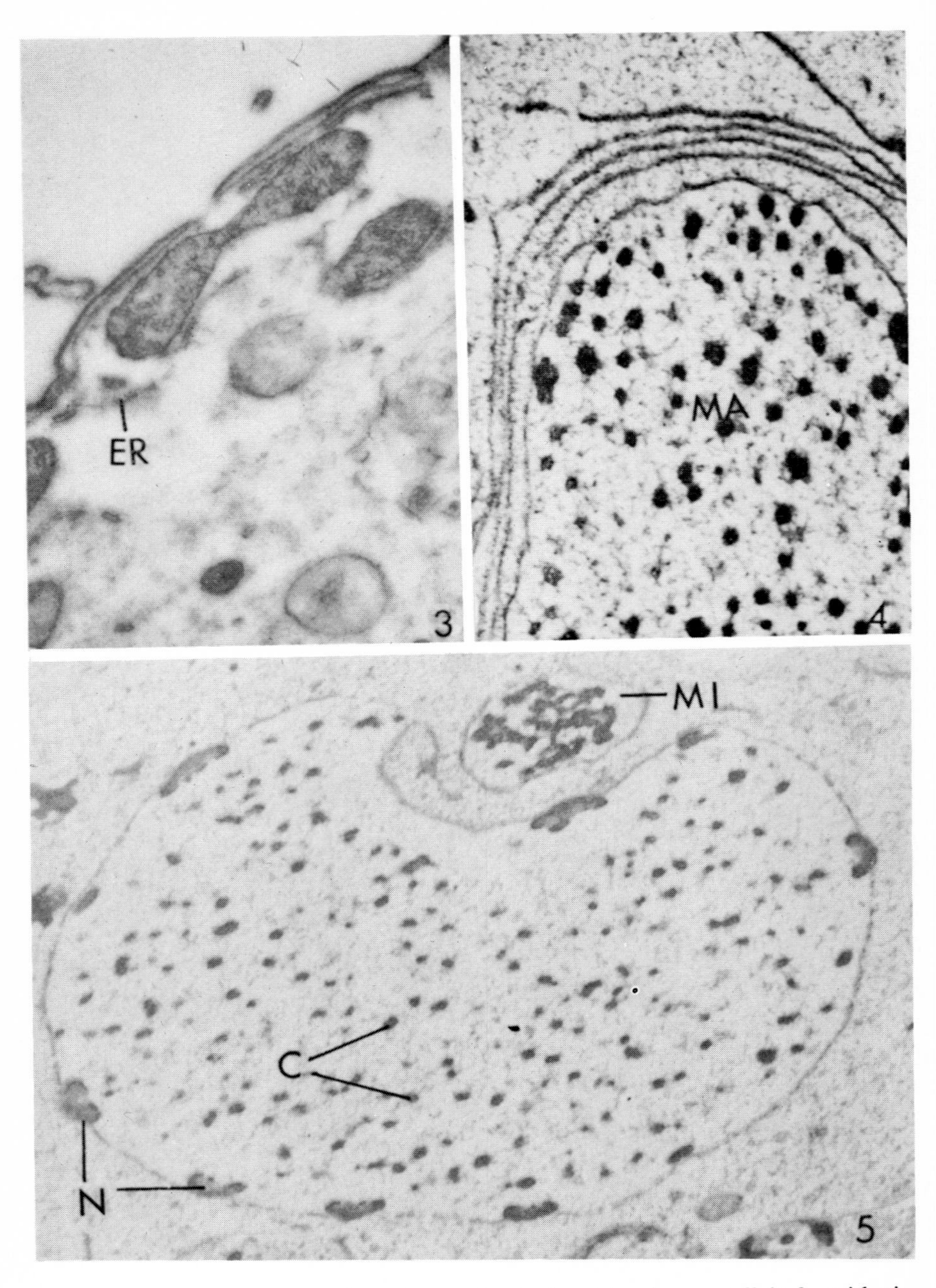

FIG. 3. High power view of an elongated mitochondrion from a cell in logarithmic growth. Note the associated endoplasmic reticulum (ER). $\times$ 25,000.

FIG. 4. The endoplasmic reticulum is conspicuous in strains from another variety

T. pyriformis, like all ciliates, is binucleate. Its large macronucleus (10 microns) consists of a porous double membrane, numerous chromatin bodies, and peripherally located Feulgen-negative bodies (Fig. 5). Both types of intranuclear bodies have been identified with the light microscope and more recently with the electron microscope (Roth and Minick, 1961; Sedar and Rudzinska, 1956). In sectioned interphase cells the former type consists of dense bodies evenly distributed throughout the nucleoplasm which are interpreted as chromatin material and may be sectioned chromosomes. When subjected to the Feulgen schedule they show a positive reaction and are destroyed by deoxyribonuclease (DNase) (Elliott *et al.,* 1961). The second type of intranuclear body lies in close proximity to, and evenly distributed along, the innner membrane of the macronuclear envelope. They lose their stainability when treated with ribonuclease (RNase) but retain their morphology indicating that they are composed in a large part of protein. They have been identified as nucleoli by other workers (Roth and Minick, 1961; Sedar and Rudzinska, 1956), and the present studies confirm their observations. The nucleoli consist of a dense outer shell and a less dense inner core (Fig. 6). The particles of which they are composed are remarkably similar to the cytoplasmic ribosomes. A conspicuous break in the shell can be observed in most sections, which gives them the appearance of crescents. The opening in the shell is usually oriented toward the center of the nucleus.

The interphase micronucleus is seen as a homogeneous Feulgen-positive bead (2–3 microns) with the light microscope, lying in an indentation in the macronucleus. Electron micrographs, however, reveal a double nuclear envelope containing thready chromosomes (Fig. 7). No structure that could be identified as a nucleolus has been seen, other than the Feulgen-negative bodies of the macronucleus.

III. Vegetative Division

A. Cortical Structures

Most morphological studies of vegetative division in ciliates have centered around the ectoplasmic cortex, involving the self-duplicating

---

(9) where they appear in the vicinity of the macronucleus (MA). Note the attached as well as the dispersed granules. × 7000.

Fig. 5. Section of the macronucleus with the micronucleus (MI) lying in an indentation. Both the position of the micronucleus and the nature of its chromosomes indicates that it is in prophase. Within the macronuclear envelope are the peripherally located crescent-shaped nucleoli (N) and the evenly distributed chromatin bodies (C). Note that the opening in the nucleoli is oriented toward the center of the macronucleus. × 8000.

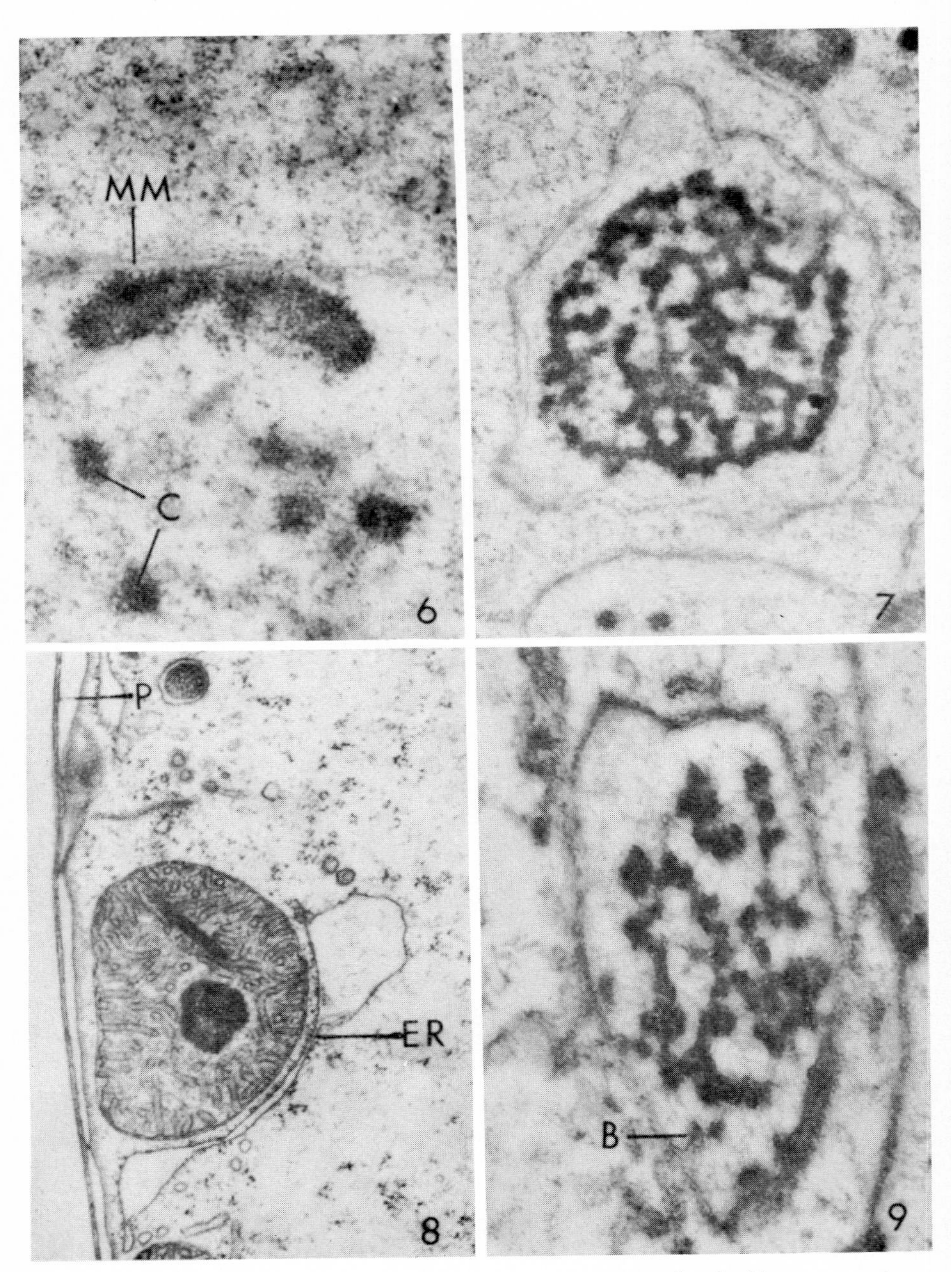

Fig. 6. High magnification of a nucleolus lying close to the double macronuclear envelope (MM). Note its granular composition. Chromatin bodies (C) may also be seen in detail. Embedded in Epon. × 30,000.

Fig. 7. Sectioned micronucleus in interphase. The nuclear envelope is clearly double and the chromosomes are uncoiled. Embedded in Epon. × 20,000.

nature of the kinetosomes and their derivatives (Lwoff, 1950). Such studies have been reported for *T. pyriformis* with normal dividing (Chatton *et al.*, 1931) and synchronized cells (Williams and Scherbaum, 1959; Williams, 1962).

At cytokinesis the ciliate is bisected transversely so that each daughter cell receives one-half of each of the eighteen kineties which then increase in length by the addition of new kinetosomes, cilia, and fibers. Williams and Scherbaum (1959) suggest that lengthening of the kinety takes place throughout cytokinesis as well as during the initial part of the cell cycle before the new oral primordium appears.

Considerable effort has been devoted to the origin of kinetosomes. The view generally accepted today is that kinetosomes are self-duplicating entities with the capacity to give rise to cilia and fibers (Lwoff, 1950). Chatton and Brachon (1935) obtained evidence that in *Glaucoma piriformis* (*T. pyriformis*) the kinetosomes divide into two daughter kinetosomes, one retaining the old cilium. The two daughters then separate and the nonciliated kinetosome sprouts a new cilium. Williams (1962) has confirmed all stages in this pattern of kinety growth from Protargol preparations. Unfortunately, dividing kinetosomes have not as yet been reported by electron microscopists, which may mean that they do not divide. Williams (1962) favors the "generative" model for self-reproduction of intracellular bodies as proposed by Mazia *et al.* (1960). This proposal suggests that the intracellular body first forms as a "seed" or "germ" which later develops into the mature body. In support of this hypothesis Williams and Anderson (personal communication) have demonstrated that small bodies do appear alongside kinetosomes in the oral primordium which may be the "seeds" or "germs" required by the "generative" model. The parent kinetosome would thus convey genetic continuity to the "offspring" in some form other than the mature kinetosome. Obviously the existing kinetosome influences the formation of the new kinetosome, but just what factors are involved are unknown.

No information is presently available on the generation of cilia or kinetodesmal segments. Since the kinetosomes always appear first it is highly likely that they somehow influence the development of both cilia and kinetodesmal segments.

FIG. 8. Mitochondria lying near the pellicle (P) often contain a dense mass of material. The endoplasmic reticulum sometimes forms large vesicles (ER). × 15,000.

FIG. 9. A micronucleus at prophase showing the chromosomes condensed. They appear to coil slightly. Paired intranuclear bodies (B) appear at this stage as well as later (Fig. 10). × 18,000.

B. Stomatogenesis

Our present knowledge of stomatogenesis in tetrahymena-like ciliates is based on the work of Chatton *et al.* (1931), which has recently been confirmed by Williams and Scherbaum (1959) in synchronized cultures of *T. pyriformis.* According to these authors the oral primordium (anarchic field) appears in the right post-oral meridian near the midregion of the cell. This is formed as a result of repeated divisions of kinetosomes in this region. The kinetosomes are identical to those in the kineties and their appearance seems to result from factors extrinsic to the kinetosomes themselves.

Shortly following the formation of the anarchic field the kinetosomes become oriented into the architecture of the future oral apparatus. The basal regions of the three membranelles and undulating membrane appear, after which the kinetosomes give rise to the cilia.

The cilia appear to grow from the kinetosomes fully formed, merely increasing in length (Williams, 1962). Since ciliary morphology is strikingly different from that of the kinetosomes, the morphogenesis is even more impressive. Just how this synthetic process occurs is completely unknown. The buccal cavity forms by invagination of the anarchic region after which the buccal ridges appear under the influence of unknown forces.

C. Mitochondria

We have found no evidence for the *de novo* origin of mitochondria since no "small" precursor bodies have been seen. All efforts to identify nuclear extruded bodies that might become mitochondria have failed. The remaining explanation for their increase in number seems to be by fragmentation and/or division. What appears to be dividing mitochondria is encountered often but this has obvious limitations as confirmatory evidence. Sato (1960) reports that mitochondria in strain W gradually increase from 600–800 during interphase to 1200–1500 just before fission. Our studies seem to indicate that the numbers of mitochondria are correlated with the nutritional state of the cell rather than the stage in division.

Roth and Minick (1961) claim that structural changes occur in mitochondria during division. They report the appearance of dense intramitochondrial masses during macronuclear division. We have observed mitochondria containing these dense masses during division but they also occur in stationary growth interphase cells (Fig. 8).

D. Nuclear Events

Very little is known about the nuclear events during vegetative fission in *T. pyriformis,* probably because the process is far less spectacular than conjugation where chromosomes can be easily seen and all the steps in meiosis delineated (Elliott and Hayes, 1953). The detailed cytology of conjugation has been most lucidly described by Ray, (1956).

The nuclear events during vegetative fission described here are based on a series of studies of synchronized cells. Mass cultures of *T. pyriformis*, strain WH_6, were prepared for study, following the procedure of Holz *et al.* (1957). Four-day-old stationary phase cells, grown in 2% proteose-peptone at 35° C, were transferred to fresh media where they were alternated between 35° ± 0.5° and 42.8° ± 0.01° C, at 30-minute intervals for 4½ hours (5 heat shocks). Over 80% of the cells showed cleavage furrows 1 hour following the last heat shock. Cells were withdrawn from the flask at 30-minute intervals (starting with the initial heat shock and continuing for 8½ hours) and fixed for both light and electron microscope studies. For the former, the fixative was Carnoy's (formula 2) and several stains were used. For the latter the fixative was 2% osmium tetroxide and, with the exceptions noted, the embedding medium was methacrylate. The usual cytochemical techniques were employed. Details of the procedures have been described in another report (Elliott *et al.*, 1962).

1. *The Micronucleus*

The dividing micronucleus of ciliates follows the pattern of metazoan cells in that a metaphase plate is formed. However, in other respects it seems to be quite different. Because of its small size, protozoan cytologists have had difficulty in interpreting what is seen with the light microscope. Kidder and Diller (1934), following binary fission in several small ciliates closely related to *T. pyriformis*, observed that the first sign of the approach of fission is the swelling of the micronucleus and its subsequent movement away from the macronucleus. Very small thready chromosomes become visible at metaphase, but the number could not be determined, owing to their small size. The chromosomes separate, move to the poles, and daughter micronuclei are subsequently formed which move apart, condense, and become located in the prospective daughter cells near the elongating macronucleus.

We have been able to add nothing to this sequence of events based on electron microscopy. However, the details of the micronucleus in various stages of division have been observed. At prophase the micronucleus swells and the chromosomes become more dense and appear to coil (Fig. 9). The movement of the chromosomes during metaphase and anaphase follow a conventional pattern, although in our preparations no spindle fibers, asters, or centrioles were seen. The nucleoplasm appears homogeneous and the nuclear envelope remains intact throughout mitosis (Fig. 10). Small paired intranuclear bodies observed during prophase

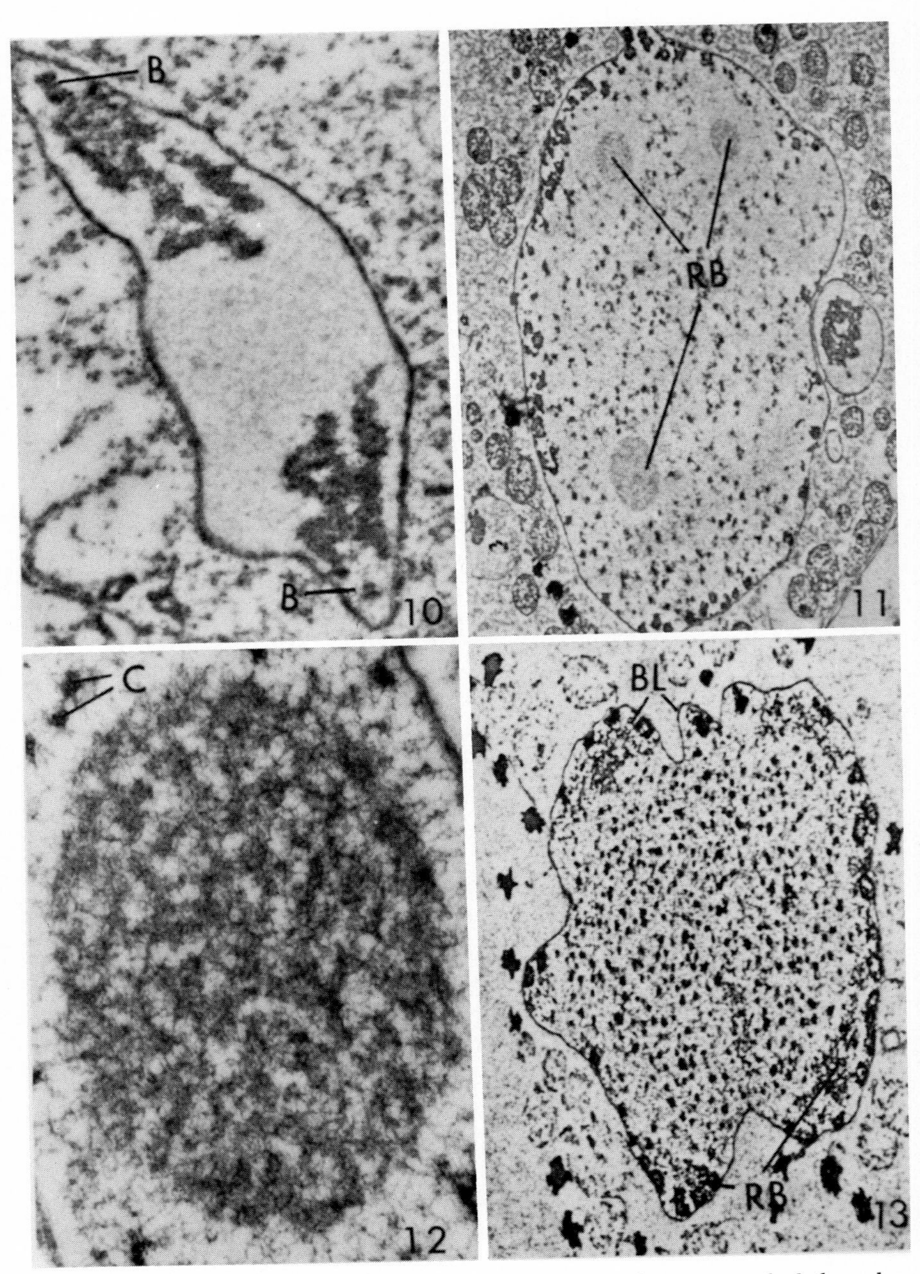

FIG. 10. A micronucleus at anaphase. The chromosomes have approached the poles where the paired bodies (B) noted in Fig. 9, are again visible. The micronuclear envelope remains intact throughout mitosis. Note the lack of chromosomal or spindle fibers. × 14,000.

FIG. 11. A sectioned macronucleus during the third heat shock (approximately 2

(Fig. 9) and at the poles during anaphase (Fig. 10) may be interpreted as centrioles although their intranuclear location seems to rule this out.

2. *The Macronucleus*

The macronucleus of ciliates functions not only in the metabolic processes of the cell but also in regeneration which has been beautifully demonstrated in Stentor (Tartar, 1961). Its genetic function has stimulated interest in its detailed morphology.

Feulgen preparations of the interphase macronucleus of *T. pyriformis* show a homogeneous distribution of Feulgen-positive material. Following the amitotic pattern for all ciliates during division the macronucleus pulls out and pinches into two equal parts. At this time many small bodies are evident, which have been called subnuclei each of which are presumed to contain a diploid set of chromosomes. This condition is thought to occur generally among ciliates (Sonneborn, 1947). Little more can be gained from light microscope studies, hence the fine structure of the dividing macronucleus has recently been investigated (Elliott *et al.,* 1962).

The first obvious change in the macronucleus of synchronized cells occurs during and following the second heat shock at which time discrete, oval, membraneless bodies appear near the center. At first they are only slightly more osmophilic than the nucleoplasm but they soon become clearly defined structures (Fig. 11). From one to four appear in each section, and they are composed of coiled fibers (Fig. 12). Since they are completely destroyed with RNase, and react with RNA stains, they have been identified as RNA bodies (Elliott *et al.,* 1962). They cannot be confused with the nucleoli for several reasons. First, their morphology is entirely different; second, they completely disappear with RNase treatment whereas the nucleoli are only partially destroyed; third, they appear only at specific periods in the division cycle whereas the nucleoli can be identified throughout the division cycle. It would seem that the RNA bodies are composed almost entirely of RNA whereas the nucleoli are mostly protein. The RNA body is surrounded by a clear area through which fibers pass to the chromatin bodies. These possibly function in the transfer of RNA from the chromatin bodies to the RNA body.

---

hours from the beginning of the initial heat shock) showing three RNA bodies (RB). × 8000.

FIG. 12. High magnification of the RNA body showing its fibrous composition. Note the fine lines of communication between the RNA body and the chromatin bodies (C). × 31,000.

FIG. 13. Macronucleus following the third heat shock. The RNA bodies (RB) are near the periphery and are disorganized. The chromatin bodies have doubled in number (compare with Fig. 11). Blebs (BL) containing nucleoli are beginning to form. × 5000.

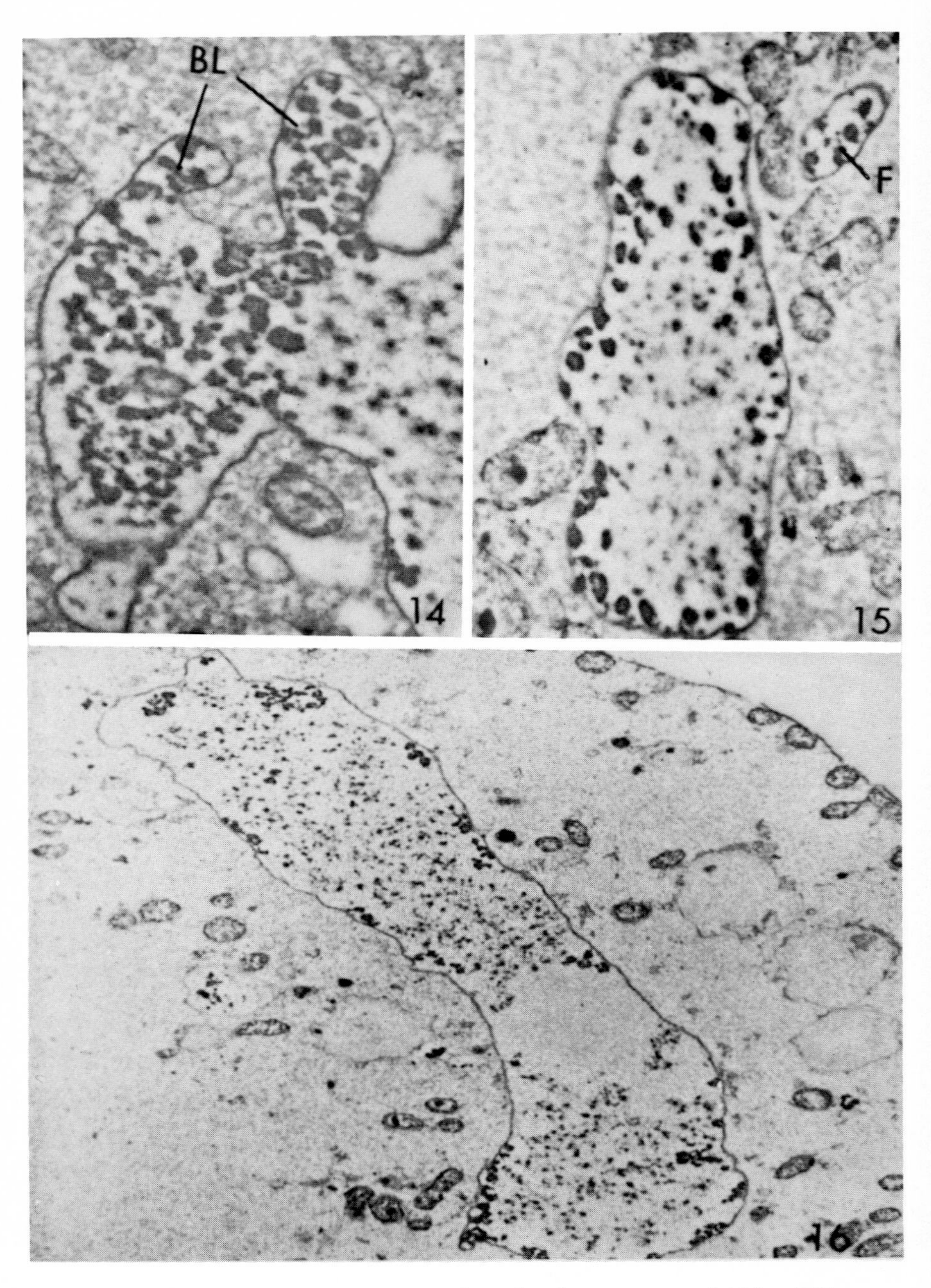

FIG. 14. The portion of the macronucleus showing pronounced blebs (BL) containing nucleoli. These extend into the cytoplasm and appear to pinch off (Fig. 15). × 15,000.

FIG. 15. The blebs shown in Fig. 14 appear to pinch off leaving fragments (F) in

The RNA bodies can be seen for approximately 2 hours from the time they can first be recognized until they disappear. They seem to wax and wane in a rhythmic fashion which is correlated with a specific stage in division. At the time of RNA body disintegration chromatin bodies can be seen among the fibers (Fig. 13).

Once the RNA bodies disappear the peripheral nucleoli increase in density and number. They also aggregate into masses interconnected by a network of fibers. These aggregated nucleoli then protrude into the cytoplasm forming blebs (Fig. 14). Cytokinesis does not occur and the nuclear cycle is initiated again. In spite of the failure of cytokinesis the number of chromatin bodies double at about this time (compare Fig. 11 and 13). Just how this is accomplished is not clear.

The RNA bodies reappear at 4 hours and ½ hour later blebs form on the nuclear envelope once again, this time extending farther into the cytoplasm. Occasionally, a bleb will contain RNA body material, as well as nucleoli. When sections at this stage are treated with RNase and examined electronoptically, complete destruction of the RNA bodies is observed (Elliott *et al.*, 1962).

The chemical nature of the bleb contents can be demonstrated with whole mounts, examined under the light microscope. Azure II-stained sections taken from the same blocks prepared for electron microscopy show heavily stained blebs (Elliott *et al.*, 1962). Following RNase treatment only the clear bleb outlines remain indicating that they contain RNA. The blebs are not stained by the Feulgen reaction following treatment with either DNase or RNase. From this cytochemical evidence one can assume that they contain RNA. The electron micrographs show little destruction of the nucleoli following RNase treatment, as indicated earlier, which demonstrates that RNA is more tightly bound to protein than is the case with the RNA bodies.

The cycle of bleb formation and disappearance requires about 30 minutes, beginning as slight evaginations of the macronuclear envelope in the neighborhood of the aggregated nucleoli, then becoming full-sized blebs a few minutes later. They extend some distance into the cytoplasm after which they appear to pinch off as fragments and shortly disappear (Fig. 15).

Several reports have appeared which demonstrate that RNA moves from the nucleus into the cytoplasm (Sirlin, 1960). Most of the evidence

---

the cytoplasm. Note that the fragment contains nucleoli. × 6000.

FIG. 16. A macronucleus during the early stages of amitosis at 5½ hours from the beginning of the initial heat shock. × 4000.

for this transfer is based on tracer studies, very little from morphological observations. Macronuclear chromatin elimination in ciliates has been reported (Diller, 1954; Kidder and Diller, 1934; Scherbaum *et al.,* 1958) but since these fragments are Feulgen-positive, it would appear that they are primarily DNA. Moreover, since they form during cleavage of the macronucleus, they probably are not identical with the blebs seen in these studies. Nucleolar extrusion from the macronucleus of *Paramecium bursaria* has been reported by Ehret and Powers (1955). This occurs during both conjugation and vegetative fission. They also suggest that mitochondria may originate from the macronucleus, an observation which we could not confirm in *T. pyriformis.*

One might ask whether or not the RNA body cycle and the bleb formation is an artifact resulting from the heat-shock treatment. During the course of other studies we have examined many cells in log growth cultures and have on occasion seen nucleoli and RNA bodies inside the blebs. Hence, we are quite certain that the observations reported here occur in normal division of *T. pyriformis.* The heat-shock treatment perhaps accentuates the process.

Approximately 5½ hours after the first heat shock the macronucleus undergoes fission accompanied by cytokinesis (Fig. 16). The nucleoli tend to cluster at the ends of the elongated macronucleus and no RNA bodies can be seen. Fibers near the center of the cleaving macronucleus reported by Roth and Minick (1961) were not seen which may be due to a difference in technique. No evidence was found for fibers which might take part in the formation of the membrane joining the two cells. Once the daughter macronuclei form, RNA bodies can be seen almost immediately. As one would expect, the chromatin bodies have been reduced to one-half and they appear smaller and less dense (Elliott *et al.,* 1962). They soon increase in size but not in number. By 6½ hours from the first heat shock the first cell division has been completed. Within the next 2 hours all of the events involving the RNA bodies and bleb formation are repeated.

These observations may be interpreted to mean that RNA is produced by the chromatin bodies and accumulates in the RNA bodies. The machinery for the transportation of RNA to the RNA body could be through the fine communicating fibers that extend from the chromatin bodies to the RNA body. The RNA bodies with their load of RNA then move to the periphery of the macronucleus and transfer it to the nucleoli which in turn carry it to the cytoplasm via the blebs.

Only fragmentary details are included in this account. Much remains to be accomplished before all the events occurring during division in *T. pyriformis* are understood.

ACKNOWLEDGMENT

The author is grateful to Il Jin Bak and J. R. Kennedy for technical assistance.

REFERENCES

Chatton, E., and Brachon, S. (1935). *Compt. rend. soc. biol.* **118**, 399.
Chatton, E., Lwoff, A., Lwoff, M., and Monod, J. L. (1931). *Compt. rend. soc. biol.* **107**, 540.
Corliss, J. O. (1953). *Parasitology* **43**, 49-81.
Diller, W. F. (1954). *J. Protozool.* **1**, 60-70.
Ehret, C. F., and Powers, E. E. (1955). *Exptl. Cell Research* **9**, 241-257.
Elliott, A. M., and Hayes, R. E. (1953). *Biol. Bull.* **105**, 269.
Elliott, A. M., Kennedy, J. R., and Bak, I. J. (1962). *J. Cell Biol.* **12** (3), 515-531.
Furgason, W. H. (1940). *Arch. Protistenk.* **94**, 244.
Holz, G. G., Scherbaum, O. H., and Williams, N. E. (1957). *Exptl. Cell Research* **13**, 618.
Kidder, G. W., and Diller, W. F. (1934). *Biol. Bull.* **67**, 201-219.
Lwoff, A. (1950). "Problems of Morphogenesis in Ciliates." Wiley, New York.
Mazia, D., Harris, P. J., and Bibring, T. (1960). *J. Biophys. Biochem. Cytol.* **7**, 1.
Metz, C. B., and Westfall, J. A. (1954). *Biol. Bull.* **107**, 106.
Pitelka, D. R. (1961). *J. Protozool.* **8**, 75.
Ray, C., Jr. (1956). *J. Protozool.* **3**, 88-96.
Roth, L. E., and Minick, O. T. (1961). *J. Protozool.* **8**, 12-21.
Sato, H. (1960). *Anat. Record* **138** (3), 381.
Scherbaum, O., and Zeuthen, E. (1953). *Exptl. Cell. Research* **6**, 221-227.
Scherbaum, O. H., Louderback, A. L., and Jahn, T. L. (1958). *Biol. Bull.* **115**, 269-275.
Sedar, A. W., and Rudzinska, M. A. (1956). *J. Biophys. Biochem. Cytol.* **2** (4), Part 2, 331-384.
Sirlin, J. L. (1960). *In* "The Cell Nucleus" (J. S. Mitchell, ed.), pp. 35-48. Academic Press, New York.
Sonneborn, T. M. (1947). *Advances in Genet.* **1**, 258-263.
Tartar, V. (1961). "The Biology of Stentor." Pergamon, New York.
Williams, N. E. (1962). *In* "Synchrony in Cell Division and Growth" (E. Zeuthen, ed.) Wiley (Interscience), New York, Press.
Williams, N. E., and Scherbaum, O. H. (1959). *J. Embryol. Exptl. Morphol.* **7**, 241-256.
Williams, N. E., Anderson, E., Kessel, R., and Beams, H. E. (1960). *J. Protozool.* **7**, Suppl. 27.

Discussion by C. Ray

Emory University, Atlanta, Georgia

Yesterday Dr. Elliott and I looked over his electron-micrographs of *Tetrahymena*, comparing the observations he has made with those I have been making using light microscopy. We looked chiefly at nuclear behavior as found in *Tetrahymena*.

My own observations of *Tetrahymena* have come from the use of fluorescent microscopy and phase microscopy of living animals and of light microscopy of fixed and stained animals. Observations using these three techniques have been most rewarding in a study of the micronucleus during conjugation. The chromosomes are easily seen in prophase of the first meiotic division — particularly at diplotene and diakinesis.

The small number of chromosomes (5 pairs) is an advantage. Moreover, individual chromosomes can be distinguished. One pair is longer than others in the complement and one pair is shorter than all others in the complement. The other three pairs can be distinguished, with some difficulty, by the position of the centromere. The chromosomes can be observed through both meiotic divisions and through the post-meiotic division, which is a mitosis of the haploid complement.

I would like to make a few observations about prophase of the first meiotic division. At the beginning of meiosis the micronucleus moves away from its position near the macronucleus and enlarges. It then becomes ovoid. In the pointed end of the nucleus is a mass of deeply staining heterochromatic portions of the chromosomes. Spinning out from the heterochromatic area are the thin threads of uncoiled, uncondensed chromosomes. The micronucleus next elongates, stretching throughout the length of the animal, producing what has been called the "crescent" stage of the micronucleus. The nuclear envelope is present, as it is throughout division of the micronucleus. The chromatin lies through the central part of this elongated micronucleus as what appears to be a continuous thread (or threads). In some preparations there may be unstained regions along the thread, but we have not been able to satisfy ourselves that these unstained regions are consistent in number or in position along the chromatin material. It appears that the uncoiled chromosomes are disposed end-to-end through the length of the elongated micronucleus. I suggest that the end-to-end arrangement is not a consequence of specific pairing but rather is a nonspecific attraction between the heterochromatic regions at the end of the chromosomes. This is suggested by the fact that in early diplotene the chromosomes are all highly heterochromatic at their ends. There is also a heterochromatic region near the centromeres. Even more pertinent is the position of the chromosomes relative to each other along the elongated nucleus. When the chromosomes first become distinguishable in early diplotene, we can compare different nuclei to determine whether the long chromosome and the short chromosome have a constant positional relation to each other in the linear arrangement. They do not. The long chromosome may be at the end or in an intermediate position in the linear arrangement. The long chromosome is next to the short one in some nuclei but next to a medium one in other nuclei.

In contrast to the detail which can be followed in meiosis, we have been disappointed in what we have been able to make out in the micronucleus during mitosis at fission. We can report little about prophase of the micronucleus during fission aside from the fact that we have been unable to demonstrate any nucleolar material by any of various staining and cytochemical tests.

In especially favorable squash preparations of mitoses during fission we can see chromosomes in late metaphase and in anaphase. In these stages the chromosomes are more slender and somewhat longer than they are during conjugation. The chromosomes are not joined in an end-to-end arrangement. They are crowded together within the nuclear envelope, making it difficult to determine precisely their number. But, the number approximates that of the diploid ten.

Evidence available from a number of sources indicates that the macronucleus controls the metabolic activities of the animal. No strains are known which lack a macronucleus, although strains lacking a micronucleus are common in nature and in the laboratory. Prescott's study (1962, *J. Histochem. and Cytochem.* in press) of site of RNA synthesis in *Tetrahymena,* using radioactive isotopes, indicates that most, if not all, RNA synthesis is in the macronucleus and that the RNA moves out from the macronucleus into the cytoplasm. Elliott's electron-micrographs show a number of nearly spherical bodies distributed around the periphery of the macronucleus. These bodies are fewer in number and larger than the chromatin bodies distributed throughout the macronucleus. Moreover, the peripheral bodies are not homogeneous; they are surrounded by an electron dense shell (protein?) which is not apparent in the chromatin bodies. The pictures suggest that material is budded from or discharged from these bodies into the cytoplasm. Elliott suggests that these bodies are nucleolar materials. We, too, with the light microscope have been observing small bodies distributed around the periphery of the macronucleus in vegetative cells. These are larger than the chromatin bodies and with different staining affinities. On the basis of staining reactions and on the basis of sensitivity of these bodies to ribonuclease treatment, we decided that these bodies were nucleolar materials. Comparing size, position, and number of the bodies in our preparations and those Elliott observes in electron microscope pictures, we believe that we have been observing the same bodies and that they are indeed RNA-containing.

The division of the macronucleus during vegetative reproduction of the animals poses an intriguing set of problems. What mechanism assures the distribution of balanced genetic information to the two daughter macronuclei? At fission the macronucleus pinches into two approximately equal parts. Each daughter cell receives one of these parts. During this division no spindles or no discrete, condensed chromosomes have been seen in light microscope preparations. And yet, a constancy in genetic information must be maintained from generation to generation even in the absence of the usual mitotic process. Elliott has an amicronucleate strain which he has kept in culture since about 1932. During this time there has been no detectable change in physiological traits or morphological characteristics. Therefore some mechanism must assure the distribution of complete and balanced genetic information at the divisions.

During macronuclear division chromatin material is aggregated into numerous small masses. These masses as seen in the light microscope are Feulgen-positive bodies. I have been following the behavior of these bodies in living dividing animals by means of fluorescence microscopy following the use of very dilute (1/100,000) solutions of acridine dyes that are specific for DNA. The number of these bodies distributed through the macronucleus is not precisely constant from cell to cell even in the same preparation. The number seems to be of the order of 40–60 at about the time when the macronucleus is dividing. The distribution of these masses to the two daughter macronuclei is not always equal. The bodies may be parceled out 20–40, 25–35, or other. In most cells some of the DNA bodies fail to get into either daughter macronucleus and are subsequently eliminated.

What can be said about the mechanism of macronuclear division from a study of Elliott's electron micrographs? A study of his pictures would seem to place restrictions upon the possible interpretations accounting for distribution of the chromatin material. In his pictures the chromatin bodies are distributed more or less evenly throughout the macronucleus. No membranes can be detected around these chromatin masses. The masses are of slightly different sizes; their outlines are irregular. Thus, it is unlikely that the macronucleus is a structure containing a group of membrane-bounded diploid micronuclei. In none of the many sections of the macronucleus has there been any evidence of numerous small isolated spindles. There has been no evidence for fibers running through the macronucleus as a whole. Looking closely at the chromatin bodies shown in Elliott's pictures, some thin chromatinic threads projecting from the dense chromatin bodies are visible.

Perhaps each chromatin mass as seen in the electron micrograph is a fusion of heterochromatic parts of a diploid complement (or even of several complements) and the thin threads projecting from the dense masses are the uncoiled, uncondensed euchromatic chromosome portions. In the light microscope the masses are somewhat larger and fewer in number during division than those seen in the electron micrograph. Perhaps at the time of division some separate chromatin masses fuse. The distribution of these masses to the two daughter nuclei would then result in the transmission of complete balanced diploid complements to each of the two daughter nuclei. This suggests further that the euchromatic portions of the chromosomes never coil and become condensed. Recall that in the first meiotic prophase of the micronucleus (during conjugation) the heterochromatin of the chromosomes is fused into a mass at one side of the micronucleus and that the euchromatic portions of the chromosomes project from the heterochromatic masses. The chromosomes appear to remain attached to each other until the time (diplotene) when the chromosome begin to condense. Is the same binding together of a diploid complement taking place in the macronucleus with a failure of chromosome coiling and a consequent persistence of heterochromatic fusion?

Although this suggestion raises questions about the mechanism of replication of the genetic material and duplication of the chromatin bodies, the model does suggest a mechanism whereby distribution of complete diploid subunits to daughter macronuclei would be assured.

Chemical Prerequisites for Cell Division[1]

O. H. Scherbaum

Department of Zoology, University of California, Los Angeles, California

I. Introduction

In the life cycle of a growing cell we can distinguish two phases: one is mitosis, the subject of this symposium. The other phase — less spectacular cytologically speaking, but equally important — is the preparation for mitosis, which may occupy the whole interphase between two subsequent divisions.

During interphase, the cell must complete certain prerequisites for division; superficially it would appear that the cell has to duplicate its structures and metabolic machinery. However, it is not my intention here to consider all these aspects involved in the study of the interphase cell. We will restrict our discussion to a problem which can be summarized in the question: "When does a cell enter mitosis?" or more specifically: "What determines the length of the interphase between two subsequent divisions under optimum conditions?" This question might seem naive if we consider the biochemical and structural complexity of the system, but today I will give you an example of how we can control (or induce) cell division in a microbial mass culture and predict exactly when the cells have to divide. I will also show you how this system can be used in the search for an answer to the question: What is the *last* prerequisite which normally controls the duration of the interphase under optimum conditions?

II. Approach to the Problem

If we search for dividing cells in a microbial mass culture we find invariably that the proliferating cells are the larger ones. It was probably this fact which led to the hypothesis that the cell size plays an important role for the initiation of mitosis. However, a critical comparison of the

[1] Unpublished work from the author's laboratory has been supported by the following grants: (1) Cancer Funds of the University of California, (2) Grant #1626 from the University of California, (3) Grant #G-9082 and #G-18554 from the National Science Foundation, (4) #RG-6461 from the United States Public Health Service.

sizes of cells entering fission in mass cultures of protozoa shows great variability (Scherbaum and Rasch, 1957). It is not surprising then that parameters other than the size in general, such as the nucleo-cytoplasmic ratio and the surface-volume ratio of the cell have been utilized for plausible explanations. A discussion of the problem in question along these lines seems to be fruitless and we shall start from the more general view that the larger cells in a mass culture enter fission with greater probability than the smaller ones. There is a good reason for this phenomenon. A cell which has just been formed by mitosis duplicates itself and when it has reached approximately twice its original size it may enter mitosis. The duration of this interphase — the time the cell spends between two subsequent divisions — is determined partly by chemical and physical factors in the environment of the cell. Under optimum conditions for multiplication this time is the shortest, well reproducible an expression of the capacity of the system for duplication.

If it is not the total cell size per se which provides the stimulus for division could it then be that a small fraction of a cell's mass could have such a role? For example, specific key substances which have to accumulate to a critical level or concentration in the cell could perform the task which has been accredited to the vaguely defined, global parameter "size"? In the approach to this rather complex problem we will inquire what experimental evidence we have in support of such a "trigger" hypothesis and how it may function in the living cell. And, by "trigger" we will mean simply the last prerequisite which, when fulfilled, leaves the cell no choice but to enter mitosis.

A. Experimental Evidence for the Control of Duration of the Interphase

It is common knowledge that a good sample of sea urchin eggs go through a few synchronous division cycles following simultaneous fertilization. Brief treatment of these cells with CO at various times up to a certain point between two successive cleavages causes a delay of the following division equal to the time of action of the gas. Ingenious experimentation along these lines led to a "reservoir hypothesis" (Swann, 1953; Swann, 1954): Energy-rich compounds accumulate in interphase, and when they reach a critical level the initiation of mitosis is triggered.

The importance of the internal environment of the cell as a natural synchronizer is illustrated in an impressive photomicrograph taken by Sonnenblick (1950) some 20 years ago. Upon fertilization of a *Drosophila* egg, mitotic nuclear division occurs in perfect synchrony for as many as twelve successive mitotic cycles. In about 2 hours, more than two thousand nuclei are formed. Each synchronous step lasts only 10 minutes, one of the shortest mitotic cycles ever reported in the literature.

Unfortunately, since growth of the egg occurs in the ovary it is negligible during the early cleavage stages after fertilization. Such a system is therefore not well suited for the study of the subtle interplay between cell growth and mitosis.

As I mentioned previously we are interested in the final prerequisite, which, when fulfilled, leaves the cell no choice but to divide. For the argument's sake we may assume that this prerequisite consists of an "ac-

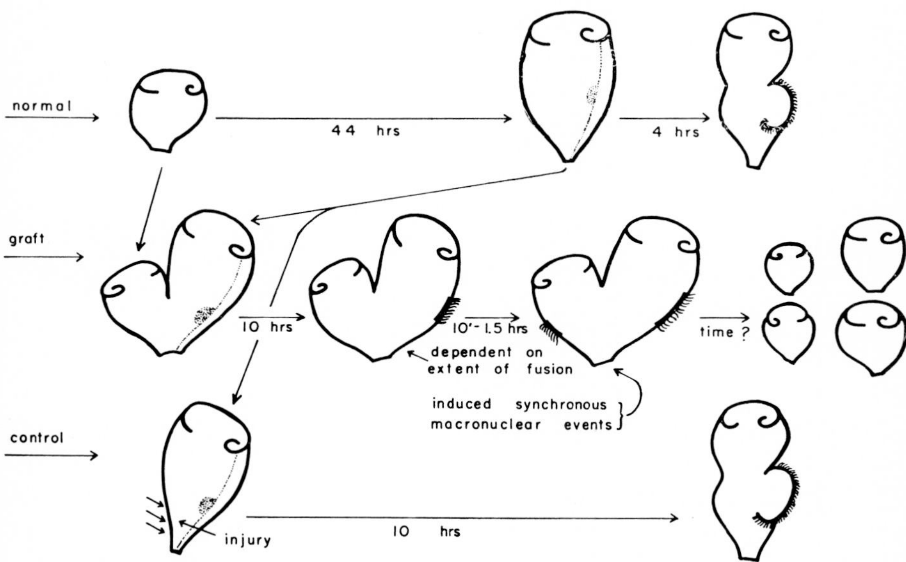

FIG. 1. Life cycle and graft experiment in *Stentor coeruleus*. The normal life cycle lasts about 48 hours (top row). If a normal post-division animal is grafted onto a pre-division animal, further development of adoral primordium and macronucleus in the small cell becomes synchronized with the development in the large pre-division cell (middle row). In a control experiment, injury alone causes a delay similar to that in the grafts (bottom row). Experiments by Weisz (1956); figures redrawn from Weisz (1956).

cumulation" of a certain key substance or substances which must be present at a threshold concentration before mitosis can begin. It follows that more of this material has to be present in a large cell just prior to division than in a young one right after division. Is it then possible to induce division in a young cell prematurely by fusion with a mature cell? Such experiments have been reported by Weisz and show clearly that division can indeed be induced prematurely in a young cell of *Stentor coeroleus* (Weisz, 1956). This is shown in Fig. 1. Mature individuals in the predivision stage show accumulation of kinetosomes in the anarchic field along kinety #1. If such a cell is fused by microsurgery to a young, post-division cell, the following morphogenetic events become synchronized: formation of the adoral membranelles, macronuclear changes, and finally simultaneous cleavage. Such experiments leave no doubt as to the importance of the internal environment in the role of initiation of mitosis. We may then postulate that induction of premature mitosis in *Stentor* is caused by the presence of a key compound or compounds. I might add that the other, more general prerequisite for division—duplication of all structures—has certainly not been fulfilled. *Stentor* may well be a fortunate exception to the observation that a daughter cell normally grows to twice its size before entering mitosis again. This is probably due to the highly developed ability for regeneration (Weisz, 1956; Tartar, 1961). We should not expect such a behavior in other systems, lacking this potential for regeneration.

B. Types of Cellular Control Mechanisms

The prerequisites of cell division consist of a multitude of synthetic reactions aimed at the duplication and separation of highly complex structures. These metabolic processes have been studied extensively in bacterial systems (for further references see Pardee, 1959). The results of these studies show a high flexibility of the cellular machinery. For example, the addition of a suitable substrate to the medium induces synthesis of the enzyme for its utilization (Induction). Conversely the presence of a product can inhibit the formation of the enzyme concerned with its formation (Repression). Not only the size of the machinery can be controlled by precursor or product, also the activity of the machinery may be regulated (feed back control). For example the presence of a surplus of products inhibits enzymatic reactions necessary for the formation of these products.

These are a few examples of the regulatory devices discovered in living cells that serve the important function of integration of the metabolic processes. These integrating devices mediate the processes of duplication of cell structure and machinery always found when the steady state of exponential multiplication under constant environmental conditions (as in a chemostat) or in a changing environment (adaptation) are studied. These control mechanisms, then, have an important function in the economy of the cell, avoiding accumulation of small molecules or the maintenance of a machinery (enzyme systems) when not required. It can be observed that this concept seems to violate the pool or threshold concept mentioned earlier because it appears not to be in the economy of the cell to synthesize and accumulate key substances during the interphase when they cannot be immediately utilized.

We have to inquire now whether it is the sum of all these reactions which determine the duration of the interphase or whether we can single out a sequence of reactions which are primarily concerned with the preparation of a cell for division. If so, what is the chemical nature of it? How does it control the timing of this discontinuous phenomenon of duplication and separation of the duplicated material? And finally, how does this system fit into the "economy concept?"

III. Experimental Separation of Cell Growth and Cell Division

If a mass culture of the ciliate *Tetrahymena pyriformis* GL is exposed to a series of seven temperature cycles (each cycle consists of half an hour at 29°C and half an hour of 34°C) division stages disappear but increase in size continues (Scherbaum and Zeuthen, 1954). After the end of the treatment the cells are kept at the optimum temperature. A lag period ("recovery period") of 1 hour and 20 minutes is followed by a burst of one or more synchronous division cycles.

Temperature cycles of similar types have been worked out for *Tetrahymena pyriformis* mating type 1, variety 1 (synonymous with *Tetrahymena* WH-6) (Holz *et al.*, 1957). It is of interest that the optimum for this strain is 35°C and the shocking level is 43°C. Repetitive temperature changes induce similar effects in other protozoa, bacteria, yeasts, and in cells of higher organisms such as HeLa cells (Scherbaum, 1960a).

Due to the limitations in available time I have to restrict myself in the following discussion to only one system, namely *Tetrahymena*.

A. Cell Growth in Synchronized Cultures of *Tetrahymena*

The increase in cell size of *Tetrahymena* during the heat treatment is an expression of synthesis of cellular structures and machinery during the synchrony-inducing temperature treatment (Scherbaum, 1960a; Scherbaum and Levy, 1961). As an example, I would like to give you some figures (Fig. 2). These values vary somewhat from experiment to experiment depending on the duration of the treatment. Size studies did not support our early assumption that the treatment brings the cells in phase. They do not become more uniform in size, just the contrary is true (Scherbaum, 1956). However, a strikingly uniform response to the temperature treatment in cortical structures and micronuclear behavior was observed. All cells just duplicate their number of kinetosomes (established by direct counts), accumulate kinetosomes in the anarchic field, and all micronuclei are blocked in an anaphase-like configuration (Holz *et al.*, 1957; Williams and Scherbaum, 1959). We may postulate that the synchrony-inducing temperature treatment interferes specifically with morphogenetic events connected with formation of a functional mouth, separation of micronuclear material, and probably with initiation of the cleavage furrow and the division process of the macronucleus.

Since we have succeeded in separating duplication of structures and machinery from specific prerequisites for mitosis proper, let us attempt to characterize this system further and see how we can explain the induction of synchrony of mitosis in *Tetrahymena*.

	A Normal log cell	B Heat treated cell
Dry weight	5.15×10^{-3} μg (1)	11.9×10^{-3} μg (2.31)
D N A	0.33 %	0.26 %
R N A	6.0 %	6.5 %
Protein	57.3 %	52.9 %
Glycogen	5.5 %	11.5 %

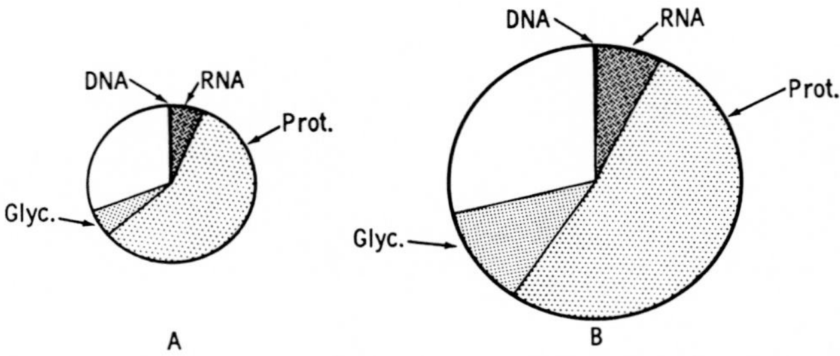

FIG. 2. Cellular constituents in a normal *Tetrahymena* cell (A) and after the temperature treatment (B). The areas of the circles represent the total dry weight of the average cell.

B. Multiplication and Recovery as a Function of Temperature

Figure 3 shows growth of *Tetrahymena* GL at various temperatures. The generation time is shortest at about 29°C and increases on either side beyond this temperature (Thormar, unpublished data and 1959). The recovery (the duration of the induced lag between the end of the treatment and first synchronous division) has an identical optimum temperature, however, it lasts only about 1 hour and 20 minutes (Scherbaum and Zeuthen, 1955). The temperature dependence of this recovery follows a similar pattern of normal multiplication.

If we set the optimum of normal multiplication and recovery equal to 100 we can compute the relative rate of these phenomena as a function of the absolute temperature (Scherbaum, 1957). This is seen in Fig. 4. We can conclude that normal multiplication and recovery have the same temperature dependence. Actually we have suggested that processes occur in the recovering period which prepare the normal cell for division, but are interfered with during the treatment (Scherbaum and Zeuthen, 1955). The optima for this temperature dependence are strain specific under comparable conditions. I have pointed out earlier that the optimum for *Tetrahymena* mating type 1, variety 1, is 35°C; for *Tetrahymena* HS it is 27°C (see Fig. 4). Quite similar temperature characteristics were found for normal multiplication of *E. coli* with an optimum around 39°C.

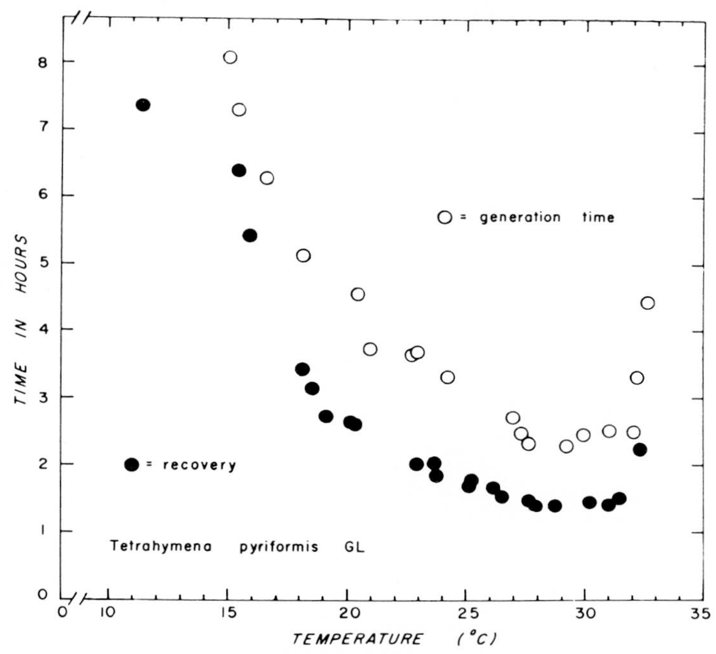

FIG. 3. Generation time (○) and time of recovery (●) of *Tetrahymena pyriformis* GL in hours as a function of temperature. "Recovery" is the time between the end of the synchrony-inducing temperature treatment and the first synchronous division.

It is important for our discussion of the final prerequisite for division that this temperature dependence can be explained on the basis of the assumption that an equilibrium between the active and inactive form of the machinery producing a key substance(s) exists, which is temperature dependent in a simple way: at optimum temperature the equilibrium is in favor of the active form of the catalyst(s), and it is shifted towards the inactive form in the temperature range above and below the optimum.

I might interject here that low temperature can also be used for induction of synchrony. For our discussion it is immaterial at the moment whether we consider a complex machinery or a single enzyme system for which the formula in Fig. 4 has been derived. This figure reveals striking similarities in the temperature-sensitive systems of *E. coli* and *Tetra-*

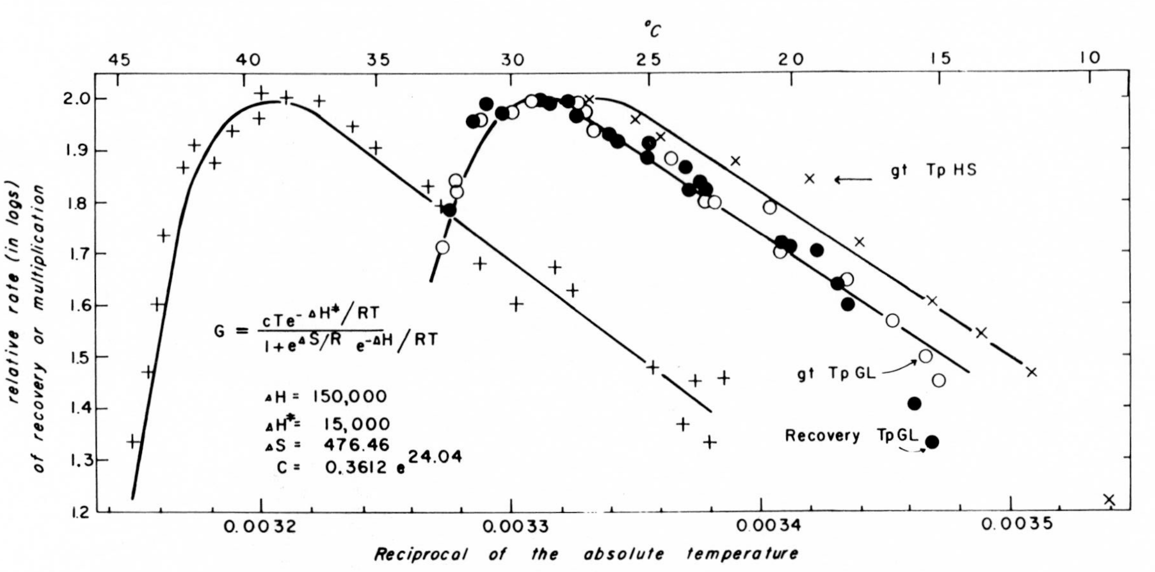

FIG. 4. Relative rates of recovery and multiplication as a function of temperature (maximum rates are set equal to 100). ○: Relative rate of recovery from the temperature treatment in mass cultures of *Tetrahymena pyriformis* GL. ●: Relative rate of multiplication in single cell cultures of *Tetrahymena pyriformis* GL (Thormar, unpublished data and 1959). ×: Relative rate of multiplication in mass cultures of *Tetrahymena pyriformis* (Tp) (Phelps, 1946). +: Relative rate of multiplication of *Escherichia coli*. The points represent experimental data and the curve has been calculated in accordance with the equation and constants given in this figure (Johnson and Lewin, 1946).

hymena. However I am not trying to sell you a "unitarian doctrine" maintaining that the temperature-sensitive systems in such diverse organisms as bacteria and protozoa are chemically identical. The important point here is that we have separated in time two events which occur simultaneously in the normal cell.

C. Temperature Sensitivity of the Interphase

If a single cell of certain age (that is at a known position in interphase) is exposed to a single temperature shock, the following division is always

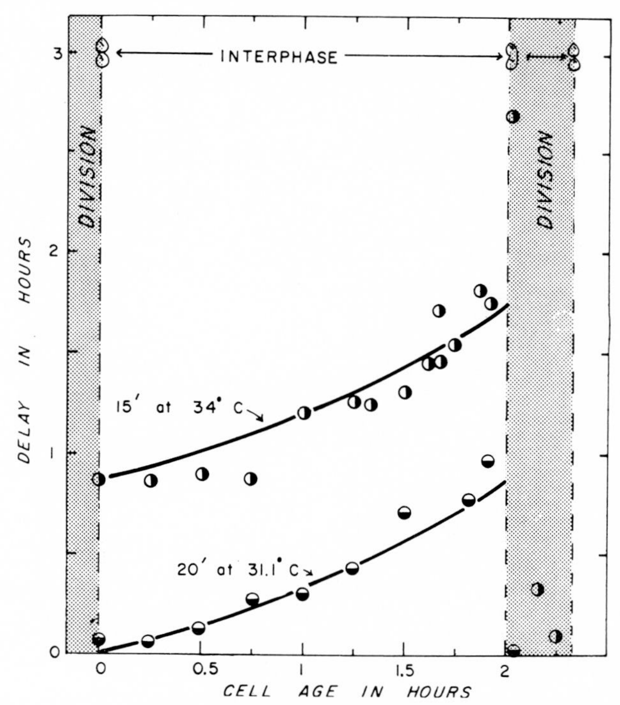

FIG. 5. Division delay of single cells as a function of their position in the interphase. Experimental data (◒, ◑) by Thormar (unpublished data and 1959). Interpreted curves by the author.

delayed (this is to be expected from the preceding discussion of the rate of multiplication of mass cultures). Furthermore this delay always increases with the age of the cell (Fig. 5). These experimental data were obtained by Thormar (1959). However, the curves here are this author's interpretation (I have to admit that these data have been interpreted already three times: Scherbaum, 1957; Zeuthen, 1958; Thormar, 1959). Exponential curves were fitted to the experimental points and we see that these curves agree fairly well with the observed data concerning the events in the interphase. Of course we take into account the technical difficulties connected with the experiments. The use of exponential delay is quite attractive and permits us further quantitation of this phenomenon in connection with induced synchrony in mass cultures.

1. *Temperature-Sensitive Interphase Processes and the Effect of a Single Temperature Shock on Multiplication in a Mass Culture*

The interphase of *T. pyriformis* GL at 29°C lasts about 2 hours. In a mass culture we find all stages, from birth of a cell at time 0 to beginning of cleavage at 2 hours. We can draw a simple diagram assum-

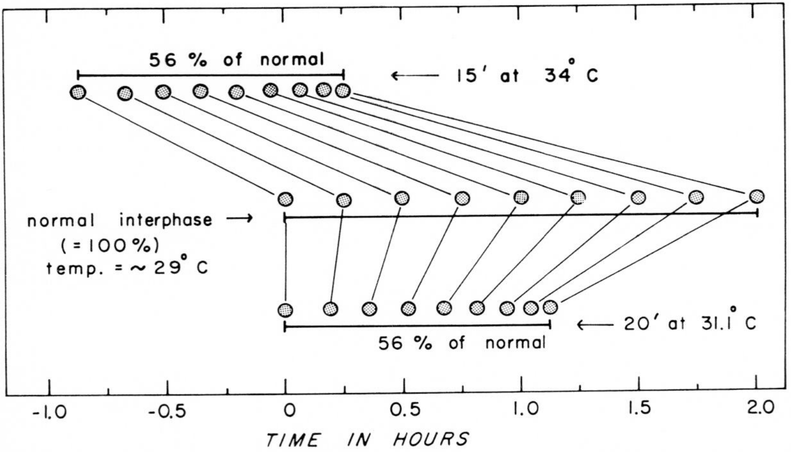

FIG. 6. The effect of single temperature shocks on *Tetrahymena* cells at various stages of the normal interphase. The time delay read from the curves in Fig. 5 was used. Time 0 and 2 hours are the beginning and the end of interphase, respectively.

ing nine cells to be equally spaced in time over this period. For each cell we know the delay caused by a single shock (Fig. 5). We can therefore plot this delay for any temperature. Figure 6 shows the result for 31.1°C and 34°C. It is interesting to note that this exponential delay causes a "compression" of the normal interphase of 2 hours to 56% of this time, irrespective of the temperature (31.1°C or 34°C) applied. There is one difference however: The cells in the compressed interphase are pushed further from division by a shock of 34°C than by 31°C.

2. *Temperature-Sensitive Interphase Processes and the Effect of Repetitive Heat Shocks*

It is known from other experiments (Scherbaum, 1957) that immediately after the heat shock the cells move towards division — now in a "compressed interphase." However, their effort meets frustration when hit by the second shock: they are set back again, but not to the same extent. For a quantitation we plot the exponential delay curve for 34°C in Fig. 5 as the logarithm of delay on the ordinate against time on the abscissa. We see that the delay is 54 minutes for a newborn cell at time 0 and 108 minutes at 2 hours (Fig. 7). From this curve we can read off the delay for any part of the interphase and also the further delay of cells already hit by one or more heat shocks. The repetition of the heat shocks at

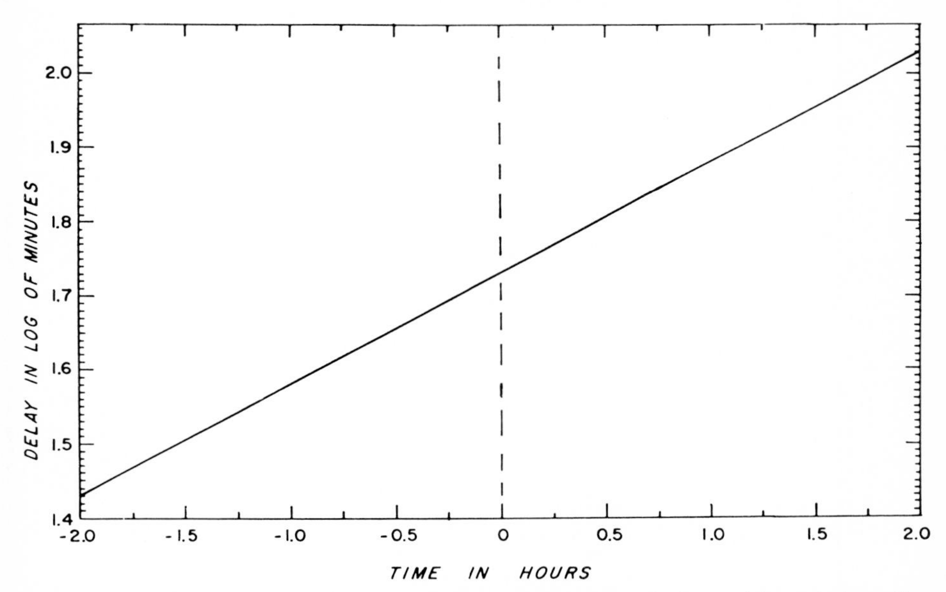

FIG. 7. Division delay (on the ordinate) as a function of the position of the cell in the interphase. The curve for the division delay in *Tetrahymena* after exposure to a single shock (15 minutes at 34°C, Fig. 5) is plotted in log of time in minutes. This plot permits discussion of division delay beyond time 0. This part of the curve (from —2.0 to 0 hours) is an extrapolation of the observed values between 0 and 2 hours.

one-half hour intervals in the standard heat treatment suppresses division but permits duplication of machinery. This is the reason why we have to correct for "increasing efficiency" of the machinery to move the cells towards division. Since at the end of the treatment the cellular machinery has doubled, it will carry out the task to bring the cells toward division twice as fast as at the beginning of the treatment. Assuming a linear increase in mass at the optimum temperature we have to add 5 minutes to each of the six half-hour periods at the optimum temperature between the heat shocks. This calculation revealed quite unexpected results. This is seen in Fig. 8 and can be summarized as follows:

(1) The cells distributed over the interphase of the normal life cycle are pushed together in time due to the different temperature delay at various ages.

(2) The reduction of the interphase follows a characteristic pattern in course of exposure to seven temperature shocks. This reduction is most pronounced during the first heat shock (reduction to 56% of normal cycle). However, during the last (seventh) shock it is only reduced to 74% of the length of interphase found prior to the seventh shock.

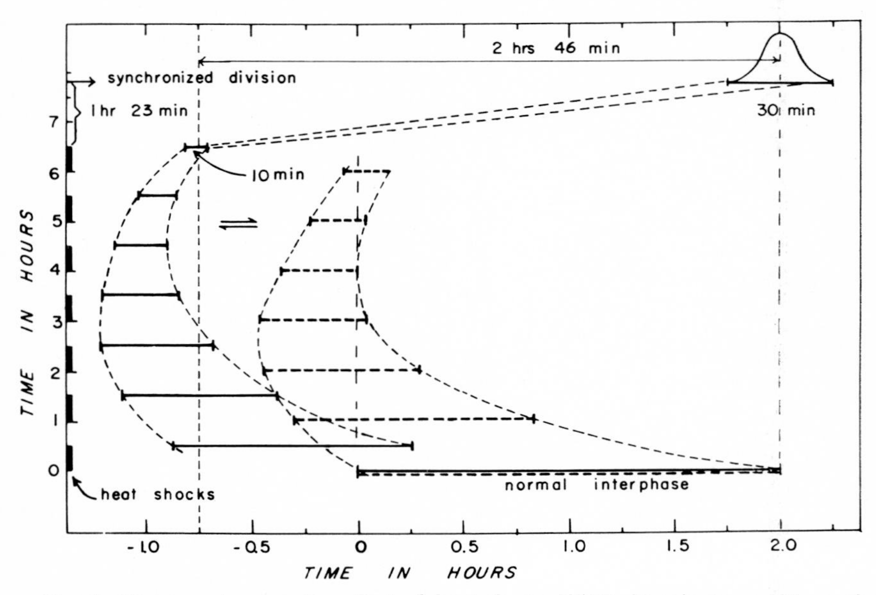

FIG. 8. Diagram showing the effect of heat shocks (34°C, 30 minutes each) on the length of interphase in a mass culture of *Tetrahymena pyriformis*. For the calculation of the division delay by the shock, Fig. 7 was used. In between two successive shocks, the cultures are kept at 29°C. During this period all cells "move" toward division at a normal rate. Further explanation in the text.

(3) At the end of the last shock the normal interphase of 2 hours has been reduced to 10 minutes! These cells, having all a doubled machinery, cover the time towards division with double velocity (as compared a normal young cell). We see that the observation checks very well with this deduction.

(4) The cells in "compressed" interphase of 10 minutes at the end of the treatment move towards division within 1 hour and 20 minutes (recovery period). At that point the interphase has "expanded" to about one-half hour, the time measured from the first to the last cell fission observed in the mass culture. This spread in time (τ-distribution) is

greater for normal cells. For example McDonald (1958) measured the generation times in five hundred fifteen cells and found the following distribution (Fig. 9). A sum curve of these data permits the calculation of the synchronization index (SI), which is 0.43 in this case but is 0.68 for the synchronized system described here (perfect synchrony would have a SI of 1.00).

(5) On the basis of the scheme in Fig. 6 we can predict that even a single shock of 34° C should produce some degree of synchrony. Since

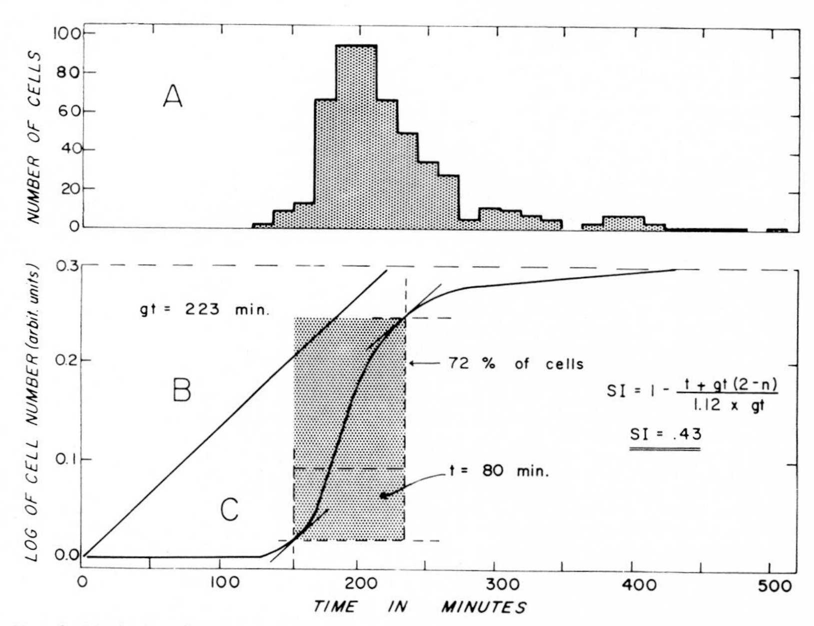

FIG. 9. Variation in generation times of *Tetrahymena pyriformis* H. Histogram (A) represents 515 generation times of isolated cells (redrawn after McDonald, 1958); (B) shows a cumulative curve of (A) and average generation time. Further details concerning the synchronization index (SI) in Scherbaum (1962a).

the cell machinery does not (or at least not significantly) increase, the division delay can be read off directly: it is $2\frac{1}{2}-3$ hours. This agrees very well with earlier observed values (Zeuthen and Scherbaum, 1954).

(6) It becomes obvious from the scheme that the temperature treatment is fully effective only when at the same time cell growth is permitted to occur: Proper aeration and a rich nutrient medium is necessary (population density should not exceed $50–100 \times 10^3$ cells per milliliter when the treatment starts).

Finally we might inquire whether a single temperature shock of longer duration than the half hour used for *Tetrahymena* would improve the degree of synchrony. Observations do not support this possibility. In fact it has been shown that a shock of 60 minutes causes the same delay as a shock of 15 minutes (Fig. 10).

D. Quantitation of the Key Substance(s) in Various Growth Stages

The diagram of Fig. 8 not only provides us with an explanation of the events in time occurring in the mass culture at the optimum temperature and during heat shocks, it permits us also to make some calculations. The result should be useful as a guide for biochemical analyses.

For such a calculation we have to make the following assumptions: (1) The delay curve of Fig. 7 represents a synthesis curve of a key substance(s) necessary for the last prerequisite for the initiation of mitosis.

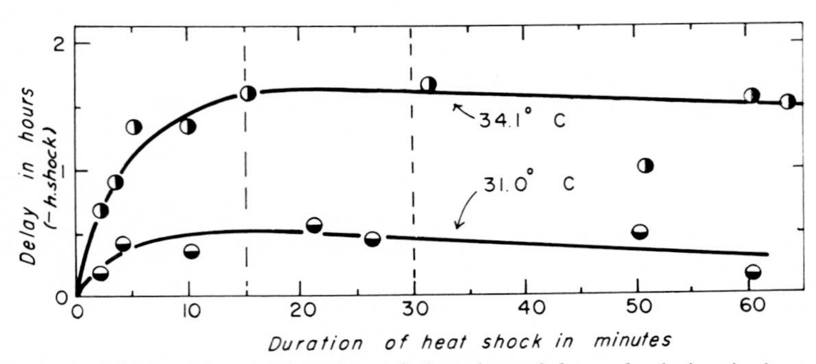

FIG. 10. Division delay as function of duration of heat shock in single cells of *Tetrahymena pyriformis* GL. Approximately 100 minutes after fission, the cells were exposed to 31°C or 34.1°C. In this experiment, the duration of the shock has been subtracted from the delay time (redrawn after Thormar, 1959).

If this is true it would indicate that the material is formed throughout the interphase at an exponential rate. This might strike us as a surprise but we have concluded on another basis that total cell mass (volume) increases almost linearly during the interphase (Scherbaum and Rasch, 1957). Furthermore, exponential synthesis for protein, nucleic acids, and enzymes was reported for synchronized bacteria (Abbo and Pardee, 1960). (2) The increased delay with increasing age is assumed to reflect removal of these key substance(s). This notion agrees with what I have said about the economy of the cellular metabolism: although the formation and accumulation of the key substance(s) occur all through the interphase, part of the pool is removed during the heat shock. At the same time the machinery normally synthesizing this material is more or less inhibited. This fits into the concept of the "absolute reaction rates," explaining the length of the generation times at various temperatures on the basis of the theory of a shift in the equilibrium between active and inactive form of the catalyst(s). We can now proceed by dividing the normal interphase into age groups and expressing the percentage of cells in these groups (age gradient). This is shown in Fig. 11. This percentage can be tabulated. The pool size of the key compound(s) can be read off on Fig. 7 for the mean of each age group and multiplied

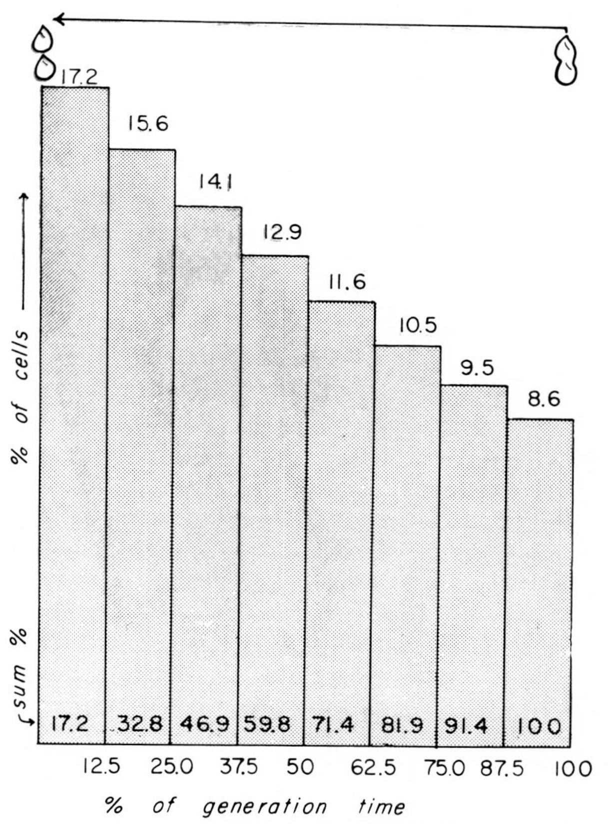

FIG. 11. The age gradient in a normal exponentially growing culture of *Tetrahymena pyriformis*. The generation time has been divided into eight parts. The fraction of the total population found in each part is given on top of the histogram.

with the percentage of cells present in each group. These values can be summed up. The resulting figure can be compared with figures obtained in the same way for other growth stages when the pool size has been affected by the heat shocks. The result of such calculations is shown in Table I.

It is indeed surprising that a key substance which has to double in amount during the normal interphase is neither completely removed by the synchrony-inducing temperature treatment nor does it have to accumulate prior to division. This might be the main reason why such key substances have not as yet been discovered. In a synchronized system

TABLE I

THEORETICAL RELATIVE AMOUNTS OF THE KEY SUBSTANCE (s) AT THREE GROWTH STAGES

Growth stage	Amount	
	Per average cell	Per protein
1. Normal multiplication	1.00	1.00
2. End of seventh heat shock	0.58	0.29
3. Prior to division	1.45	0.73

this normal interplay between cell growth and cell division has been experimentally affected. We can then reasonably assume that the calculated ratio of key substance(s) to total protein must change (Table I).

IV. What Is the Chemical Nature of the Last Prerequisite for Cell Division?

I have previously mentioned that cell growth in *Tetrahymena* cultures occurs during the synchrony-inducing temperature treatment; the cells grow abnormally large but do not become more uniform in size. However, a study of differentiation in the ectoplasmic cortex revealed that all cells are arrested in a phase of development through which an untreated cell has to pass in its normal development. After treatment all cells were found to possess anarchic fields of kinetosomes in the stomatogenic region. The cells remain in this condition for about 50 minutes after the last heat shock. During this time the macronuclear chromatin granules increase in size. At the end of this period morphogenesis is resumed in synchrony and culminates in simultaneous cleavage. All these observations, including counts of the number of kinetosomes, ciliary meridians, etc., have indicated that morphogenetic events in synchronized organisms are no different from these processes in untreated cultures.

There are several plausible explanations possible for this "arrest in synchrony":

(1) Inhibition of the formation and/or distribution of a messenger substance of nuclear origin carrying the code for the spatial arrangement of the accumulated kinetosomes into a functional mouth.

(2) Inhibition of the formation of specific proteins concerned with movement of kinetosomes, nuclear, and cytoplasmic cleavage.

(3) The energy requirements for the blocked morphogenetic and cytokinetic events are not fulfilled.

This list can certainly be extended, but I would like to restrict myself to a discussion of the following aspects: (A) Possible role of deoxyribonucleic acid (DNA), (B) the pattern of isolated microsomal particles, and (C) phosphorus metabolism.

A. DNA in Synchronized *Tetrahymena*

The DNA content in the average cell approximately doubles during the temperature treatment (Scherbaum, 1960a). A further DNA-synthesis occurs during the recovery period (Scherbaum *et al.,* 1959). At the present it does not seem likely that this additional synthesis is a prerequisite for the following synchronous division (Zeuthen, personal communication). Cytospectrophotometric estimates of DNA in single nuclei indicate that the amount of DNA increases in all nuclei during the temperature treatment. It was surprising to find that nuclei which had already an amount of DNA characteristic for normal cell entering division, synthesized "excessive" DNA (Scherbaum *et al.,* 1959)

Results of column chromatography of DNA indicate, however, that some molecular changes might occur during the recovery period (T. Mita and Scherbaum, unpublished data). DNA was isolated by the method of Kay, Simmons, and Dounce (1952) from *Tetrahymena* cells grown in mass culture of seven liters. Approximately 3 mg of DNA were adsorbed onto columns of ECTEOLA and eluted with NaCl of changing concentration (Bendich *et al.,* 1958). The results are shown in Fig. 12. A considerable increase occurs at the end of the temperature treatment (stage 2) in the fraction eluted with 0.5 *M* NaCl. One hour later, just prior to synchronous division, the elution profile is similar to the pattern found in normal cells prior to the treatment. At the present the "homogeneity" of these samples in sedimentation analyses in the analytical ultracentrifuge is under study. I have suggested elsewhere that the temperature treatment might interfere with the formation of high molecular DNA, only eluted by higher salt concentration at a basic pH (Scherbaum, 1962b).

B. Microsomal Particles

It has been repeatedly shown that the sedimentation patterns of microsomal particles depend on the growth stage of the cells. These reports prompted us to attack this problem in *Tetrahymena* (J. Byfield and Scherbaum, unpublished data). First a fractionation scheme for their isolation had to be worked out (Fig. 13). It can be seen that five fractions were obtained. Fraction #3 was pelleted out in a Spinco preparatory centrifuge, resuspended in $10^{-3}M$ phosphate buffer, the concentration of the material was adjusted to 280–300 DU at $\lambda 280$ in a Beckman spectrophotometer, and run in a Spinco analytical ultracentrifuge at a speed of 50,740 rpm. The bar angle was 50. Some preliminary stability tests

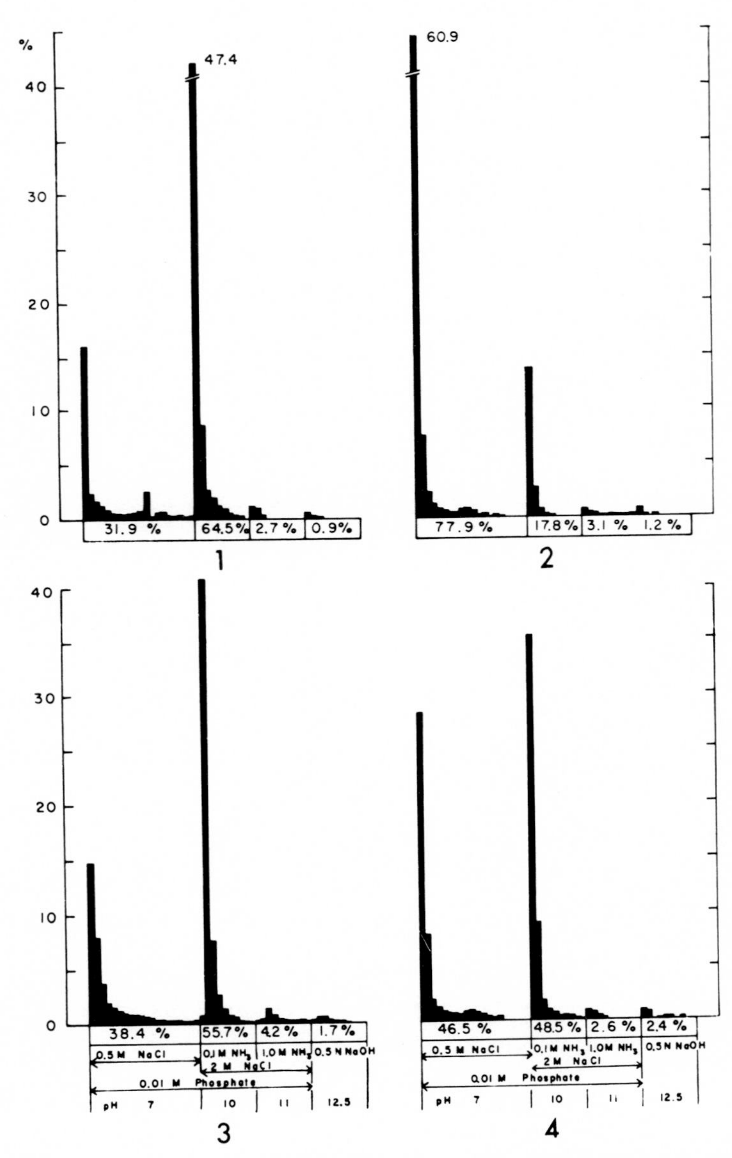

FIG. 12. Elution profiles of *Tetrahymena* DNA chromatographed on columns of ECTEOLA (0.9 × 6 cm; 0.29 m-meq/gram; flow rate about 5 ml/hour of approximately 3 mg DNA for each of the following stages: (1) exponential multiplication, (2) end of the heat treatment, (3) prior to synchronous division, (4) maximum stationary phase.

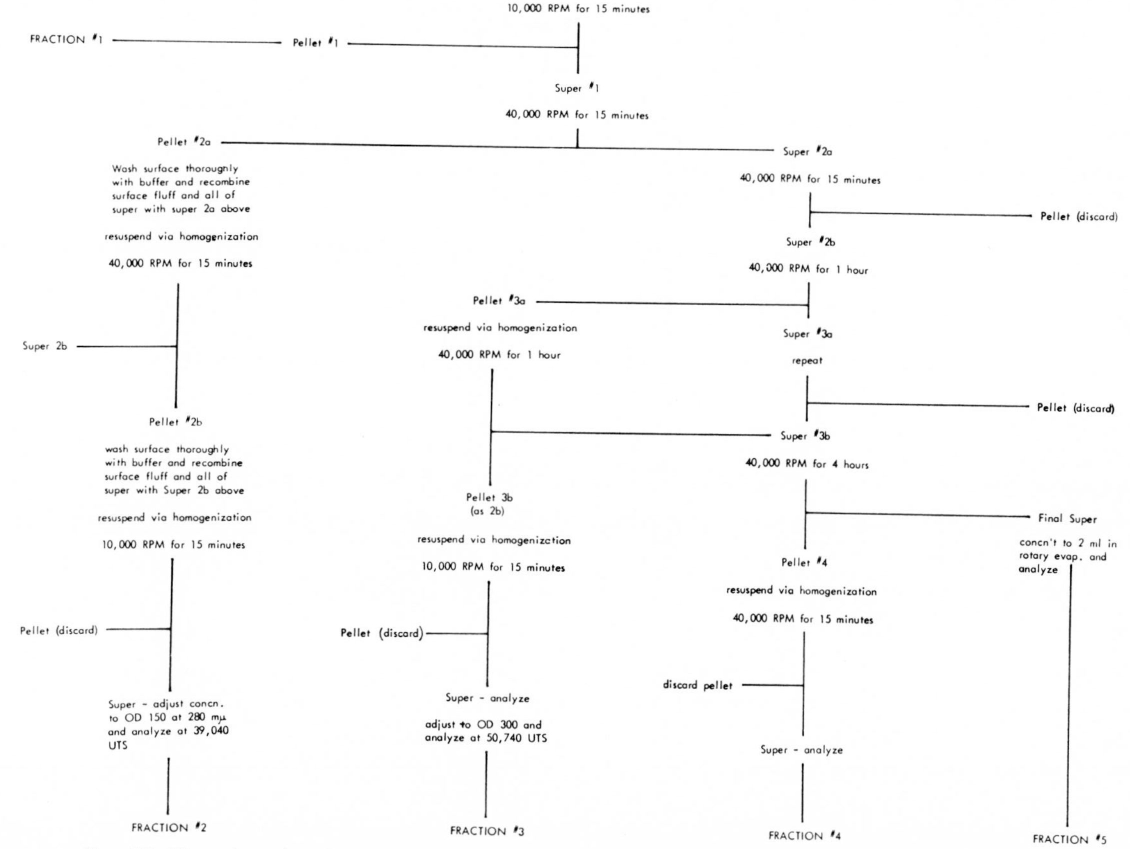

FIG. 13. Flow sheet for the isolation of a microsomal fraction (#3) of *Tetrahymena pyriformis* GL.

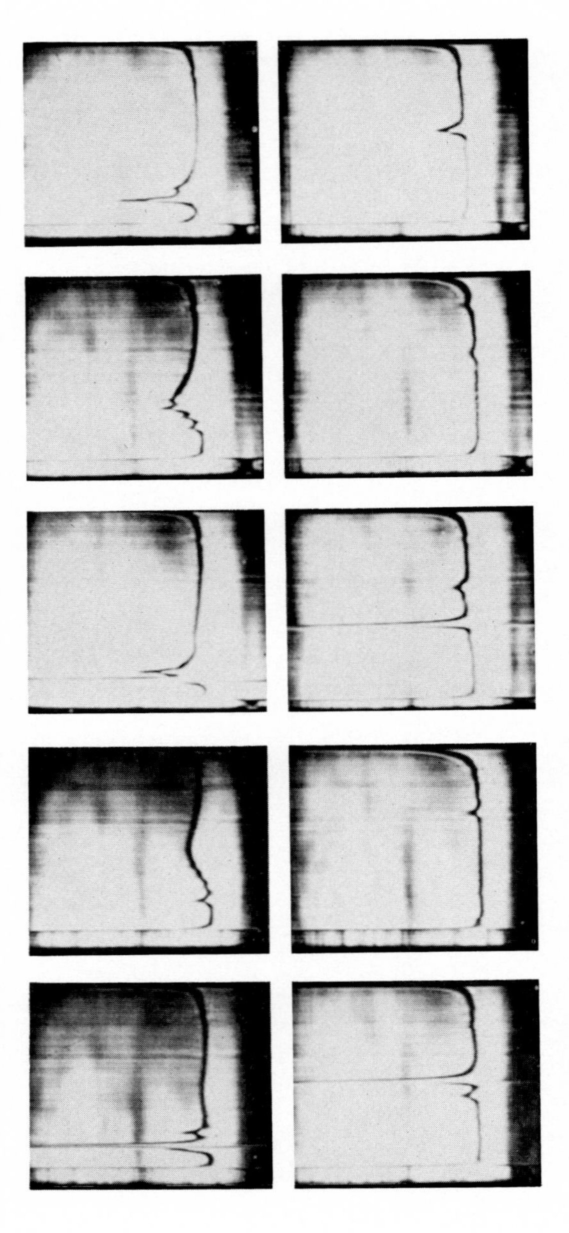

Fig. 14. Sedimentation pattern of microsomal particles isolated from *Tetrahymena* cells in the maximum stationary phase of growth. Top row: pictures taken 2 minutes after UTS at 50,740 rpm. Bottom row: pictures taken 8 minutes after UTS. The sample was divided into 5 parts and treated as follows: (1) normal control; (2) incubation at 37°C for 30 minutes; (3) storage at +5°C for 24 hours; (4) storage at +5°C for 48 hours; (5) storage in a frozen state for 48 hours. Direction of sedimentation is from left to right. Further explanation in the text.

are shown in Fig. 14. A microsomal fraction isolated from stationary phase cells was divided into five aliquots: top row: pictures taken 2 minutes after UTS; bottom row: 8 minutes after UTS. Number 1, Normal untreated sample, no. 2, sample was heated 37°C for 30 minutes. The normal pattern has disappeared obviously due to aggregation of the material; storage of the sample at +5°C for 24 hours shows pattern in no. 3 and storage at +5°C for 48 hours indicates aggregation of the material (no. 4).

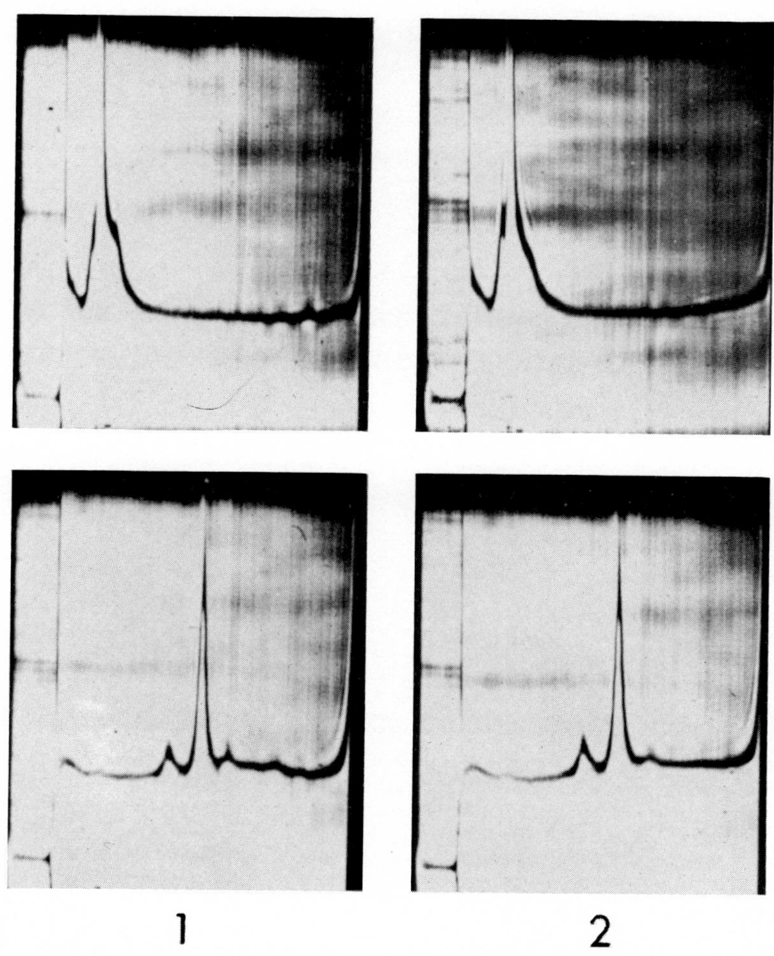

FIG. 15. Sedimentation pattern of microsomal particles isolated from *Tetrahymena* cells in the maximum stationary phase of growth. Top row: pictures taken 2 minutes after UTS at 50,740 rpm. Bottom row: pictures taken 8 minutes after UTS. The sample was divided into two parts (1) $MgCl_2$ was added to make final concentration 10^{-4} *M*; (2) control without $MgCl_2$. Further explanation in the text.

Storage of the sample in frozen state for 48 hours shows a marked decrease in concentration of the material originally present (no. 5). Figure 15 shows the effect of Mg ions. Homogenization and all following resuspensions were carried out in 10^{-4} M $MgCl_2$ in one-half of the sample while a control in normal 10^{-3} phosphate buffer comprised the other. The

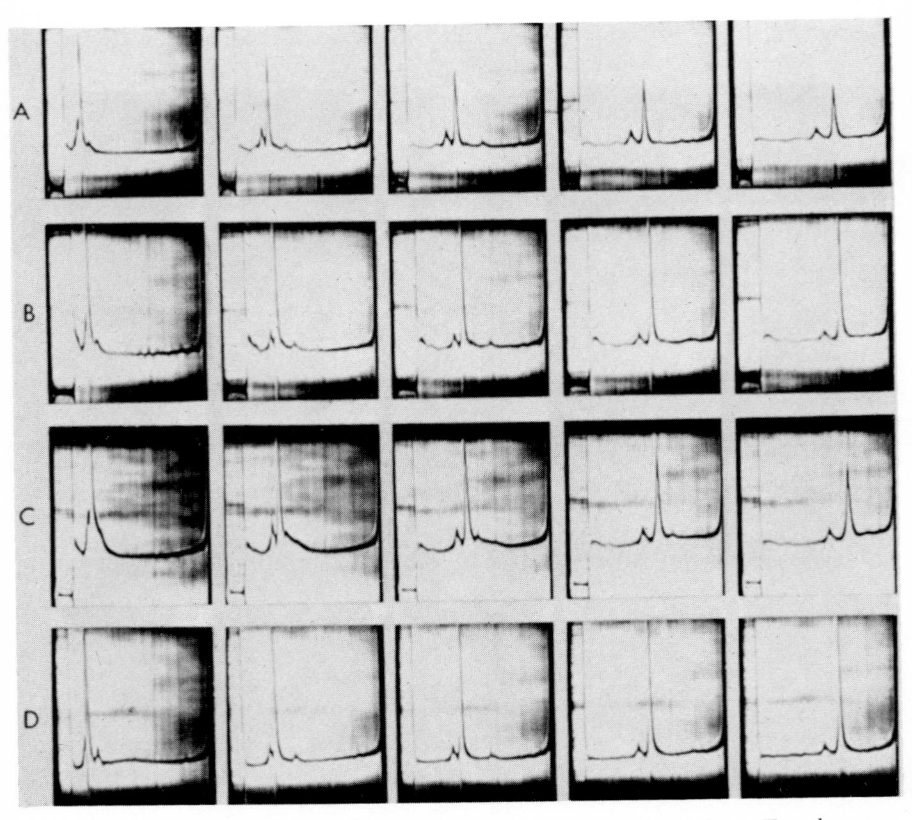

FIG. 16. Sedimentation pattern of microsomal particles isolated from *Tetrahymena* in the following growth stages: (A) normal exponential multiplication, (B) after the end of the heat treatment, (C) 1 hour later, prior to synchronous division, (D) maximum stationary phase, 48 hours after inoculation. The first pictures at the left were taken at 2 minutes after UTS. The following 4 pictures were taken at 2-minute intervals. Direction of sedimentation is from left to right. Further explanation in the text.

pictures on top were taken 2 minutes after UTS, pictures on bottom were taken 6 minutes later. No appreciable difference in the sedimentation pattern could be detected.

Microsomal fractions were prepared from *Tetrahymena* cells harvested at four growth stages (Fig. 16): (A) exponential multiplication; (B) at the end of the synchrony-inducing temperature treatment; (C) 1 hour later, just prior to synchronous fission, and (D) in the maximum stationary phase. In all four samples three microsomal peaks could be observed with the following (approximate) sedimentation constants: 125 S, 65 S, and 50 S. It is surprising indeed, how similar these patterns are, irrespective of the growth stages which they supposedly represent. At the present we attempt to prepare pellets from these peaks for a study of the ultrastructural elements present. This project is under study now with Dr. Sjöstrand's group in our department. I might add that—independent of our work—the effect of ATP on microsomal particles isolated from synchronized *Tetrahymena* has been reported by Plesner (1962).

C. Phosphorus Metabolism

Although the energy requirements of dividing cells have been discussed often, the details are still obscure. Formation and utilization of energy-rich compounds in the cell cycle play an important role in the reservoir concept mentioned above. Plesner reported a detailed account of the nucleoside triphosphate content of synchronized *Tetrahymena* (Zeuthen, 1958; Plesner, 1958a,b, 1959, 1962). Our group was interested in the adenosine triphosphate (ATP) content in relation to the temperature shocks. We have to admit that our early reports (Scherbaum, 1960b; Byfield *et al.*, 1960) are not in keeping with the latest findings. This is due to the presence of a strong ATPase activity in the neutralized acid extracts. In the method finally adopted, the cells are extracted with 2.5% perchloric acid and small aliquots (200μl or less) are added to the buffered luciferase solution together with an internal ATP standard. The emission of light is recorded within 20 seconds (McElroy and Strehler, 1949). Typical analyses are shown in Fig. 17. The difference between duplicate determinations is indicated by the small bars in Experiment A. From the results shown we concluded that inhibition of cell division during the temperature treatment and in the maximum stationary phase does not seem to be accompanied or caused by a decrease in the total cellular ATP content. Obviously the data do not fit the theoretical values calculated on the basis of the pool concept developed above. Another problem concerns the specificity of the enzyme assay for ATP. Other nucleoside triphosphates may contribute to the light emission measured.

The report that a dihydrodiphosphopyridine nucleotide (DPNH) oxidase in *Tetrahymena* homogenates could be inhibited by heating for 10 minutes at 40°C, prompted our studies on synchronized cells (Eichel, 1959). I had suggested elsewhere that such a system might conceivably be an effective regulator in a cell's metabolism (Scherbaum, 1960a). There is some indication that the cellular DPNH content increases during the

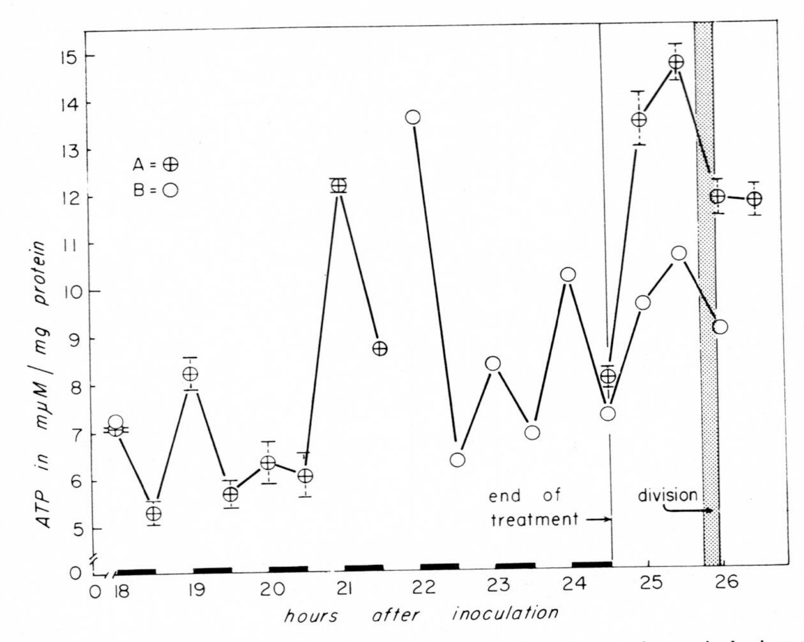

FIG. 17. ATP content of *Tetrahymena* during and after the synchrony-inducing temperature treatment. Further explanation in the text.

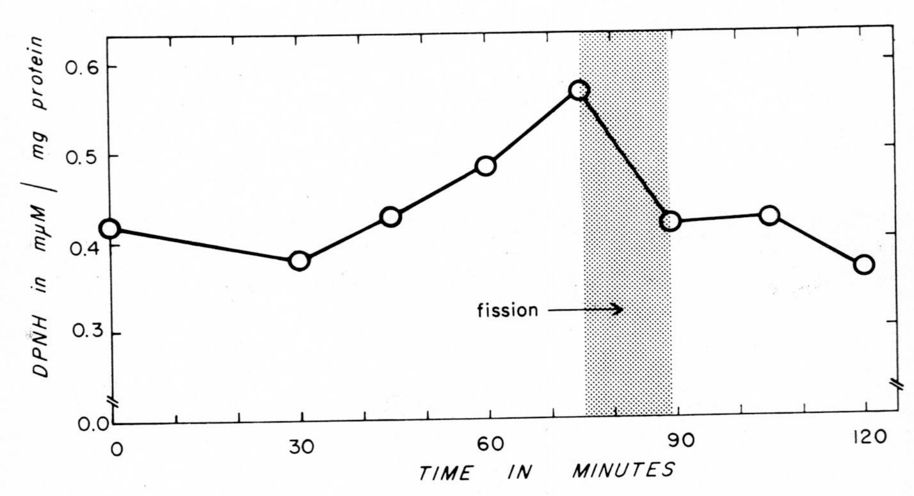

FIG. 18. DPNH content of *Tetrahymena* between the end of the temperature treatment (time 0) and the first synchronous division (dotted area). Further explanation in the text.

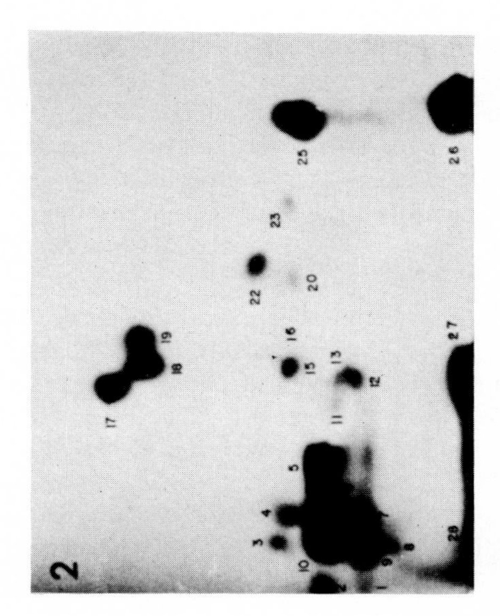

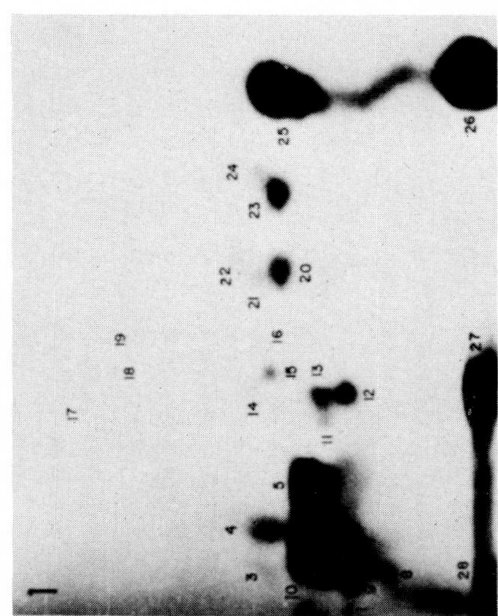

EBEL'S ACID SOLVENT 2nd →

EBEL'S ALKALINE SOLVENT 1st →

Fig. 19. Autographs on X-ray film of paper chromatograms showing the activity of compounds labeled with P^{32}. The following solvents were used: first dimension (isopropanol, isobutanol, NH_4OH, water—40:20:1:39); second dimension (isopropanol, trichloroacetic acid, NH_4OH, water—75:5:0.3:25, Ebel). TCA extracts of *Tetrahymena* corresponding to 10 µg P were used for each chromatogram. 1: Normal exponential multiplication; 2: end of the synchrony-inducing temperature treatment.

recovery period prior to synchronous division (Fig. 18). However, I am hesitant to interpret these findings at the present. Other data have to be scrutinized. For example, the DPNH oxidase activity was assayed in *Tetrahymena* extracts prepared from various growth states of synchrony (A. Nishi and Scherbaum, unpublished data). The activity was found to be within 1.3 and 2.6 mμ moles of DPNH oxidized/minute/mg protein. It appears that the DPNH oxidase is not inhibited in extracts prepared from temperature treated cells. Results of this nature illustrate how difficult it is to extrapolate results obtained from the study with extracts to the intact living cell.

In view of the importance of small phosphorus-containing molecules in the energy metabolism of cells, we have made attempts to separate such compounds occurring in the acid-soluble extracts of *Tetrahymena* (S. C. Chou and O. H. Scherbaum, to be published). The cells were cultured overnight in a proteose-peptone medium containing $P^{32}O_4$ (0.3 μC/ml). The pictures in Fig. 19 show paper chromatograms of phosphorus compounds in acid extracts of cells harvested during normal exponential multiplication (1, Fig. 19) and at the end of the temperature treatment (2, Fig. 19). The method is reproduced very well and permits separation of 28 compounds. For example, spots 17, 18, 19 — although they are not visible in the picture — appear very faint on chromatogram 1 (exponential multiplication); however a marked increase in the amounts of these compounds was found at the end of the heat treatment (chromatogram 2, Fig. 19). This accumulation occurs *only* in the extract of the heat-treated cells; it disappears after the first burst of synchronized cell division; nor can any label be detected in spots 17, 18, or 19 in cells of the early stationary phase of growth. The chemical nature of these compounds is not yet known.

In conclusion we may glance back at the theories and observations concerning the last prerequisite for division which I had outlined in this presentation: the approach is clear, but it will take considerable efforts to get insight into the molecular events controlling and preparing a *Tetrahymena* cell for mitosis. Only then we will see where our guiding hypothesis was right or wrong.

References

Abbo, F. E., and Pardee, A. B. (1960). *Biochim. et Biophys. Acta* **39**, 478.

Ashikawa, J. (1958). *In* "Microsomal Particles and Protein Synthesis" (R. Roberts, ed.). Pergamon, New York.

Bendich, A., Pahl, H. B., Korngold, G. C., Rosenkranz, H. S., and Fresco, J. R. (1958). *J. Am. Chem. Soc.* **80**, 3949.

Byfield, J., Seraydarian, K. H., and Scherbaum, O. H. (1960). *Physiologist* **3**, 35.

Ebel, J. P. (1953). *Bull. soc. chim. France* **20**, 991.

Eichel, H. J. (1959). *Biochim. et Biophys. Acta* **34**, 589.
Holz, G. G., Scherbaum, O. H., and Williams, N. (1957). *Exptl. Cell Research* **13**, 618.
Johnson, F. H., and Lewin, J. (1946). *J. Cellular Comp. Physiol.* **28**, 47.
Kay, E. R. M., Simmons, N. S. and Dounce, A. L. (1952). *J. Am. Chem. Soc.* **74**, 1724.
McDonald, B. B. (1958). *Biol. Bull.* **114**, 71.
McElroy, W. D., and Strehler, B. L. (1949). *Arch. Biochem.* **22**, 420.
Pardee, A. B. (1959). *In* "Ciba Foundation Symposium Regulation of Cell Metabolism" (G. E. W. Wolstenholme, and C. M. O'Connor, eds.), p. 295, Little, Brown, Boston, Massachusetts.
Phelps, A. J. (1946). J. *Exptl. Zool.* **102**, 277.
Plesner, P. (1958a). *Biochim. et Biophys. Acta* **29**, 462.
Plesner, P. (1958b). *Intern. Abstr. Biol. Sci., Suppl.*
Plesner, P. (1959). *Proc. 4th Intern. Congr. Biochem., Vienna, 1958.*
Plesner, P. (1962). *Cold Spring Harbor Symposia Quant. Biol.* in press.
Scherbaum, O. H. (1956). *Exptl. Cell Research* **11**, 464.
Scherbaum, O. H. (1957). *Exptl. Cell Research* **13**, 11.
Scherbaum, O. H. (1960a). *Ann. Rev. Microbiol.* **14**, 283.
Scherbaum, O. H. (1960b). *Ann. N. Y. Acad. Sci.* **90**, 565.
Scherbaum, O. H. (1962a). *J. Protozool.* **9**, 61.
Scherbaum, O. H. (1962b). The Univ. of Michigan 1961 Summer Biological Symposium, in press.
Scherbaum, O. H., and Levy, M. (1961). *Pathol. et Biol.* **9**, 514.
Scherbaum, O. H., and Rasch, G. (1957). *Acta Pathol. Microbiol. Scand.* **41**, 161.
Scherbaum, O. H., and Zeuthen, E. (1954). *Exptl. Cell Research* **6**, 221.
Scherbaum, O. H., and Zeuthen, E. (1955). *Exptl. Cell Research* Suppl. **3**, 312.
Scherbaum, O. H., Louderback, A., and Jahn, T. L. (1959). *Exptl. Cell Research* **18**, 150.
Sonnenblick, B. P. (1950). *In* "Biology of Drosophila" (M. Demerec, ed.), p. 62. Wiley, New York.
Swann, M. M. (1953). *Quart. J. Microscop. Sci.* **94**, 369.
Swann, M. M. (1954). *In* "Recent Developments in Cell Physiology" (J. H. Kitching, ed.), p. 185. Butterworths, London.
Tartar, V. (1961). "The Biology of Stentor." Pergamon, New York.
Thormar, H. (1959). *Compt. ren. Trav. Lab. Carlsberg* **31**, 207.
Weisz, P. B. (1956). *J. Exptl. Zool.* **131**, 137.
Williams, N. E., and Scherbaum, O. H. (1959). *J. Embryol. Exptl. Morphol.* I, 241.
Zeuthen, E. (1958). *Advances in Biol. and Med. Phys.* **6**, 37.
Zeuthen, E., and Scherbaum, O. H. (1954). *In* "Recent Developments in Cell Physiology" (J. A. Kitching, ed.), p. 141. Butterworths, London.

Discussion by Seymour Gelfant

Department of Zoology, Syracuse University, Syracuse, New York

Dr. S. Gelfant (Syracuse University): The main point I would like to bring up for discussion involves the relationship of cell division in synchronized *Tetrahymena* cells as compared to the generalized process of cell division in higher organisms.

Dr. Scherbaum focuses on the period of interphase in his system of synchronized *Tetrahymena* and asks, "What determines the length of interphase between two subsequent divisions, and what is the last prerequisite which controls the duration of interphase?"

Let us relate Dr. Scherbaum's focus of attention in *Tetrahymena* to the generalized cycle of cell division in metazoan plant and animal cell types. The cell division cycle as defined by autoradiographic methods consists of three interphase periods plus the period of mitosis. DNA is synthesized during a specific and delimited period in interphase known as the S-period. This period is preceded and followed by two non-DNA synthetic periods of gaps, called G_1 and G_2, respectively. These are the three interphase periods ($G_1 \rightarrow S \rightarrow G_2$), and if we place the period of mitosis (M) in between the pre-mitotic gap (G_2) and the post mitotic gap (G_1) periods, the cell division cycle is completed.

The first question I would like to ask Dr. Scherbaum is whether he visualizes the same sort of situation in *Tetrahymena* cells and whether the lag period he is studying in *Tetrahymena* may be comparable to the G_2 period, as just described.

The second question I am concerned with is the point of inhibition of cell division in the heat synchronized *Tetrahymena* cells. To begin with, the micronucleus is presumably at anaphase at the time the cells are blocked by heat treatments — which is pretty far along in terms of the process of cell division. As Dr. Elliott had shown yesterday, the macronucleus has already undergone DNA synthesis — which would mean that the macronucleus is in G_2. Moreover, the morphogenetic events associated with division in *Tetrahymena* are fairly well completed, except for a few of the events involved in mouth formation. I would, therefore, like to ask Dr. Scherbaum how he relates his system in *Tetrahymena* to the concept of interphase and to the generalized system of cell division, as known for most other cells. The only major cell division event that has not occurred in the heat-blocked *Tetrahymena* cells is cytoplasmic cleavage (aside from the completion of morphogenesis and macronuclear division). Is it possible that you are studying the biochemical prerequisites for cytoplasmic cleavage?

Now with regard to the "last prerequisite for division" I would like to ask Dr. Scherbaum if this last prerequisite is some sort of a protein, wouldn't this protein have the same sensitivity to heat treatment as other proteins? Is the synthetic machinery not of the same nature in both cases? And how do you explain the difference in temperature sensitivities?

And finally I would like to ask how your views differ from Zeuthen's concerning the heat-shock effect producing synchrony in *Tetrahymena*.

General Discussion

Dr. O. Scherbaum (University of California, Los Angeles): It is difficult to make a comparison of the interphase in *Tetrahymena* with the interesting scheme of distinct interphase periods which Dr. Gelfant has just shown us. His scheme is based on observations that the time of DNA synthesis in mammalian systems is restricted to a well-defined part of the interphase. This may not necessarily be the case in micro-organisms. DNA synthesis in macronuclei and micronuclei of *Paramecium caudatum* is restricted to approximately 25% of the late interphase. Here we have, certainly, a G_1 period and an S-period, but no G_2 period. Such a distinction cannot be made in *Tetrahymena pyriformis* for DNA synthesis in the macronucleus: DNA synthesis occurs throughout the whole interphase at a constant rate (Walker, P. M. B., and Mitchison, J. M., *Exptl. Cell Research* **13**, 167, 1957).

In synchronously dividing cells of *Escherichia coli*, the synthesis of DNA and protein is exponential, and the rate of this synthesis apparently is not changed during the stepwise increase in cell number (Abbo, F. E., and Pardee, A. B., *Biochim. et Biophys. Acta* **39**, 478, 1960).

In the synchronization experiments of *Tetrahymena*, we see that during the treatment the cells synthesized enough DNA, RNA, and protein to divide once in inorganic medium (Hamburger, K., and Zeuthen, E., *Exptl. Cell Research* **13**, 443, 1957). Despite this excessive accumulation of cellular material during the treatment, further synthesis of proteins (Christensson, E., *Acta Physiol. Scand* **45**, 339, 1959), and DNA (Scherbaum, O. H., Louderback, A. L., and Jahn, T. L., *Exptl. Cell Research* **18**, 150, 1959) can occur prior to synchronous division. The synthesis of cellular protein is also expressed in dry weight and volume of the cells. Extensive volume distribution studies have shown that the temperature treatment does not block in the increase in cell size (bulk synthesis) at a specific step in interphase (Scherbaum, O. H., *Exptl. Cell Research* **11**, 464, 1956). Considering DNA synthesis once more, we have estimated microspectrophotometrically the amount of DNA in individual nuclei of *Tetrahymena* cells prior to and after the treatment. There is evidence that DNA synthesis occurs during the treatment, irrespective of whether the cell has enough DNA for division or not (Scherbaum, Louderback and Jahn, *op. cit.*). All these observations do not permit us to suggest that we study the G_2 period mentioned by Dr. Gelfant. In *Tetrahymena* the emphasis is not so much on *sequential* steps (such as the well-defined periods of Dr. Gelfant) but rather on *parallel* events. This brings us to the discussion of your second question: What is the nature and significance of the last prerequisite?

We know that the number of kinetosomes is strictly duplicated during the synchrony-inducing temperature treatment, while the development of the functional mouth by proper arrangement of the kinetosomes is arrested. Other events which are not so obvious as the formation of the mouth, but which are equally important for mitosis may be similarly arrested. For the sake of this discussion, we may assume that a specific protein is concerned with the spatial arrangement of cellular structures within the cell prior to mitosis. We have made this suggestion earlier (Holz, G. G., Scherbaum, O. H., and Williams, N., *Exptl. Cell Research* **13**, 618, 1957).

The work by Zeuthen and his group shows clearly that *p*-fluorophenylalanine interferes with the "recovery" of the heat treated cells (Zeuthen, E., *in* "Growth in Living Systems," M. X. Zarrow, ed. Basic Books, New York, 1961). This means that protein synthesis must occur between the release from the temperature treatment and the first synchronous division, although the cells are already abnormally large. Consequently, the temperature must have interfered with the formation of this protein (or proteins) which is necessary for the following division.

The significance of this "last prerequisite," then, becomes obvious — at least in a speculative model. The cell has a very sensitive regulatory mechanism which can only be slowed down as we change the conditions (food, temperature) from the optimum for division. It has often been observed (for example: James, T. W., and Reed, C. P., *Exptl. Cell Research* **13**, 510, 1957) that cells are larger at lower temperatures, when the generation time is longer. It appears that growth in general, that is, duplication of most of the structures, is less sensitive to temperature changes than the synthesis of this specific division-linked protein. Experimental data suggest that this protein (if it is a protein) is not formed *after* the other prerequisites are fulfilled, just prior to division, but *throughout* the interphase.

Another important question concerning the formation of the "last prerequisite" *throughout* the interphase is the following. Let us assume that a *Tetrahymena* cell has successfully matured and is ready to enter division. At this moment it is exposed to the cyclic temperature treatment. We know that it grows further but does not divide. A sort of quasi-stationary phase has been induced. The question is then what has happened to the postulated division-linked protein? Zeuthen has suggested that this material might be rendered useless by temperature-induced changes in its tertiary structure (Zeuthen, E., *Advances in Biol. and Med. Phys.* **6**, 37, 1958). The evaluation of the experiments concerning the temperature effects on single *Tetrahymena* and on mass cultures of this ciliate leads us to believe that the material in question is broken down during the treatment. It is known from work on bacterial systems that there is a turnover of proteins in *E. coli* during the maximum stationary phase, (for example: Halvorson, H. O., Paper presented at *Conf. on Free Amino Acids, City of Hope, Duarte, California, 1961*) . If we assume that the situation is similar in *Tetrahymena* during the induced lag phase of the temperature treatment, we can speculate along the following lines: If some proteins synthesized during the interphase of the normal cell cycle are concerned with spatial arrangement of the structures duplicated in bulk growth, we can illustrate then the situation in the following scheme:

A: Preparation for division $$AA \underset{k_2}{\overset{k_1}{\rightleftharpoons}} P$$

B: Growth $$AA \underset{k_2}{\overset{k_1}{\rightleftharpoons}} P$$

If the temperature treatment shifts the equilibrium in favor of k_2 in pathway A, the amino acids released from protein degradation could be used for other, unblocked processes in pathway B, such as synthesis of permanent structures. If this reasoning is correct, an old cell prior to division has more "division protein" than a young cell, and the cyclic temperature treatment makes this material available in the form of amino acids for unblocked processes. Actually, we have found that larger cells do grow more than smaller cells when the culture is exposed to the synchrony-inducing temperature treatment.

Now to Dr. Gelfant's last question. Why should the synthesis of structural proteins be less affected by the temperature treatment than the synthesis of the last prerequisite, which we assumed to be a protein also? Zeuthen has suggested that the synchrony-inducing temperature treatment for *Tetrahymena* renders certain structures useless. For example, the temperature shocks might break H-bonds of some macromolecules necessary for division (Zeuthen, E., *loc. cit.*) . In comparing his view with what I have said before, we see that he stresses the change in configuration of the division-linked macromolecule. If these macromolecules are proteins, it follows that the temperature treatment has a quite *specific* effect on these. So here we have one possibility: the last prerequisite (a protein) , once formed, is more sensitive to temperature than similar molecules with other functions. Another possibility we can think of is the following: the *product* is not inactivated, but the *machinery* forming the product or last prerequisite is blocked by the temperature treatment. As a result, little or no product is formed.

There is also a third possibility, which we favor at the present. Under optimum environmental conditions for multiplication, the generation time is minimum, and it cannot be shortened more by increasing the food concentration. Under these conditions, the enzyme systems of the small machinery producing the last prerequisite are always saturated and work at full capacity, while the enzyme systems of the bigger machinery concerned with bulk synthesis are not saturated. If the temperature treatment causes partial inactivation of all systems, it is obvious that the output of the small saturated system is changed noticeably but not so in the larger systems. You see that despite the *identical* temperature sensitivity of the synthetic machineries, there is a *differential* effect on the processes of over-all growth (synthesis of bulk proteins) and of the preparation for cell division. This effect is what we actually can observe. It is another question whether our speculations are correct.

Dr. D. Block (University of Texas) : I was wondering if Dr. Gelfant's proposition could really be eliminated on the basis that the stages move back prior to the previous division. I wonder if still you could not consider synthesis as some prerequisite at a specified stage during the interphase. However, normally this substance might be made not for the next division but for the division succeeding the next division; in these treatments you might be destroying this and the cell might respond to the stress by moving up the time of synthesis of substance X.

Dr. O. Scherbaum: The proposed scheme still allows for the possibility that some prerequisite for division can be formed at a specific stage during the interphase, as long as it is within the system for bulk synthesis. However, the "last prerequisite" is assumed to be formed *throughout* the interphase. Perhaps "last" is the wrong word. It is not called last because it is formed last, but because the rate of its formation is readily affected by temperature changes and thus its formation may fall behind the formation of other material. If this "last prerequisite" or let's say compound X is normally formed not for the first division but for the second division, then the temperature treatment should *not* interfere with the first forthcoming division. Furthermore, if there were a specific stage during the interphase at which compound X would be formed, we have to assume then that this period is the target of the temperature treatment. From Thormar's singly cell-single shock experiments, we see that *all* phases of the cycle are affected by the temperature treatment. For example, a small *Tetrahymena* cell after division is set back 54 minutes by a single shock of 34° C. Dr. Block, how would you account for this setback?

Dr. Block: We could account for it by saying that you have completely eliminated substance X. Normally there is enough, that is, a threshold value of substance X. In, let's say, the G_2 stage you have more synthesized normally. You have enough for division one and in the G_2 stage you have more accumulated which will be for division two, so that if you do heat shock and remove all substance X, you would have to start all over again. This is a little bit contrived, but I think at least it's a possibility and Dr. Gelfant's suggestion couldn't be dismissed because of it.

Dr. O. Scherbaum: In constructing a model, we should consider the following observations: (1) the duration of the interphase of *Tetrahymena* as a function of temperature under normal conditions; (2) the division delay as a function of cell age (i.e., a cell's position in interphase) when exposed to a single shock; (3) the length of the induced lag after a mass culture of *Tetrahymena* has been exposed to a single shock; and (4) the length of the recovery period in synchronized populations of *Tetrahymena*. I cannot see at the moment how these observations can be explained by Dr. Block's suggestion.

Dr. H. Ris (University of Wisconsin) : One of the difficulties in understanding the physiology of mitosis in the past has been, in my view, that it was always treated as a single problem and in a formalistic way. One looked for the trigger of the mitosis. I am getting a bit frustrated that this same view is dominating the discussion today. In modern cell biology the integration of physiological and structural aspects of the cell has become fruitful — and I would like to see more of this in the study of cell division. I think we should pay more attention to cell organization and to differences in the complexity of organization in different types of cells. For instance, we have seen today a tendency to pull in one thing from bacteria, another from *Tetrahymena,* another from Hela cells and so on. I think this approach is misleading. There is quite a difference between *Tetrahymena* and bacterial cells. If you are disturbed by the different timing of DNA synthesis in bacteria you must remember that the genetic system in bacteria is quite different from the chromosomal system in higher cells. Of course they have certain things in common. Both have DNA. But in one case it may be just DNA. In the other it is organized in a complex way together with different kinds of proteins. And so the replication of a chromosome is not the same as replication of DNA in bacteria. Again *Tetrahymena* differs from a vertebrate cell. The presence of a micro- and macronucleus will cause differences and their replication need not be similar. The fact that we have a complex cytoplasmic organization in addition, such as kinetosomes causes further differences. I also heard a statement that cytoplasmic structures replicate before cell division. This again is not universally true. Let me just mention one example. Many algae contain two chloroplasts which are segregated through cell division, so that each daughter cell has only one. They divide afterwards, in early interphase. We have thus a number of replicating systems which may be synchronized or which may be staggered in their replication. All these things complicate the process and so perhaps you have studied just one aspect of division which comes at the end. Before this many essential components of mitosis have already taken place and all these and their complex interrelation must be taken into account in a discussion of the physiological aspects of mitosis.

DR. SCHERBAUM: I agree with Dr. Ris that the rapid advances in the elucidation of cellular ultrastructures and the integration of structure and function have been rewarding to cell physiology. As an example that this approach has not been neglected with *Tetrahymena,* I would like to draw attention to Dr. Elliott's interesting electron micrographs of *Tetrahymena.* The photographs showed cells in various physiological stages such as logarithmic phase, stationary phase, during and after the synchrony-inducing temperature treatment. In our laboratory we have spent considerable time to estimate the amount of DNA in individual nuclei cytospectrophotometrically and we have related the data to the physiological stage from which the cells were taken (Scherbaum, Louderback, and Jahn, *Exptl. Cell Research* **18**, 150, 1959). We have also correlated the cortical events in *Tetrahymena* with the time of the synchrony-inducing temperature treatment. At the present, a project is under way which is the joint effort of our group and Dr. Sjöstrand's laboratory in our department. The sedimentation analyses of the microsomal fraction which I have shown you before are being done with a thin membrane inserted in the sedimentation cell of the ultracentrifuge. We hope that the material pelleted out on this membrane might reveal a stratification in the electron microscope. Our hope is to correlate the observed sedimentation pattern presented here with a stratification of structural material.

As to Dr. Ris's comment on the comparison of different biological systems such as protozoa, bacteria, and mammalian cells, I certainly did not want to sell a unitarian doctrine or a trigger hypothesis. Although we have to take into consideration the structural and metabolic differences of a bacterium and a mammalian cell, we are allowed, I think, to compare these systems in one or the other aspect, such as the processes of the intermediary metabolism, presence or absence of enzymes of certain pathways, etc. Similarly, we can study the temperature effect on the generation time and compare the kinetic data so obtained. Temperature is an ecological factor and any cell, regardless of its complexity, has to cope with its environment. Even in the simplest type there must be a choice for alternative pathways so that the system can "decide" whether to divide or not. The mode and the time of duplication of cellular structures are then questions which we would like to know, but which are secondary in importance to the problem of learning something about metabolic control in connection with division. We have seen in *Tetrahymena* that these two processes —

duplication of the cellular structures and preparation for division — can experimentally be separated.

The similarity of the temperature dependence of over-all growth and replication in microbial systems, the response to single temperature shocks, the distribution of the generation times of individual cells in a mass culture, and the induced synchrony described for *Tetrahymena* have led us to suggest that we are not studying the *last* step in a series of events, but rather processes occurring *during* the whole interphase and distinct from duplication of the bulk of the cellular material. Similar kinetic data for multiplication in different microorganisms should not lead us to conclude that this is an expression of the participation of the *same* chemical pathways in different systems. For example, if the energy metabolism is involved in all systems and a certain energy demand for division must be met, it could be supplied as a polyphosphate in a bacterium or as phosphocreatine in a mammalian cell. We are not concerned with a search for a universal trigger, but specifically with a study of the biochemical systems in *Tetrahymena* which are responsible for temperature-induced division synchrony.

DR. GELFANT: I just have a short comment. There is really one universal thing that you can look for which is not a specific chemical prerequisite. As you showed in *Stentor,* the injury, is a phenomenon that will really kick a wide variety of plant and animal cell types off into mitosis. So as an experiment tool, it might be interesting to make some studies to relate chemical analysis to an injury stimulus.

DR. SCHERBAUM: This is a very interesting suggestion. I may add that in the elegant experiment by Weisz on Stentor, which I mentioned earlier, injury control experiments were also done. Injury which was comparable to the injury done during grafting caused a division delay comparable to the delay caused by grafting, that is, about 10 hours.

DR. I. A. BERNSTEIN (University of Michigan): I didn't catch whether you indicated that the compounds that incorporate P^{32} were inorganic or organic. If organic, were you, able to determine whether they had any ultraviolet absorption?

DR. O. SCHERBAUM: All we can say at the present is that the compounds are acid soluble. The chromatograms were run in the second dimension with TCA as the solvent. Since TCA absorbs UV, we tried to remove it with ether. Under these conditions we could not locate the material at $\lambda 260$. I might add that biochemical analyses yielded a phosphorus containing fraction, precipitable with barium salt at pH 2.5 and 4.5. This finding indicates that inorganic polyphosphates are present in *Tetrahymena.*

Chemical Aspects of the Isolated Mitotic Apparatus[1]

Arthur M. Zimmerman

Department of Pharmacology, State University of New York, Downstate Medical Center, Brooklyn, New York

I. Introduction

Interest in mitosis, especially the mechanism, has led to extensive investigation of the cell organelles which are directly responsible for the separation of the chromosomes. The mitotic spindles and asters, whose primary function is to carry to each daughter cell its proper complement of hereditary factors, probably play a pivotal role in the control of cell division. However, the actual controlling mechanism of mitosis remains elusive; in fact, we are gathered at this symposium in an attempt to elucidate some of the complex problems associated with cell division.

Extensive cytochemical studies have been conducted on dividing cells and these studies have been most informative in helping to elucidate the nature of mitotic structures. Another approach in studying the character of the organelles responsible for mitosis is that of isolating the structure which is to be studied from the surrounding cell matter and subjecting the isolated material to analysis. In this manner it is possible to treat the isolated structure as though we were dealing with a homogeneous biochemical material and thus analyze it accordingly.

In the last decade methods have been devised which permit one to isolate the mitotic complex from dividing sea urchin eggs in sufficient quantities for biochemical and physiochemical analysis. Mazia and Dan, in 1952, found that if sea urchin eggs in metaphase are treated with cold ethanol (−10°C), and the mitotic apparatus subsequently stabilized with H_2O_2, the spindle-aster-chromosome complex (mitotic apparatus) may be isolated from the cell by solubilizing the surrounding cytoplasm with Duponol. Shortly afterward, in 1954, Dr. Mazia reported that hydrogen peroxide was not necessary for stabilizing the mitotic apparatus and more gentle methods of isolation could be obtained by "selectively

[1] The work described in this paper was supported in part by a grant from the National Institutes of Health, United States Public Health Service.

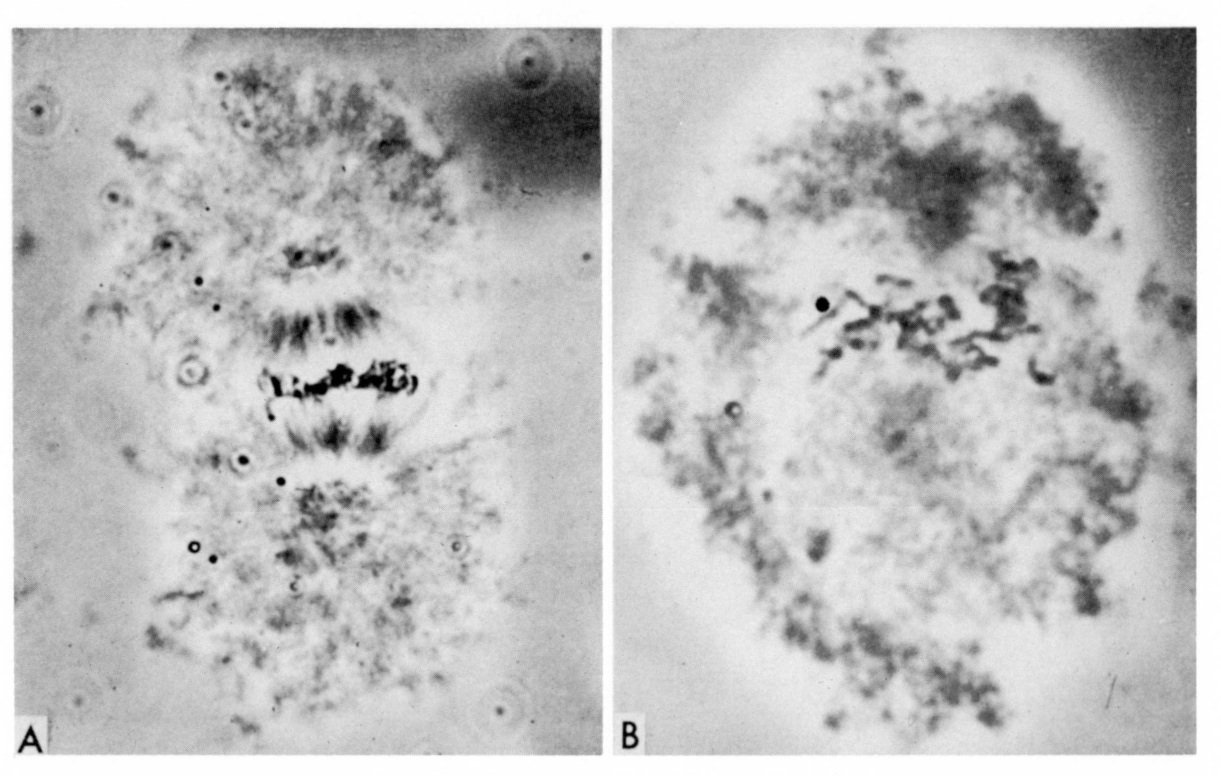

Fig. 1. Mitotic apparatus isolated from eggs of *Strongylocentrotus purpuratus*. A. Mitotic apparatus isolated 75 minutes after insemination at 18°C. B. Mitotic apparatus isolated after 10 minute exposure to 0.075 *M* mercaptoethanol; this photo illustrates the extensive structural disorganization of the mitotic apparatus after brief exposure to mercaptoethanol.

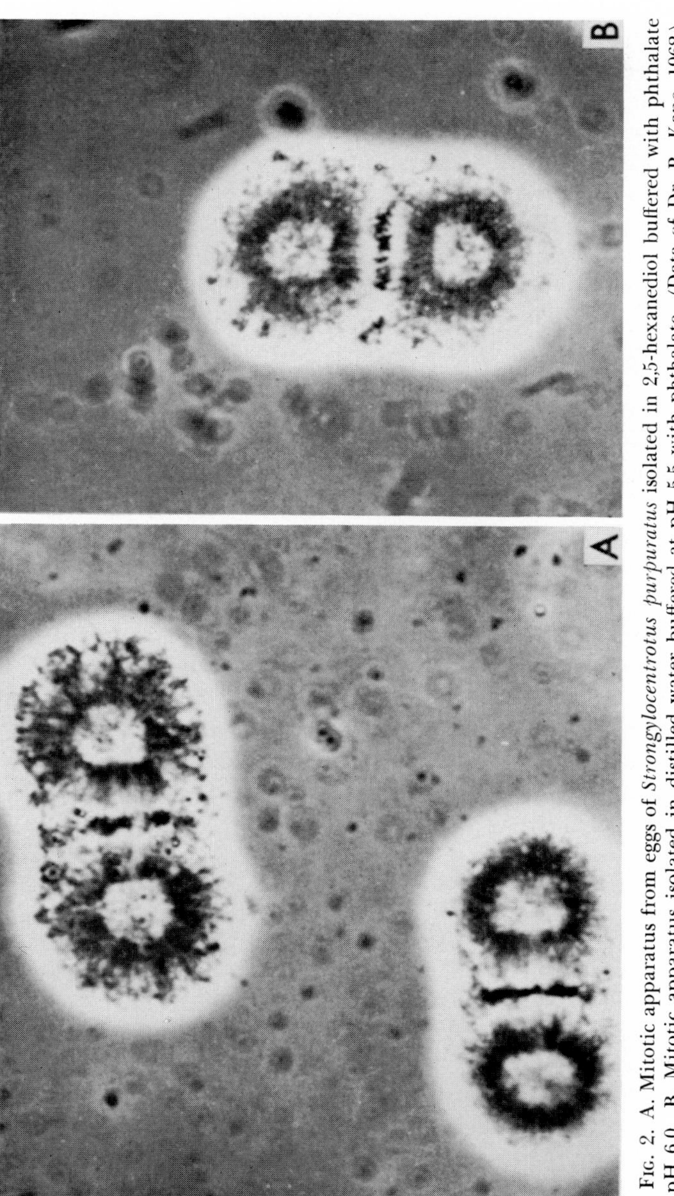

FIG. 2. A. Mitotic apparatus from eggs of *Strongylocentrotus purpuratus* isolated in 2,5-hexanediol buffered with phthalate at pH 6.0. B. Mitotic apparatus isolated in distilled water buffered at pH 5.5 with phthalate. (Data of Dr. R. Kane, 1962.)

solubilizing" the surrounding cytoplasm of the alcohol-treated eggs with a solution of 1% digitonin (see Fig. 1A).

More recently it has been possible to isolate the mitotic complex directly from living cells. Mazia (1959a,b, 1961), in collaboration with J. M. Mitchison and H. Medina (Mazia *et al.,* 1961b), reported that when demembranated and dehyalinated sea urchin eggs are placed in a dextrose-versene-dithiodiglycol medium at pH 6.2–6.3 and briskly shaken, the cytoplasm surrounding the mitotic apparatus disperses into a smooth particulate suspension and the intact mitotic apparatus is released (called the DTDG method). Probably, the dithiodiglycol stabilizes the mitotic apparatus by oxidizing the thiol groups of the mitotic apparatus, or alternately the existing disulfide groups are stabilized by supplying an excess of S-S compound. The versene (ethylenediaminetetraacetic acid) is necessary for removal of the remaining calcium thus preventing aggregation of the dispersed cytoplasmic particles. The dextrose was found to be a favorable medium for the easy dispersal of the cytoplasm.

Robert Kane (1962) also reported the isolation of mitotic apparatus directly from living cells. By using 1 *M* 2,5-hexanediol, buffered at pH 6.0–6.5 with 0.005 *M* phthalate, mitotic apparatus are isolated from the surrounding cytoplasm (see Fig. 2A). These isolations are pH sensitive. Furthermore, Dr. Kane has been able to isolate mitotic apparatus directly by placing demembranated eggs into distilled water, buffered to pH 5.5 with 0.005 *M* phthalate (see Fig. 2B). However, it is more difficult to disrupt the cells, and the isolated mitotic apparatus are contaminated with yolk and cytoplasmic granules. Dr. Kane has found that the mitotic apparatus isolated either by hexanediol or distilled water display similar solubility characteristics.

II. Biochemical Composition

A. Stability of Isolated Mitotic Apparatus

In general the alcohol-digitonin isolated mitotic apparatus is fairly stable in aqueous media. However, when the pH is increased to 12 or when alkaline thioglycolate is used at pH 11.5, the mitotic apparatus is readily solubilized. On the other hand, when the isolated mitotic apparatus is placed into a solution of Salyrgan (mersalyl acid), at pH 8.5–9.0, there is almost instantaneous solubilization of the alcohol–digitonin isolated mitotic apparatus (Zimmerman, 1958). Similarly, *p*-chloromercuribenzoate will also dissolve the mitotic apparatus. It is of extreme interest that both of these agents, which readily solubilize the mitotic apparatus, have a marked affinity for association to thiol groups. It was established that there is an inverse relationship between the effective solubilizing concentration of these agents and the time for dissolution of the mitotic apparatus (see Fig. 3). In comparing critically the ability of these agents to solubilize the mitotic apparatus, we find that Salyrgan is a more effective solubilizing agent than *p*-chloromercuribenzoate. Although the critical concentration for dissolution of the mitotic apparatus is the same for both agents, it is seen in Table I that Salyrgan is almost twice as effective a solubilizing agent as the *p*-chloromercuribenzoate. At $2 \times 10^{-3} M$ Salyrgan the mitotic apparatus were dissolved in less than 12 minutes, in contrast to the

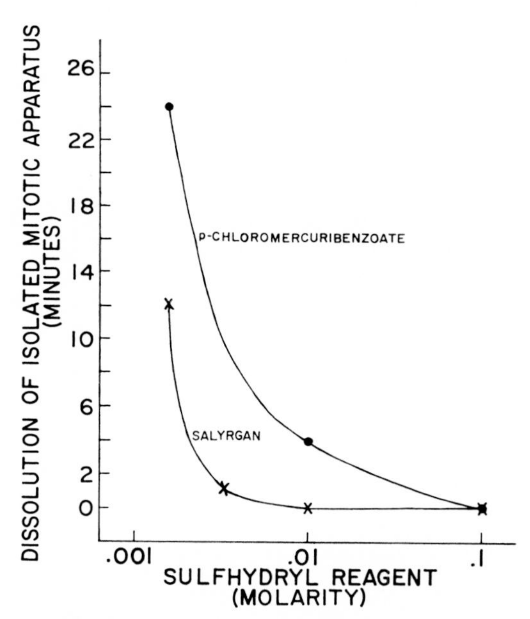

FIG. 3. Effects of sulfhydryl reagents upon dissolution of the mitotic apparatus. The minimum time required to dissolve the mitotic apparatus is plotted as a function of sulfhydryl concentration for both Salyrgan and *p*-chloromercuribenzoate.

p-chloromercuribenzoate which required a little over 25 minutes for complete solubilization of mitotic apparatus. However, with both agents a concentration of $10^{-3} M$ was not sufficient to dissolve the mitotic apparatus when it was placed in Salyrgan or *p*-chloromercuribenzoate.

The sulfur acting reagents, *N*-ethyl maleimide, mercuric acetate (pH 5), and sodium arsenite, were ineffectual as solubilizing agents of the isolated mitotic apparatus.

Since it has been well established that Salyrgan and *p*-chloromercuribenzoate are potent thiol inhibitors, it was not difficult to propose that these agents were acting in a similar fashion on dissolving the alcohol-digitonin mitotic apparatus. Possibly these agents react by breaking the relatively labile hydrogen bonds and thus permitting the agent to react with the free sulfhydryl (SH) groups. Certainly, this supposition would be supported by the observations of Madsen and Cori (1956) who reported that *p*-chloromercuribenzoate is capable of breaking the enzyme α-phosphorylase into smaller fractions, a reaction which is reversible by the addition of cysteine. It is very possible that the sulfhydryl reagent may be acting in a similar manner dissolving the isolated mitotic apparatus. Amberson and colleagues (Amberson *et al.,* 1957; White *et al.,* 1957) have demonstrated that Salyrgan may act on Δ-myosin and thus separate Δ-protein from the Δ-myosin complex. Of course, this does not exclude the possibility that these sulfhydryl reagents are mediating their action through their effect on disulfide bonds. Under mild alkaline conditions disulfide

TABLE I
A Comparison of the Effects of Salyrgan and *p*-Chloromercuribenzoate Solubilization on the Isolated Mitotic Apparatus

Concentration of agent (molarity)	Sulfhydryl reagent	
	Salyrgan	*p*-Chloromercuribenzoate
0.1	Immediate	Immediate
0.01	$<$ 15 seconds	$<$ 4 minutes
0.005	$<$ 60 seconds	—
0.002	$<$ 12 minutes	$>$ 25 minutes
0.001	Faint but visible after 60 minutes	Slight effect

bonds are in a state of dynamic equilibrium with thiol groups (Cecil, 1950; Stricks and Kolthoff, 1953). When hydroxyl ions are available the equilibrium would shift as the frequency of SH formation increases. This reaction is favored when heavy metals are available to tie up the newly formed SH-groups.

As we mentioned earlier, the alcohol-digitonin isolated mitotic apparatus is morphologically relatively stable in aqueous solution; however, this is not quite true with respect to its biochemical nature. In our early studies we frequently stored the isolated mitotic apparatus in 30% glycerol at −10°C until we had sufficient material for analysis. We soon found that storing the mitotic apparatus modified their susceptibility to Salyrgan solubilization. Freshly prepared mitotic apparatus readily dissolve with Salyrgan and *p*-chloromercuribenzoate. However, with the passage of time, even at low temperature, it becomes progressively more difficult to dissolve the mitotic apparatus. Moreover, exposure of the mitotic apparatus to air or to oxidizing agents (ferrocyanide and hydrogen peroxide), within a relatively short interval of time, causes the mitotic apparatus to be resistant to solubilization with Salyrgan. As a result of these observations it was evident that any subsequent investigation pertaining to the structural nature of the isolated mitotic apparatus would of necessity have to be conducted exclusively on "freshly" prepared material, since any delay after isolating the mitotic apparatus would probably not give a true picture of its biochemical and structural composition.

From our discussion to date it has been suggested that hydrogen bonded sulfhydryl groups, as well as S-S (disulfide) interactions with covalent bonds, are probably instrumental in maintaining the structural integrity of the mitotic apparatus. The isolation of the mitotic apparatus and subsequent selective solubilization with Salyrgan and *p*-chloromercuribenzoate lend evidence in support of the hypothesis that SH groups are necessary for mitotic apparatus stability. Other data which we have collected points to the possibility that the integrity and assembly of the mitotic apparatus is probably also dependent upon S-S interactions. In studies by Mazia and the author (Mazia, 1958; Mazia and Zimmerman, 1958) employing the agent mercaptoethanol (monothioethylene glycol) data was collected in support of the hypothesis that the assembly of the mitotic apparatus may also involve S-S interaction. These studies were conducted to test the premise of whether S-S interaction was essential for assembly of the mitotic apparatus. Thus, if S-S interaction were a prime aspect of mitotic apparatus assembly the introduction of an appropriate SH could conceivably prevent disulfide formation, consequently interfering with mitotic apparatus assembly and/or stability. The agent selected for these studies was mercaptoethanol, since it is a rather simple thiol compound of low molecular weight which readily penetrates the cell. When fertilized eggs of *Strongylocentrotus purpuratus* were exposed to inhibiting concentrations (0.075 *M*) of mercaptoethanol at prophase or earlier, spindle assembly was blocked. Furthermore it was found that when the eggs were exposed to the SH compound at the time of mitotic apparatus formation, namely metaphase (see Fig. 1A) — 75 minutes after insemination at 18°C — the effect of the mercaptoethanol was that of structural disorganization.

The appearance of mitotic apparatus isolated from eggs treated with mercaptoethanol at 75 minutes after insemination for a period of 10 minutes is that of an amorphous mass whose spindle and astral fibers are indistinguishable (Fig. 1B). The chromosomes are rather randomly placed in the center of the structure with no apparent orientation. Evidently, these cells are extremely susceptible to mercaptoethanol blockage. Furthermore, the mitotic apparatus is extremely reactive to the thiol compound even for short durations of exposure. The time when the eggs are exposed to mercaptoethanol is very critical. If the fertilized eggs are placed into the mercaptoethanol at early metaphase, mitotic blockage is complete; however, if the cells are placed into the mercaptoethanol at late metaphase

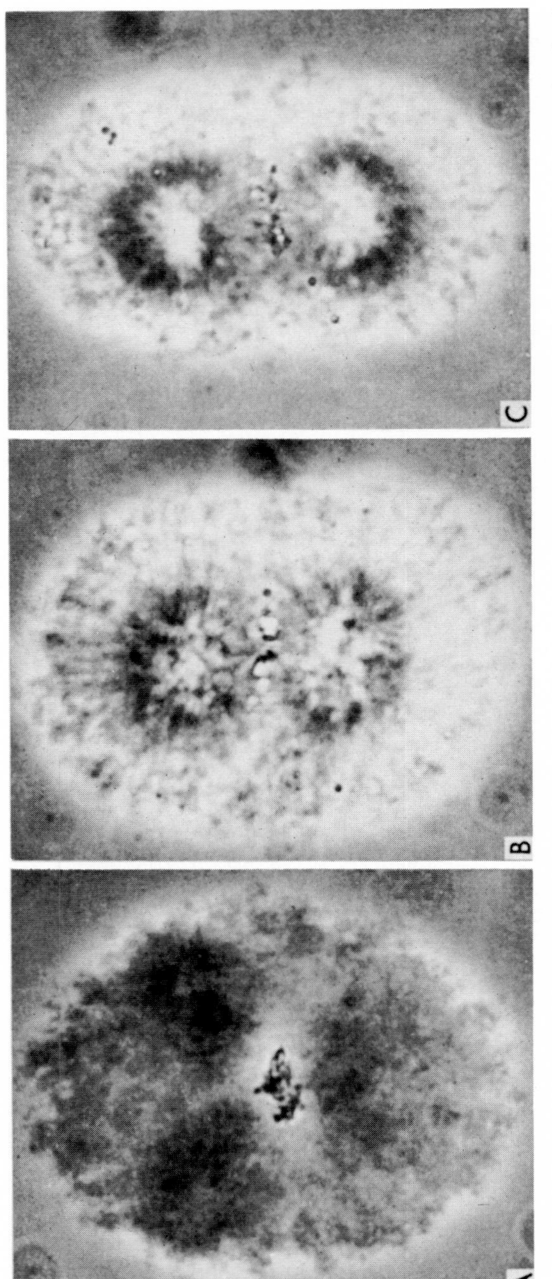

Fɪɢ. 4. Successive stages in reorganization of the mitotic apparatus after treatment with 0.075 *M* mercaptoethanol. A. Isolated mitotic apparatus after exposure to mercaptoethanol for 29 minutes. B. Mitotic apparatus isolated from eggs 4 minutes after removal of the mercaptoethanol shows that structural reorganization has already begun. C. Isolated mitotic apparatus 8 minutes after the eggs are transferred to sea water.

or anaphase there is no blockage of the mitotic apparatus and the cells cleave on schedule.

The recovery from mercaptoethanol is perhaps almost as rapid as that of the induced disorganization. In experiments where the eggs were placed in mercaptoethanol at metaphase for 29 minutes (Fig. 4A) and rapidly washed in sea water the mitotic apparatus readily reforms. As soon as 4 minutes after removal of the mercaptoethanol, recovery is evident with the asters becoming clearly defined (Fig. 4B) ; this progresses to fibrous mitotic spindle formation (Fig. 4C) .

In general these experiments tend to confirm the supposition that the protein-sulfur is probably directly implicated in mitotic formation and assembly.

Fluctuations in the thiol content as well as possible interconversion of disulfide groups to thiol groups and vice versa during mitotic events is not a "new" theory. These same ideas have been expressed by investigators more than three decades ago. The early investigators (Rapkine, 1931; Bolognari, 1952) report that there are fluctuations in the nonprotein bound SH groups during the first mitotic cycle. More recently, however, it has been demonstrated (Neufeld and Mazia, 1957; Sakai and Dan, 1959; N. Kawamura, 1960) that the quantity of nonprotein SH in the early cleavage stages remains constant, but the protein-bound form of SH fluctuates in amount. Dr. N. Kawamura (1960) has demonstrated that in general the concentration of protein SH groups during the first division cycle in sea urchin eggs is a mirror image of the S-S groups during the same period. During the cleavage cycle there is a fall in protein SH groups in early streak stage which increases with the formation of the mitotic mechanism.

Cytochemical evidence for the presence of SH groups in the mitotic apparatus of sea urchin eggs has been reported by several workers (Dan, 1956; Shimamura *et al.*, 1957; Kawamura and Dan, 1958; N. Kawamura, 1960) . In general it appears that the spindle and asters become rich in SH groups with the formation of the mitotic apparatus. N. Kawamura (1960) reported an insignificant amount of S-S groups in the asters and spindle areas, and proposed that the mitotic apparatus is composed of SH proteins which arise from S-S proteins in the egg cytoplasm. An alternate hypothesis was formulated by Shimamura *et al.*, (1957) who reported that the mitotic apparatus deeply stained for S-S groups, and therefore proposed an active conversion of SH groups to S-S groups in the astral region.

B. Protein Composition

With a successful method for isolating the mitotic apparatus firmly established, studies were begun to investigate the biochemical nature of the mitotic mechanism. Mazia and Dan (1952) reported the composition of the mitotic apparatus to be composed primarily of protein. Further studies were made by Mazia and Roslansky (1956) on the total protein content and the amino acid composition of isolated mitotic apparatus. They reported that the protein molecules of the mitotic apparatus were generally of one type. The average total protein of an unfertilized *Strongylocentrotus purpuratus* egg was 6.2×10^{-5} mg. The average protein composition of an alcohol-digitonin apparatus was found to be 0.72×10^{-5} mg. From this data it is possible to calculate what fraction of the entire cell protein is mitotic apparatus protein. Thus, it was determined that the mitotic protein was 11.6% of the entire cell protein.

A question frequently asked regarding the origin of the mitotic apparatus protein material may be well considered at this time. Does the mitotic apparatus form from the nucleus or from the cytoplasm? It is true that in certain forms the achromatic figures may arise within the nucleus as in *Cyclops* egg (Stich, 1954) or outside of the nucleus in the cytoplasm as in *Barbulanympha* (Cleveland, 1953). Similar observations reported by K. Kawamura (1960) on living cells of grasshopper neuroblasts indicate that the achromatic figure arises within the cytoplasm. However, from the biochemical data of Mazia and Roslansky (1956) it is apparent that the mitotic spindle protein is probably primarily of cytoplasmic origin, since the isolated mitotic apparatus contains over 50× as much protein as the sea urchin nucleus (Mazia, 1961).

Recently, Rustad (1959) has demonstrated with the interference microscope that during anaphase a strip of microscopically dense material forms in the center of the interzonal region of the isolated mitotic apparatus. As anaphase continues the total of this interzonal material gradually increases. Rustad has proposed that new protein is being incorporated into the spindle structure at this time.

The immunochemical studies of Went (1959, 1960) and Went and Mazia (1959) shed additional insight as to whether the mitotic apparatus is synthesized in the cell prior to division or whether the mitotic apparatus is assembled from preformed existing submicroscopic substances. Employing the Ouchterlony gel-diffusion technique, antisera were prepared against unfertilized sea urchin egg antigens and mitotic apparatus antigens. The results reveal that of the several antigens in the unfertilized egg there appears to be only one which gives a positive precipitin reaction against a solution of dissolved mitotic apparatus (precursor 1 component). A similar pattern is obtained (Went, 1959) when the mitotic apparatus were either prepared from the living cell by the dithiodiglycol method and dissolved by increase in pH or obtained by the alcohol-digitonin method and dissolved by mersalyl acid. Thus it appears that the various antigen solutions of dissolved mitotic apparatus were immunochemically indistinguishable from one another when challenged with antisera from unfertilized eggs. Furthermore, when antisera prepared from isolated mitotic apparatus was challenged with unfertilized egg antigens, two bands were evident. The precursor 1 component was seen plus a second band which was called the precursor 2 component. Went (1960) reports that these two bands were seen routinely in the dithiodiglycol prepared mitotic apparatus, but only occasionally in the alcohol-digitonin prepared mitotic apparatus. One possible explanation for the immunochemical difference namely, the loss of precursor 2 component, may be a loss of antigen-producing substances in the alcohol-digitonin isolation procedures. However, Went believes that the loss in antigenicity more probably may be caused by the increase in pH necessary to dissolve the alcohol-digitonin mitotic apparatus. In general these immunochemical studies lead one to support the hypothesis that the mitotic apparatus is assembled from performed macromolecular components.

The formation of the mitotic apparatus may involve more than just two precursor components. By using serum of higher activity, two additional components have been found in the mitotic apparatus isolated by the DTDG method and the alcohol-digitonin method. Mazia (1961) states that H. Sauaia (unpublished) has demonstrated the presence of four antigens in the isolated mitotic apparatus.

C. Nucleotides of the Mitotic Apparatus

1. *Whole Cells*

The question of the amount and the presence of RNA in the mitotic apparatus is important in view of its cytochemical identification in a wide variety of cell types. Cytochemically RNA has been found in the mitotic apparatus of both plant and animal cells. In a fine review, Shimamura and Ota (1956) reported that the metaphase spindle in *Lilium* contained an abundance of RNA clearly distinguishable from cytoplasmic RNA. Furthermore, RNA has also been identified cytochemically in the mitotic spindles of *Vicia, Allium Tradescantia,* and *Pinus.* Shimamura and Ota (1956) conclude that generally in plant cells, as the chromosomes separate, the interzonal region between the chromosomes has a staining reaction for RNA at about the same intensity as the cytoplasm. This is somewhat different from that found in animal tissue by other investigators.

Jacobson and Webb (1952) studying chick and mouse fibroblasts in tissue culture, report that the interzonal region has an increased accumulation of RNA. Further evidence by Davies (1952), employing ultraviolet (UV) microscopy, supports the observations of increased interzonal RNA *in vivo* by recording increased UV absorption at 2650Å. However, Montgomery and Bonner (1959) using flying spot and television viewing show no conspicuous increase in the interzonal RNA in newt endothelial cells in culture. Harvey and Lavin (1944) have also determined RNA in sea urchin eggs by means of UV absorption techniques. The increase in interzonal RNA is not at all a restricted observation. Boss (1955) studying newt fibroblasts in culture and Stich (1954) working with *Cyclops* also report increase in interzonal RNA.

Quantitatively the RNA content of the mitotic apparatus has not been as thoroughly investigated as the qualitative aspects because of the technical difficulties involved in analysis of such a small amount of material. However, a quantitative value for RNA in the mitotic spindle of *Cyclops* eggs has been estimated to be about 5%. Stich and McIntyre (1958) report that the mass of the mitotic spindle, as measured by X-ray absorption is reduced 4–6% after digestion with RNase. However, as these authors pointed out, their RNA measurements are within the range of technical error and their data must be considered accordingly.

It is of interest to include the work of Jan Erik Edstrom (1956), who recently reported (1960) on a technique, microphoresis, which permits purine-pyrimidine analysis of nucleoplasm and nucleoli with a sensitivity similar to macrochemical methods. Perhaps this quantitative technique will some day be applicable for analysis of the mitotic spindle apparatus.

2. *Nucleotides of Isolated Mitotic Apparatus.*

Cytochemically RNA has been reported by Rustad (1959) studying the alcohol-digitonin isolated mitotic apparatus. Staining the mitotic apparatus for RNA by gallocyanin-chrom alum and azure B, he reported the presence of RNA in the metaphase mitotic apparatus and an increase in the RNA in the interzonal region during anaphase. It is apparent, therefore, that the cytochemical activity of the alcohol-digitonin isolated mitotic mechanism is in good agreement with the cytochemical studies of intact mitotic apparatus in other cells.

From the early studies (Mazia, 1955) it appeared that the isolated mitotic apparatus contained about 2–3% nucleic acid. The nature of this nucleic acid was not understood, but was presumed to be RNA since it was extractable with cold 5% perchloric acid. However, it was not until recently that nucleic acid associated with isolated mitotic apparatus was chemically identified as RNA and the quantitative relationship to the mitotic apparatus established (Zimmerman, 1960).

The mitotic apparatus isolated from sea urchin eggs were found to contain between 5–6% ribonucleic acid by weight. The ribonucleotide levels were identified by hydrolysis of the isolated mitotic apparatus with piperidine, which was removed by lyophilization. The protein was precipitated out by adjusting the pH to 4.7 (isoelectric point of mitotic apparatus protein) and centrifuging. The nucleotides in the supernatant solution were separated and identified by two-dimensional chromatography and paper electrophoresis.

The identity of four nucleotides may be established from the hydrolyzed RNA of the mitotic apparatus namely, adenylic, cytidylic, uridylic, and guanylic acids. Identification of the four nucleotides by chromatography was accomplished by employing two solvent systems. The butyric acid-ammonia solvent system separates adenylic and cytidylic acids from a mixture of uridylic and guanylic acids. The tertiary amyl alcohol-water-formic acid solvent system separates the uridylic from the guanylic acids (see Fig. 5A). It was also possible to identify the nucleotides by paper electrophoresis. Very sharp separation of the four nucleotides occurred in 2 hours at a field strength of 30 volts/cm in a formate buffer at pH 3.5, with ionic strength (μ) of 0.1. Uridylic acid migrated towards the anode ahead of guanylic acid. The relatively slow moving adenylic acid was sharply separated from cytidylic acid which scarcely migrated at all (see Fig. 5B).

Quantitative relationships of the isolated nucleotides were established by elution of the UV absorbing spots with 0.01 *N* HCl and subsequent spectrophotometric analysis. The mean nucleotide composition of the RNA from isolated mitotic apparatus was established to be 23.9% adenylic, 33.8% guanylic, 24.4% cytidylic, and 17.9% uridylic acids (see Table II).

From the above data it is possible to calculate the amount of RNA in a single mitotic apparatus. On the basis of the average protein per mitotic apparatus calculated to be 0.72×10^{-5} mg (Mazia, 1955; Mazia and Roslansky, 1956), RNA content of the mitotic apparatus would be 5×10^{-7} mg/mitotic apparatus.

In general the RNA of the mitotic apparatus conforms to the regularities in composition as proposed by Elson and Chargaff (1955). They propose that the ratio of the nucleotides having an amino group in the sixth position (adenine and cytidine) to those nucleotides having a keto group in that same position (guanine and uridine) approach unity. In the present study the ratio of the 6 amino/6 keto in the RNA of the isolated mitotic apparatus was found to be 0.94. In the unfertilized eggs the ratio of the 6 amino/6 keto was found to be 1.03. These values compare favorably with those found in the sea urchin *Paracentrotus lividus* whose ratio is 0.992 (Elson *et al.*, 1954). It is interesting to compare these ratios with those found in an RNA contaminant of muscle myosin. Mihalyi *et al.* (1957) report a ratio of 6 amino/6 keto of 0.97 (Table III).

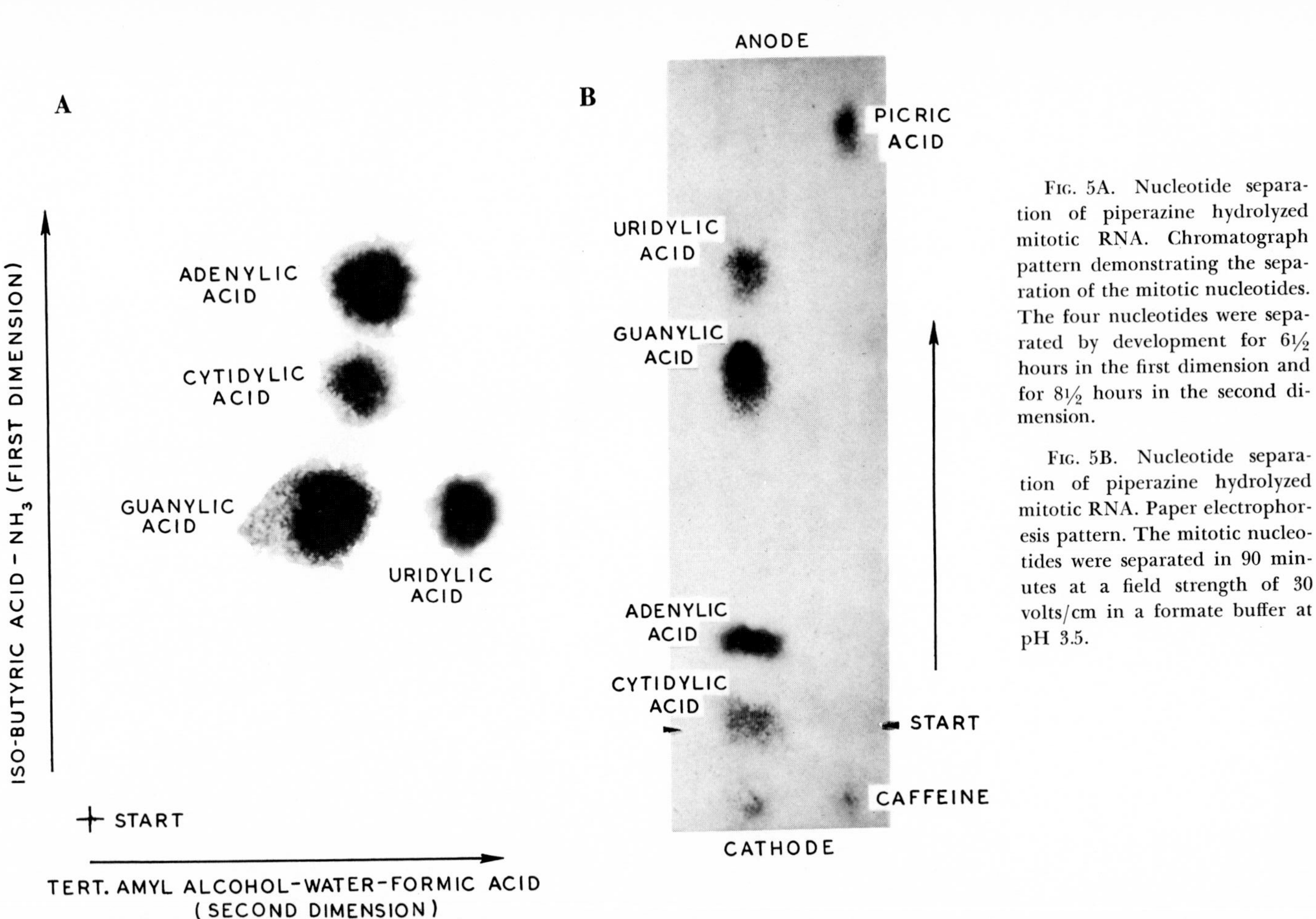

Fig. 5A. Nucleotide separation of piperazine hydrolyzed mitotic RNA. Chromatograph pattern demonstrating the separation of the mitotic nucleotides. The four nucleotides were separated by development for 6½ hours in the first dimension and for 8½ hours in the second dimension.

Fig. 5B. Nucleotide separation of piperazine hydrolyzed mitotic RNA. Paper electrophoresis pattern. The mitotic nucleotides were separated in 90 minutes at a field strength of 30 volts/cm in a formate buffer at pH 3.5.

While there is a relatively small amount of RNA in the mitotic apparatus, it is an integral part of the fibrous elements. And one may speculate that it plays a vital role in mitotic events concerned with division. Perhaps the RNA manifests its action in conjunction with contractile

TABLE II

RNA NUCLEOTIDE COMPOSITION OF ISOLATED MITOTIC APPARATUS

Method of separation	Experiment no.	Mole per cent				Mole Ratio
		Adenylic acid	Guanylic acid	Cytidylic acid	Uridylic acid	6 Amino / 6 Keto
Chromatography	1c	25.1	34.3	25.3	15.3	1.02
	2c	24.2	37.6	23.6	14.6	0.92
	3c	23.8	33.6	24.9	17.7	0.95
	4c	23.5	32.4	24.6	19.5	0.93
Average		24.1	34.5	24.6	16.8	0.95
Electrophoresis	1e	21.7	36.3	22.2	19.8	0.78
	2e	24.4	31.0	25.2	19.4	0.98
	3e	24.4	31.3	25.0	19.3	0.98
Average		23.5	32.9	24.1	19.5	0.91
Average chromatography and electrophoresis		23.9	33.8	24.4	17.9	0.94

TABLE III

COMPARISON OF NUCLEOTIDE COMPOSITION OF RNA FROM SEVERAL DIFFERENT SOURCES

Source	Adenylic	Guanylic	Cytidylic	Uridylic	6 Amino / 6 Keto
Isolated mitotic apparatus (*Strongylocentrotus*)	23.9	33.8	24.4	17.9	0.94
Unfertilized eggs (*Strongylocentrotus*)	24.8	31.3	25.9	18.0	1.03
Paracentrotus eggs and embryos[a]	22.6	29.4	27.2	20.8	0.99
Muscle (rabbit)[b]	18.3	31.9	30.9	18.9	0.97

[a] Data from Elson *et al.*, 1954.
[b] Data from Mihalyi et *al.*, 1957.

proteins of the mitotic apparatus. Mihalyi and co-workers (1957) have been able to isolate RNA up to 1% from rabbit myosin. An alternate possibility is that RNA in the mitotic apparatus is associated with the energetics of division.

Recently Mazia *et al.* (1961a) demonstrated an active and specific ATPase which was found associated with the dithiodiglycol isolated mitotic apparatus. It is reported that this ATPase is dependent upon divalent cations for activation, Mg^{++} being the most effective. Furthermore it was demonstrated that the ATPase splits inosine triphosphate, but to a lesser extent. The enzyme does not split adenosine diphosphate nor does it split cytidine triphosphate, uridine triphosphate nor guanosine triphosphate.

III. Physicochemical Characteristics

The early studies of the physicochemical characteristics of the isolated mitotic apparatus were hindered by the problem of solubilizing the mitotic complex. We assumed that strong solubilizing agents such as alkali would in all probability modify or possibly destroy the characteristics of the mitotic apparatus. Prior to the use of the Salyrgan method of solubilization it was not possible to dissolve the mitotic apparatus without employing relatively strong agents. However, with the Salyrgan dissolution method we were able to take advantage of the solubilizing property and employ this technique for study of the physicochemical properties of the isolated mitotic apparatus. The methods chosen for analysis were Tiselius electrophoresis and analytical ultracentrifugation.

A. Electrophoresis

The isolated mitotic apparatus were dissolved in 0.1 *M* Salyrgan and dialyzed against a phosphate buffer at pH 7.5 or in a veronal-HCl buffer at pH 8.5 $\mu = 0.1$. Two peaks were observed, a major peak which migrated slowly, and a minor peak which migrated rapidly. At pH 7.5 the mobility of the major peak is 5.4×10^{-5} cm^2/volt sec; the minor peak has a mobility of 10.7×10^{-4} cm^2/volt sec.

The electrophoretic study clearly illustrates that the composition of the mitotic apparatus is not homogeneous and that it consists of two molecular components possessing different electrical charges.

It is interesting to note that the patterns observed in these experiments are very similar to those obtained by Mazia (1955) who employed alkaline reagents to solubilize the mitotic apparatus. Mazia (1955) demonstrated spectrophotometrically that when the minor faster moving peak was permitted to migrate out of the electrophoretic cell, the remaining material lost most of its 260 mμ UV absorbing properties. Presumably, therefore, the minor peak contained nucleic acid. In the present study, where Salyrgan was employed as the solubilizing agent, the slower moving component accounts for over 90% of the total material and the minor peak accounts for less than 10%. Assuming that the biochemical analysis of the isolated mitotic apparatus discussed earlier is valid, then it is reasonable to propose that the minor peak found in the electrophoretic patterns of mitotic apparatus solubilized with Salyrgan also contains nucleic acids.

B. Analytical Ultracentrifugation

The ultracentrifugal studies of the Salyrgan solubilized mitotic apparatus indicates that the composition of the mitotic apparatus is not homogeneous, but consists of two distinct molecular species with different sedimentation characteristics. When isolated mitotic apparatus, repeatedly washed at 4°C in distilled water by differential centrifugation, were dissolved in O.1 *M* Salyrgan at a pH between 8.5 and 9.5 and subsequently subjected to analytical ultracentrifugation, two peaks were recorded (see Fig. 6). The major portion of the dissolved mitotic apparatus material is found located in the slower peak which has a s_{20} of 3.7. A smaller fraction of the material is found in the faster moving peak whose s_{20} is 8.6.

The identification of two peaks is found only if the material is not dialyzed prior to ultracentrifugal analysis. When the Salyrgan dissolved mitotic apparatus was dialyzed for 24 hours at pH 7.5–9.0, only one peak was observed. At pH 7.5 the dialyzed mitotic apparatus has a s_{20} of 3.2. The presence of a single peak after dialysis suggests that the solubilized mitotic apparatus is extremely labile. One possible explanation is that there is a breakdown, or possibly a breakdown followed by a reaggregation of molecular components, which gives a different sedimentation pattern. It is interesting to note that in the earlier work of Mazia and Dan (1952) they report a s_{20} of 4.0 for the dissolved mitotic apparatus; however, they employed a different method for solubilizing the isolated mitotic apparatus. Moreover, it is quite significant that the present author and Mazia and Dan (1952) both find only a single peak after dialysis.

The molecular weight of the alcohol-digitonin isolated mitotic apparatus was determined in the analytical ultracentrifuge by means of the procedure involving the "approach to sedimentation equilibrium." The technique employed for the molecular weight determination was similar to that proposed by Klainer and Kegeles (1955) and modified by Ginsburg *et al.* (1956). The molecular weight of the Salyrgan dissolved mitotic apparatus was calculated to be 315,000 ± 20,000. This larger value for the molecular weight differs appreciably from the earlier estimate (45,000) of the molecular weight by Mazia and Dan (1952). The apparent inconsistency is probably due to considerations of shape essential for the earlier studies where it was assumed that the particles were spherical. Actually they are fibrous in nature. In a recent study of the molecular weight by means of the "approach to sedimentation equilibrium," such assumptions of shape are not necessary in determining the value of the molecular weight.

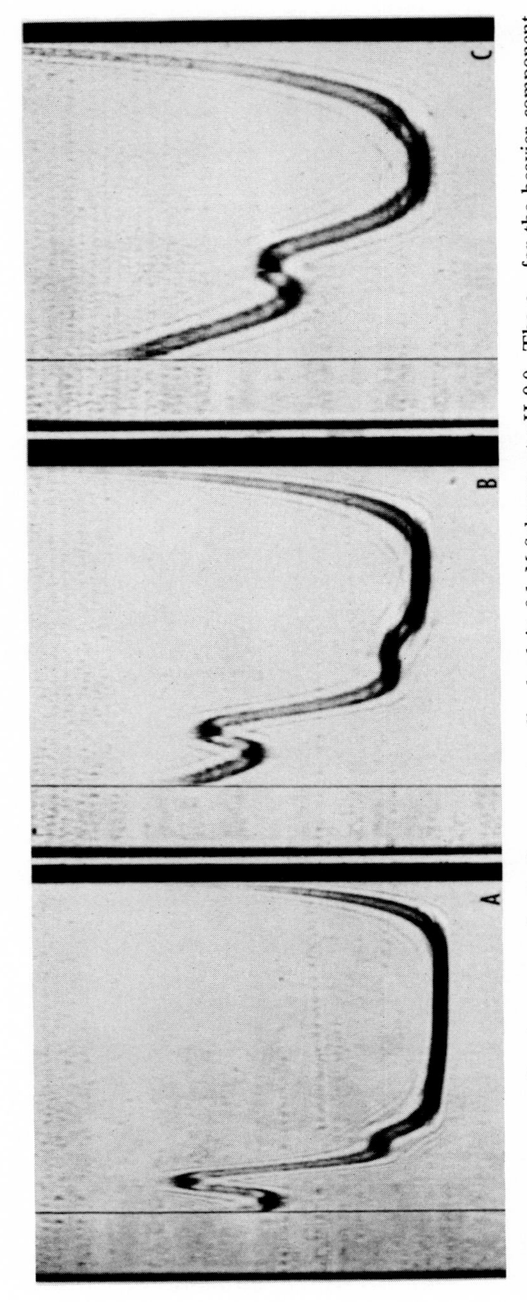

Fig. 6. Ultracentrifugal pattern of mitotic apparatus dissolved in 0.1 *M* Salyrgan at pH 9.0. The s_{20} for the heavier component is 8.6 and for the lighter component 3.7. Centrifuge speed 59,780 rpm. A. 16 minutes, bar angle 60°; B. 32 minutes, bar angle 50°; C. 56 minutes, bar angle 40°.

IV. Conclusion

In general, analysis of the isolated mitotic apparatus has afforded us insight into the complex physiological events of cell division. Biochemical analysis has established that the mitotic apparatus consists primarily of protein intimately associated with ribonucleic acid. The total nucleotide analysis indicates that the isolated mitotic apparatus contains about 5–6% ribonucleic acid by weight. The protein analysis indicates that the isolated mitotic apparatus contains about 11.6% of the total cell protein, suggesting that the spindle-aster complex is probably of cytoplasmic origin. The immunological studies demonstrate the presence of two or possibly four antigens in the isolated mitotic apparatus. The presence of a specific ATPase has been found to be associated with the isolated mitotic apparatus. The Salyrgan dissolved mitotic apparatus consists of at least two electrically charged components as demonstrated by Tiselius electrophoresis. Ultracentrifugal studies indicate the presence of at least two components with different sedimentation characteristics. Furthermore, the molecular weight of the isolated mitotic apparatus has been calculated to be in the range of 315,000.

From these studies, therefore, we can see that the isolated mitotic apparatus is not homogeneous, but is a complex heterogeneous structure. Since the mitotic apparatus is so complex in both its chemical structure and its physiological activities, the interrelationships between the chemical structure and cellular function are difficult to establish.

In the evaluation and analysis of biochemical and physicochemical measurements on the mitotic structure it is also essential to consider the interrelationships of the various cellular components. Cell division is a complex phenomenon which entails both karyokinesis and cytokinesis. Although these phenomena are closely related and probably under similar controlling mechanisms, there is an accumulation of data which illustrates that mitosis and cytokinesis, under various experimental procedures, exhibit a certain degree of autonomy (see Zimmerman and Marsland, 1960). Mitotic events may proceed with the absence of subsequent cytoplasmic division; also cytoplasmic division may occur without preliminary mitotic events. In studying any single aspect of cell division it is most fruitful to consider the data with respect to the over-all phenomena of division. Separate studies are of most value when they can be interrelated so as to help elucidate the complex pattern of division.

References

Amberson, W. R., White, J. I., Bensusan, H. B., Himmelfarb, S., and Blankenhorn, B. E. (1957). *Am. J. Physiol.* **188**, 205.

Bolognari, A. (1952). *Arch. Sci. Biol. Ital.* **36**, 40.

Boss, J. (1955). *Exptl. Cell Research* **8**, 181.
Cecil, R. (1950). *Biochem. J.* **47**, 572.
Cleveland, L. R. (1953). *Trans. Am. Phil. Soc.* [N. S.] **43**, 809.
Dan, K. (1957). *Cytologia (Tokyo)* **22** (Suppl., Proc. Intern. Genetics Symposia, 1956), p. 216.
Davies, H. G. (1952). *Exptl. Cell Research* **3**, 453.
Edstrom, J. E. (1956). *Biochim. et Biophys. Acta* **22**, 378.
Edstrom, J. E. (1960). *J. Biophys. Biochem. Cytol.* **8**, 39.
Elson, D., and Chargaff, E. (1955). *Biochim. et Biophys. Acta* **17**, 367.
Elson, D., Gustafson, T., and Chargaff, E. (1954). *J. Biol. Chem.* **209**, 285.
Ginsburg, A., Appel, P., and Schachman, H. K. (1956). *Arch. Biochim. Biophys.* **65**, 545.
Harvey, E. B., and Lavin, G. I. (1944). *Biol. Bull.* **86**, 163.
Jacobson, W., and Webb, M. (1952). *Exptl. Cell Research* **3**, 163.
Kane, R. E. (1962). *J. Cell Biol.* **12**, 47.
Kawamura, K. (1960). *Exptl. Cell Research* **21**, 1.
Kawamura, N. (1960). *Exptl. Cell Research* **20**, 127.
Kawamura, N., and Dan, K. (1958). *J. Biophys. Biochem. Cytol.* **4**, 615.
Klainer, S. M., and Kegeles, G. (1955). *J. Phys. Chem.* **59**, 952.
Madsen, N. B., and Cori, C. F. (1956). *J. Biol. Chem.* **223**, 1055.
Mazia, D. (1954). *In* "Glutathione" (S. Colowick *et al.*, eds.), p. 209. Academic Press, New York.
Mazia, D. (1955). *Symposia Soc. Exptl. Biol.* **9**, 335.
Mazia, D. (1958). *Biol. Bull.* **114**, 247.
Mazia, D. (1959a). *Harvey Lectures, 1957-1958* **53**, 130.
Mazia, D. (1959b). *In* "Sulfur in Proteins" (R. Benesch *et al.*, eds.), p. 367. Academic Press, New York.
Mazia, D. (1961). *In* "The Cell" (J. Brachet and A. E. Mirsky, eds.), Vol. III, p. 77. Academic Press, New York.
Mazia, D., Chaffee, R. R., and Iverson, R. M. (1961a). *Proc. Natl. Acad. Sci. U.S.* **47**, 788.
Mazia, D., and Dan, K. (1952). *Proc. Natl. Acad. Sci. U.S.* **38**, 826.
Mazia, D., Mitchison, J. M., Medina, H., and Harris, P. (1961b). *J. Biophys. Biochem. Cytol.* **10**, 467.
Mazia, D., and Roslansky, J. D. (1956). *Protoplasma* **46**, 528.
Mazia, D., and Zimmerman, A. M. (1958). *Exptl. Cell Research* **15**, 138.
Mihalyi, E., Bradley, D. F., and Knoller, M. I. (1957). *J. Am. Chem. Soc.* **79**, 6387.
Montgomery, P. O. B., and Bonner, W. A. (1959). *Exptl. Cell Research* **17**, 378.
Neufeld, E. F., and Mazia, D. (1957). *Exptl. Cell Research* **13**, 622.
Rapkine, L. (1931). *Ann. Physiol. Physicochim. Biol.* **7**, 382.
Rustad, R. C. (1959). *Exptl. Cell Research* **16**, 575.
Sakai, H., and Dan, K. (1959). *Exptl. Cell Research* **16**, 24.
Shimamura, T., and Ota, T. (1956). *Exptl. Cell Research* **11**, 346.
Shimamura, T., Ota, T., and Hishida, T. (1957). *Symposia Soc. Cellular Chem. (Tokyo)* **6**, 21.
Stich, H. F. (1954). *Chromosoma* **6**, 199.
Stich, H. F., and McIntyre, J. (1958). *Exptl. Cell Research* **14**, 635.
Stricks, W. and Kolthoff, I. M. (1953). *Anal. Chem.* **25**, 1050.
Went, H. A. (1959). *J. Biophys. Biochem. Cytol.* **6**, 447.
Went, H. A. (1960). *Ann. N. Y. Acad. Sci.* **90**, 422.
Went, H. A., and Mazia, D. (1959). *Exptl. Cell Research,* Suppl. 7, 200.

White, J. I., Bensusan, H. B., Himmelfarb, S., Blankenhorn, B. E., and Amberson, W. R. (1957). *Am. J. Physiol.* **188**, 212.
Zimmerman, A. M. (1958). *Federation Proc.* **17**, 174.
Zimmerman, A. M. (1960). *Exptl. Cell Research* **20**, 529.
Zimmerman, A. M., and Marsland, D. M. (1960). *Ann. N. Y. Acad. Sci.* **90**, 470.

Discussion by Ronald C. Rustad

Florida State University, Tallahassee, Florida

Today is close to the tenth anniversary of the first large-scale preparation of isolated mitotic apparatus by Mazia and Dan (1952). Dr. Zimmerman has provided an effective summary of the work on the chemistry of the isolated mitotic apparatus. He has reviewed the history and discussed the contemporary work on several aspects of mitotic apparatus biochemistry. Therefore, the most useful function that I can perform is to reassemble some of the problems and open the topic for group discussion.

Initially, it must be stressed that all of the methods for mitotic apparatus isolation are empirical methods. Hence, the chemical analyses must be performed on whatever substances the isolation method yields. The remarkable fact is that the original methods have yielded the protein backbone of the mitotic figure plus some ribonucleic acid. One of the newer methods also retains some of the lipids (Mazia, 1961).

Earlier studies and Dr. Zimmerman's recent work indicate that the low temperature alcohol-digitonin method yields a structure containing about 95% protein and 5% ribonucleic acid. Although the 260 mμ absorbing material behaved cytochemically like RNA (Rustad, 1959), its identity was not certain until Dr. Zimmerman was able to determine the base ratios (Zimmerman, 1960). Other components, such as DNA and polysaccharides, are present in the mitotic figure, but these have not been demonstrated in the isolated mitotic apparatus by gross biochemical methods.

Electrophoresis and analytical ultracentrifugation have revealed the presence of two distinct proteins in Salyrgan dissolved mitotic apparatus. Using the approach to sedimentation equilibrium method, Dr. Zimmerman (1960) has shown that the molecular weight of the major protein component is of the order of 3.15×10^5. It is interesting to note that the major soluble protein in the sea urchin egg has a molecular weight of the order of 3.5×10^5 (Kane and Hersh, 1959).

Dr. Went (1959, 1960) demonstrated the presence of two antigenically distinct proteins. More recent antibody studies indicate the presence of four different proteins (Sauaia, 1961). Obviously, if the methods can be refined we may find hundreds of different proteins present in small quantities. Some of these may be extracted by accident and some may represent substances of special physiological significance in the mitotic process.

At present we must concentrate on considering the roles of the major protein components and possibly of the RNA. A useful framework for considering possible chemical changes may consist of four major biochemical problems of mitosis.

The first problem is to synthesize the material needed for the mitotic apparatus. In growing cells this must be done every generation. In the case of the sea urchin egg, Dr. Went (1959, 1960) has shown that the two precursor proteins are already present in the unfertilized egg. We have no information indicating whether or not enough of these proteins are present to permit mitosis. It is equally uncertain whether or not other proteins and other mitotic figure components, such as RNA and polysaccharides, must be synthesized after fertilization.

Dr. Zimmerman has directed considerable attention to the second problem: the assembly of the mitotic apparatus. Various solubility studies suggest that the isolated mitotic apparatus may be held together by disulfide bonds. Studies with mercaptoethanol *in vivo* indicate the possibility that the structure of the spindle and asters is dependent on a proper balance of sulfhydryl containing materials (Mazia, 1958; Mazia and Zimmerman, 1958). Changes in protein sulfhydryl and disulfide groups have been demonstrated both biochemically and cytochemically (Kawamura, 1960; Kawamura and Dan, 1958; Neufeld and Mazia, 1957; Sakai and Dan. 1959). Nonetheless, the roles of sulfur-containing groups remain obscure (Mazia, 1961).

How does RNA become a part of the mitotic apparatus? There have been numerous suggestions that part of it is nucleolar or chromosomal in origin. The elegant electron microscope studies in Dr. Porter's laboratory show that considerable amounts of RNA migrate into the spindle from the cytoplasm in the ribonucleoprotein granules of the invading endoplasmic reticulum (Porter, 1962; Porter and Machado, 1960). There are no data concerning the synthesis of this RNA, and there are no experiments indicating whether or not it plays an active role in mitosis.

A third major question is: How does the mitotic apparatus work? Most of the papers at this symposium are devoted to some aspect of this problem. Chemical analyses of the isolated mitotic apparatus have provided two exciting pieces of information for speculation and future study. The first is the superficial similarity between the amino acid ratios in the proteins of the mitotic apparatus and of muscle actin (Roslansky, 1957). The second is the recent demonstration of what appears to be a specific magnesium-activated ATPase (Mazia *et al.*, 1961). Certainly this information calls for a careful consideration of the possible similarities and differences between the fibrils of the spindle and of muscle.

The final problem is to take the mitotic apparatus apart. The molecules of the mitotic apparatus have been assembled in a very complex array for a rather particular purpose. When mitosis is complete, this structure must be disassembled. Perhaps the chemical mechanism is the mirror image of the assembly process, but there do not seem to be any experiments dealing with this question.

Studies on the chemistry of the isolated mitotic apparatus have provided key information for speculating about some of the major biochemical problems of mitosis. While my organization of these problems has been somewhat oversimplified, this seems to be an appropriate time to turn the program back to the chairman, Dr. Went.

References

Kane, R. E., and Hersh, R. T. (1959). *Exptl. Cell Research* **16**, 59.

Kawamura, N. (1960). *Exptl. Cell Research* **20**, 127.

Kawamura, N., and Dan, K. (1958). *J. Biophys. Biochem. Cytol.* **4**, 615.

Mazia, D. (1958). *Exptl. Cell Research* **14**, 486.

Mazia, D. (1961). *In* "The Cell" (J. Brachet and A. E. Mirsky, eds.), Vol. III, p. 77. Academic Press, New York.

Mazia, D., and Dan, K. (1952). *Proc. Natl. Acad. Sci. U.S.* **38**, 826.

Mazia, D., and Zimmerman, A. M. (1958). *Exptl. Cell Research* **15**, 138.

Mazia, D., Chaffee, R. R., and Iverson, R. M. (1961). *Proc. Natl. Acad. Sci. U.S.* **47**, 788.

Neufeld, E. F., and Mazia, D. (1957). *Exptl. Cell Research* **13**, 622.

Porter, K. R., and Machado, R. D. (1960). *J. Biophys. Biochem. Cytol.* **7**, 167.

Roslansky, J. D. (1957). Ph.D. Thesis, Univ. of California (Berkeley).
Rustad, R. C. (1959). *Exptl. Cell Research* **16**, 575.
Sakai, H., and Dan, K. (1959). *Exptl. Cell Research* **16**, 24.
Sauaia, H., Unpublished work, cited in Mazia (1961).
Went, H. A. (1959). *J. Biophys. Biochem. Cytol.* **6**, 447.
Went, H. A. (1960). *Ann. N. Y. Acad. Sci.* **90**, 422.
Zimmerman, A. M. (1960). *Exptl. Cell Research* **20**, 529.

General Discussion

DR. S. GELFANT (Syracuse University): The question I have are based upon four words: recovery, reversal, *in vitro,* and *in vivo.* I think that Dr. Zimmerman is describing a recovery or recuperation and not a reversal when he speaks of the reformation of the mitotic apparatus after pressure treatment or after removal of mercaptoethanol. The distinction here is significant because a reversal would imply that you could reverse the effects of mercaptoethanol on the mitotic apparatus by some agent in the presence of mercaptoethanol. And you are not doing this.

The second point of interest involves the differential effect of mercaptoethanol *in vivo* and *in vitro.* It works on the mitotic apparatus in the *in vivo* egg but not on the isolated *in vitro* mitotic apparatus. And yet your main thesis is that the isolated mitotic apparatus has the same structure, chemical composition, and physiology as the mitotic apparatus *in vivo.* For example, you used Dr. Rustad's slides to illustrate that the staining characteristics of the isolated and the *in vivo* mitotic apparatus are the same, and therefore you are dealing with the same structure *in vivo* and *in vitro.* How do you explain the differential mercaptoethanol results which would indicate that there is a difference between the physiological response of the mitotic apparatus *in vitro* and *in vivo?*

DR. A. ZIMMERMAN (State University of New York): With respect to the first question pertaining to the reversal and recovery aspect, we found that when fertilized *Arbacia* eggs were subjected to high hydrostatic pressure and the mitotic apparatus then isolated, there was extensive disorganization. Other sister cells which were pressurized but whose mitotic apparatus were not isolated, were permitted to recover. At intervals of 5 and 10 minutes after pressurization, these eggs were prepared for mitotic isolation. In addition, some of the eggs were not treated for mitotic isolation but were permitted to go on to division. Thus we had a whole series for observation and analysis. We found that when we had disorganized mitotic apparatus by high pressures, we prevented cleavage formation. When the mitotic apparatus reorganized, the cells continued on their way to division. We were able to relate the mitotic disorganization to the delay in the first division. In other words, if it took 10 minutes for reorganization of the mitotic apparatus after a pressure treatment, there was a corresponding 10 minute delay in furrowing. To illustrate this point, in *Arbacia* at 20°C it takes 60 minutes for the first division. If we subject the eggs to a pressure of 10,000 lb/sq. in. for 1 minute, and then released the pressure, it required 70 minutes for the eggs to divide. Thus we could relate delay in division to the structural disorganization, namely the morphology of the mitotic apparatus. These pressure-treated eggs went on to form normal plutei. I hope this clarifies the first point.

With regard to the mercaptoethanol experiments, we found a somewhat different picture. When the cells were placed into the mercaptoethanol at metaphase for a period of 10, 20, or 30 minutes and then removed and permitted to recover, Dr. Mazia and I noticed that the delay in first division was related to the length of time the cells were in the mercaptoethanol. In other words, if it took 90 minutes for a cell to divide, and if we put the cell into the mercaptoethanol for 10 minutes and then removed it, the cell divided in 100 minutes. If we allowed the cell to remain in the mercaptoethanol for 20 minutes, it took 110 minutes.

In answer to your second question, it would have been more fortunate had the isolated mitotic apparatus been dissolved in the mercaptoethanol. But this really has no bearing per se. What we were trying to illustrate was that the isolated mitotic apparatus has a labile structure and certain dynamic characteristics, and by isolating this "organelle" from the cell, we were hoping to characterize some of its structural components by chemical analysis. Furthermore, I attempted to compare some of the cytochemical relationships of the isolated preparation with the mitotic apparatus found in whole cells. Dr. Plato Telaporus has conducted sulfhydryl studies on the isolated mitotic apparatus. Dr. Rustad has been conducting cytochemical studies on the isolated mitotic apparatus and on the mitotic apparatus in sectioned eggs. In addition to what has been shown in the studies I have conducted, it appears that there is a close cytochemical relationship between the isolated mitotic preparation and that which has been studied in the intact sectioned eggs. I hope this sheds some light on the problem.

Dr. H. Went (Washington State University): Have you observed the pressure-treated cells immediately after removal from the pressure chamber while still alive under phase microscopy? Does what you see in the isolated mitotic figures corroborate what you see in the living material? Namely, do you see any fibrous orientation in the spindle or astral regions?

Dr. A. Zimmerman: *Arbacia* is a rather difficult egg in which to study mitotic events. It contains a large amount of coarse cytoplasmic granules and echinochrome pigment vacuoles. When looking at this egg under phase, you do not see too much of the structural organization of the mitotic spindles or asters. You have to squash the egg to see these structures under phase. In a few cases in which I did look at the mitotic structures after pressurization, I could not see any organization. All I could see was an amorphous mass in the general shape of a mitotic structure, but I could not see any linear orientation of the spindle. I have seen "spindle fibers," however, in nonpressurized eggs, but this takes a bit of manipulation to find them each time. After pressure we were usually in a rush to place the eggs into the cold ethanol solution before the mitotic structures returned.

Dr. J. Kay (University of Rochester): I wonder if you could make any estimate of the amount of contamination of RNA in the isolated spindle? That is, how much is real, and how much is contamination?

Dr. A. Zimmerman: I doubt that there was any contamination in the preparation of the isolated mitotic apparatus; however, this cannot be said with absolute certainty. In our biochemical studies on mitotic RNA, we washed the isolated mitotic apparatus by differential centrifugation at 4°C at least 10 times. In this manner we were able to remove a good deal of the residual cytoplasmic residue.

I have some additional data which I have not presented, and I should like to take this opportunity to show it. We were concerned about the contribution that cytoplasmic residue would have on patterns obtainable in the analytical centrifuge. In order to clarify this point, we isolated mitotic apparatus from the same batch of fertilized eggs, and they were washed repeatedly and separated from the cellular residue. The mitotic apparatus was then divided into two groups. The first group consisted of the clean mitotic apparatus; the second group consisted of the clean mitotic apparatus plus the insoluble cellular residue which was collected in the preparation of the clean mitotic apparatus. We then subjected both groups to dissolution in 0.1

molar Salyrgan, centrifuged them at 125,000 *g* for 7 minutes in the preparative ultracentrifuge. A large amount of cellular residue, comparable to that which was originally added to the second group was sedimented. Only a trace of residue material was found in the bottom of the centrifuge tube which contained the clean mitotic apparatus. A sample of supernatant, the dissolved mitotic apparatus, from each tube was subjected to analysis in the analytical ultracentrifuge. It is seen from the accompanying lantern slide that the patterns obtained in the analytical ultracentrifuge for the clean mitotic apparatus and for those of the mitotic apparatus plus cellular residue are very similar. The s_{20} of the major peak for the clean mitotic apparatus and those of the mitotic apparatus plus cellular residue are 3.67 and 3.40, respectively; the s_{20} for the minor peaks are 8.56 and 8.59, respectively. It appears from the visual, as well as the analytical studies, that there is hardly any difference between the two groups. This suggested to us that the contaminants contributed very little if any to the patterns obtained in the analytical ultracentrifuge.

DR. H. RIS (University of Wisconsin) : I am puzzled by two terms that were combined here. One was "molecular weight" and the other "apparatus." Apparatus suggests to me something complex, and in the light microscope, it looks somewhat complicated. You mentioned centrioles, astral rays, spindle fibers, and chromosomes. If one looks in the electron microscope, and I have seen some electron micrographs taken by Miss Harris, who works with Dr. Mazia, the mechanism appears very complex. When it is cleaned and all the "dirt" (cytoplasmic residue) taken out, some significant "dirt" (cytoplasmic residue) might be removed. The electron micrographs show a complex system of membranes and ribosomes. Actually not much else is seen. Of course, there are also chromosomes, but the predominant material seems to be membrane systems and RNP particles. It is interesting that these micrographs look very much like the pictures that Dr. Porter showed, if you would call that structure in onion root-tip, where there are no apparent centrioles or asters, the mitotic structure. But certainly most of it is something that we do not associate with any unit molecule of which you could determine the molecular weight.

DR. A. ZIMMERMAN: I used the term, mitotic apparatus; perhaps I should use the terms chromatic and achromatic figures, which might be more applicable. It is just that we have become more familiar with the term, mitotic apparatus. It is true that the mitotic apparatus is a complex system. There is no question about it. However, what we wanted was some indication of the size of the structural components constituting the mitotic apparatus. From the earlier studies by Mazia, the molecular weight was estimated to be about 45,000 for the thiogylcolate-dissolved mitotic apparatus. This determination was based on the assumption that the dissolved particles were spherically shaped with 30% hydration. As we learned more recently, these assumptions were incorrect. By employing the Archibald technique, the "approach of sedimentation equilibrium," no assumptions of this nature were necessary to determine an estimate of the molecular weight. What we have given is an average molecular weight for the particles which constitute the mitotic apparatus.

Our values for the dissolved mitotic apparatus were in the range of 300,000. It is rather enticing to suppose that these structural components might be related to muscle, since the molecular weights of myosin are in the range of 400,000, and since we feel, in a manner of speaking, that both have a similar physiological function.

DR. A. W. BURKE (Rhode Island Hospital) : It is regrettable that Dr. Zimmerman was backed into a corner by Dr. Ris about terminology. As he first used the term "mitotic apparatus" he was using it correctly. It is an unfortunate term, but is a term, I understand, that was coined to represent the chromatic apparatus, chromosomes and centromeres, and the achromatic apparatus or that part generated by the centriole (astral rays, central spindle, centrosome). It would seem to me that a term other than mitotic apparatus might be used, like mitotic figure or mitotic anything, to avoid the confusion which arises with achromatic apparatus. Your abstinence from misuse of the term achromatic apparatus in the first part of your paper is to be congratulated, Dr. Zimmerman. There are so many terms already in use that we should understand thoroughly what we mean when we say mitotic apparatus, as isolated originally in your preparations. What was obtained was chromosomes with their centromeres, centrioles, asters, etc.; in fact, the whole "works," not just the achromatic apparatus. Any subsequent chemical treatment might lead to the isolation of the achromatic apparatus.

I have a question to ask concerning mercaptoethanol treatment in the urchin eggs. Is it true that there is a certain concentration or dose regimen or exposure which leads to subsequent tetrapolar divisions of the egg? If this is the case, this shows that mercaptoethanol while suppressing the achromatic-figure producing portion of the centriole or what we might call broadly, achromatic apparatus, clearly, does not affect its generative portion. Otherwise, it could not possibly lead to the formation of a tetrapolar figure, unless, of course, you wish to employ the interpretation that since its centriole is bifid or duplicate in each aster at the time of anaphase, then these are allowed to mature to function precociously in the next division.

Dr. A. Zimmerman: Just to comment on the first statement: When I referred to mitotic apparatus, I was referring to a spindle-aster-chromosome complex, the three components which constitute the mitotic apparatus.

Concerning the second question: If you block the development of the mitotic apparatus in a whole egg treated with mercaptoethanol for a specific period of time, and then place the cell back into sea water, very often the cell cleaves into four cells, one division right after the other. In other words, you get a four-cell stage directly from a one-cell stage. Dr. Mazia and I have seen that the centrioles have separated while under mercaptoethanol blockage. In a recent publication by Mazia, Harris, and Bibring (1960, *J. Biophys. Biochem. Cytol.* **7**, 1), they discuss the aspect of cellular movement in mercaptoethanol blocked cells. Although fibrous orientation of the spindles and asters of the mercaptoethanol blocked cells is not evident in the isolated mitotic apparatus, the centrioles are shown to have separated and possibly duplicated. Unfortunately I did not bring a slide with me to illustrate this aspect.

Dr. L. E. Roth (Iowa State): I understand that there has been some electron microscopy done on sea urchin eggs in the Mazia group. Can you tell us what has been contributed from this work?

Dr. A. Zimmerman: The photographs that I have seen show a profile very similar to that which you have shown of the spindle fibers in the interzonal region in *Amoeba*. In fact, if you were to substitute your electron micrographs of the amoeba spindle fibers for those of the sea urchin, it would not be easy to distinguish them. In the sea urchin there are many particles presumed to be RNP, and these are shown in the area of the mitotic apparatus. The spindles seem to be tubular. However, filaments are also seen in the interzonal region.

Perhaps Dr. Porter has seen some of the original electron micrographs when Patricia Harris was in New York recently. (Harris, P., 1961, *J. Cell Biol.* **11**, 419.)

Dr. K. R. Porter (Harvard University): In essence you are right in what you say. The spindle does contain the small tubular elements that Dr. Roth has found in *Amoeba*. And around these there is obvious condensation of what is probably your spindle apparatus or your mitotic apparatus protein. In addition there are many RNP particles around. Otherwise there is nothing special.

Dr. L. E. Roth: Was this isolated, or was this whole?

Dr. A. Zimmerman: I believe these were sections of whole eggs.

Studies on the Disruption of the Mitotic Cycle[1]

G. B. Wilson

Department of Botany and Plant Pathology and Biology Research Center, Michigan State University, East Lansing, Michigan

I. Introduction

Study of the disruption of the mitotic cycle by various chemical and physical agents has been a popular laboratory sport for a considerable number of years and has given rise to a rather extensive literature. While the motivation has sometimes been obscure, it generally seems to have been dictated by one of two primary interests: (*a*) concern for the biological effects of the agent per se or (*b*) interest in the way the system comes apart under various stimuli. Such work as my colleagues and I have done in this area has been stimulated both by our interest in the potential hazards of the chemical milieu in which plants and animals find themselves and by our interest in the general problem of cell replication. The two motivations are not necessarily incompatible so far as type of experimentation is concerned though the choice of disruption agent usually depends on the aim. Since the theme of the present symposium is "The Cell in Mitosis" I shall attempt to consider the data which I have to present in terms of the phenomena associated with cell division.

Until fairly recently most students of mitosis were primarily occupied with the problem of mitosis proper and, even more specifically, with the problem of organization of the metaphase plate and anaphase movement. Indeed E. B. Wilson (1925) suggested with some logic, that mitosis really begins with metaphase. Most of the classical observations and hypotheses on this problem have been reviewed in some detail by Franz Schrader who also notes that the problem is still with us. I suspect that it will still be here even at the end of this symposium.

During the past decade or so there has been an increasing tendency to stress the entire mitotic cycle rather than the most conspicuous part of it. This expansion of interest would appear to stem from the inescapable

[1] The work described was supported in part by NIH grant RG 4835.

fact that the wondrous event which we recognize as mitosis is determined by a whole series of prior events. What are these events? When do they occur? How are they related? Which ones are pertinent to actual nuclear division? Many formidable techniques including a number touched on in this symposium have been brought to bear on these several questions and I feel there is reason to hope that some rather good answers will emerge in a few years. Certainly, the questions have been sharpened to a marked degree.

A. The Mitotic Cycle

The mitotic cycle may be considered in terms of those pertinent events which occur between the inception of one division and the ensuing one, and can be represented in a very general way by some such diagram as shown in Fig. 1. Presumably if a daughter cell is to divide it must undergo a series of morphological, physiological, and biochemical changes which will return it to essentially the same state as the mother cell from which it arose. It would seem that if we are to understand the mitotic process, we must find out what these changes are, their order of occurrence, and interrelationships. Furthermore, we must distinguish between those changes which are pertinent to division and those which are not. For example, it has been quite clearly shown by Dr. Scherbaum (1960) among others that growth per se is at least partly separate from division preparation. The cycle is quite obviously divided into two major segments; namely, that part we commonly refer to as mitosis, and interphase. The former segment is easily subdivided on the basis of readily observable morphological shifts but the latter can only be split into stages, except in special cases, by biochemical criteria although further extension of electron micrographic studies of interphase may well provide a morphological basis for division. So far the most reliable segmentation of interphase appears to be in terms of time of synthesis of deoxyribonucleic acid (DNA) and, to a lesser extent of ribonucleic acid (RNA), (Taylor, 1960; Wimber, 1960; Prescott, 1960).

If one is interested in antimitotics, as I am, he finds himself involved in three questions: (1) the type of disruption; (2) the part of the cycle where inimicable action appears to take place; and (3) the biochemical nature of the disruption. While it is a rare case where one might hope to answer all three questions with any degree of confidence, the questions themselves should be kept in mind.

Perhaps the time has come to work out a usable taxonomy of antimitotics. A number of efforts have been made in this direction, the most notable being that of Dr. Biesele (1958), but much of the data which he had to use were not obtained with classification in mind. Presumably such a classification should be based on both morphological and biochemical criteria. Despite the very considerable work which has been carried out with chemical antimitotics, there appear to be too few comparative studies to allow us to make even much of a start at a valid classification. None the less some samples of what I have in mind can be offered. It has been shown rather convincingly by many workers (Eigsti and Dustin, 1955; Hadder and Wilson, 1958) that colchicine affects the

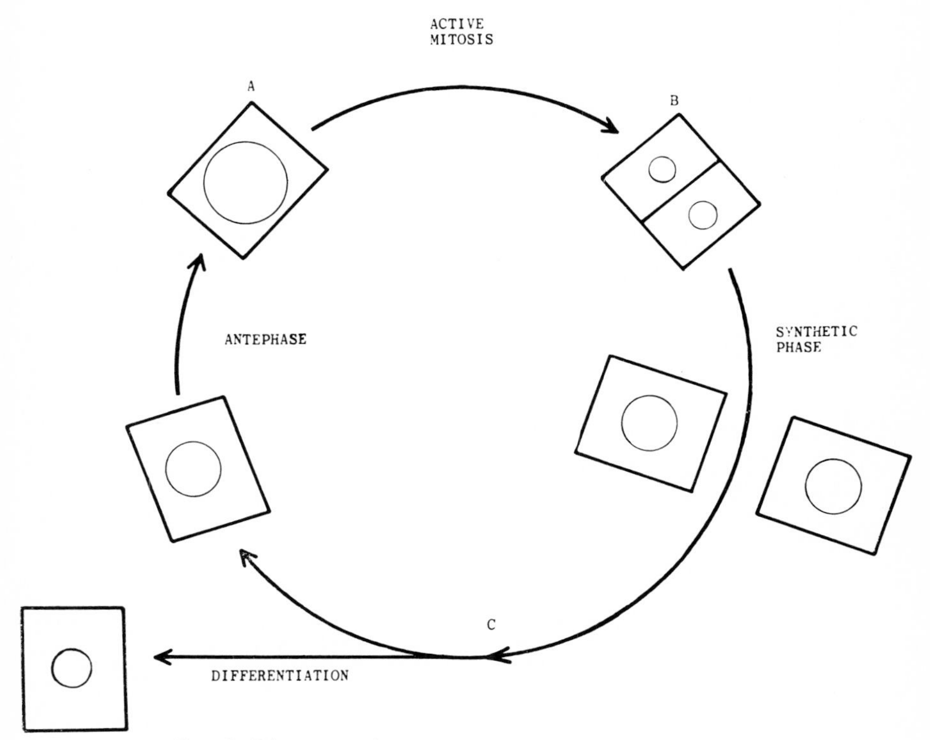

FIG. 1. Diagrammatic representation of the mitotic cycle.

movement of chromosomes from prometaphase on, apparently by preventing the organization of the spindle or destroying the organization if it has already been established. On the basis of morphological data it would appear that podophyllin and γ-cyclochlorohexane belong to the same class. In none of these cases do we really know what the biochemical mode of action is. Antimitotics such as the antibiotic, actidione, appear to have a major effect at many points in the cycle though the period from prometaphase to early interphase and preprophase seem to be unaffected or at least highly resistant (Hadder and Wilson, 1958). Again no definite biochemical basis for the antimitotic activity has been established though, for what it is worth, we can probably exclude any direct effect on respiration (Van Dreal, 1961). The substituted phenols, although similar in their effects to both colchicine and actidione, can be separated on the basis of both morphological and biochemical responses. These are examples of antimitotics which quite obviously belong to different classes on the basis of a number of criteria. Even in the present not altogether satisfactory state of the art, we could add others which can be shown one way or another to belong to still other classes. However, I think the cases cited serve to illustrate the point. The study of antimitotic systems is both tedious and expensive and, therefore, is a game that should be played according to some fairly definite rules lest it degenerate into *ad hoc* screening which is almost as tedious and expensive and less likely to have scientific validity. To quote Dr. Biesele: "The great diversity of possible initial points of attack by mitotic poisons in the broad sense results from the complicated interlocking of the metabolic features of cell life." This means in fact that different antimitotics acting in different ways may appear to have the same action if measured by one criterion. It is therefore highly desirable to measure antimitotic activity in a number of different ways.

B. Some Characteristics of the Mitotic Cycle

I would like to turn my attention now to some general characteristics of the mitotic cycle which are especially pertinent to the task of studying antimitotic activity. As noted earlier (Fig. 1) the mitotic cycle can be divided into two major segments; namely, mitosis and interphase, and each of these can be subdivided on the basis of a variety of criteria.

The method of measuring antimitotic activity depends, in part at least, on the experimental system used and the part or parts of the cycle showing the effect.

Wilson and Hyypio (1955) provided evidence that once a nucleus enters mitosis, the morphological changes (contraction and telomorphic transformation) take place independently of breakdown of the nuclear boundary, formation and functioning of the spindle or cleavage of the kinetochore. In short, any nucleus entering mitosis will sooner or later revert to interphase. This fact can be used to measure certain kinds of antimitotic effects especially those involving disruption of some phase of the mitotic process itself. Two general sorts of criteria may be used: (1) the rate of appearance of diagnostic configurations such as the "scatters" or "clumps" induced by colchicine or the stalled prophase induced by actidione, and (2) the change in ratio of one set of stages to another with time. With regard to the latter method the model shown in Fig. 2 indicates the basis of argument. The horizontal line represents normal flow from prophase to telophase, the vertical lines, flow from a disruption point to telomorphic transformation. If one plots the change in relative frequency of stages to the left of a block compared to those to the right, it is possible to provide a rather good picture both of the general nature of the block and how it is applied (Wilson and Morrison, 1958; Wilson, 1959). Even where more than one transition is blocked, this method of analysis may be used though with somewhat less precision in practice. In the model (Fig. 2) vertical and horizontal distances from a given

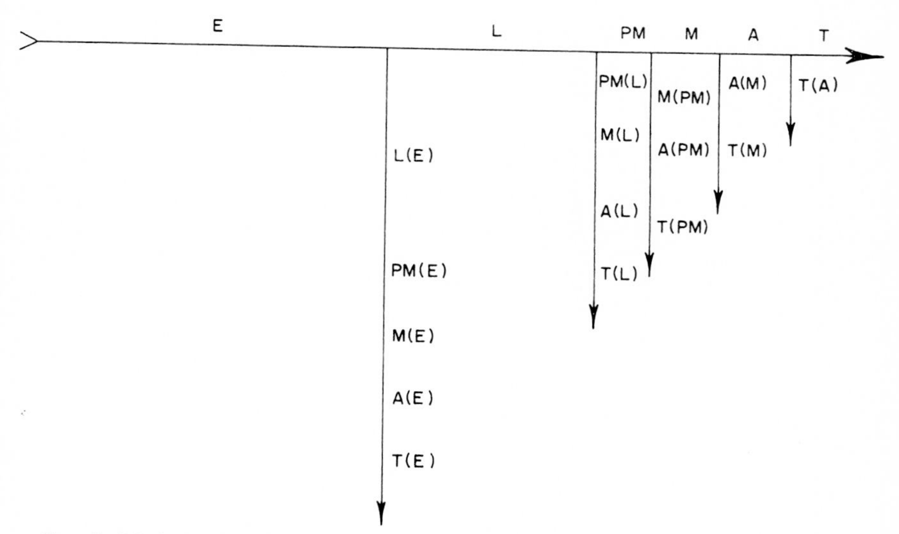

FIG. 2. Model indicating changes in movement and morphological cycles during mitosis. Horizontal arrow, normal sequence; vertical arrows, telomorphic transformation as result of interruption of chromosome movement cycle at specific stages.

point are shown as equal on the assumption that the time from initiation of prophase to interphase is the same regardless of route. It should be recognized that such an assumption is not automatically justified. For example from our work with colchicine we have concluded that a clumped prometaphase takes about 15 minutes (about 20%) longer to reach interphase than a control prometaphase. Guttman (1952) has estimated a much greater delay (about four times as long) though we have been able to find little evidence for this even by direct observation of living *Tradescantia* stamen hair cells. The measurement of antimitotic effects which are primarily centered in some portion of interphase poses somewhat more difficult problems. Plotting decrease in mitotic index with time and concentration can, of course, be used, providing suitable controls are employed to establish valid baselines. As a rule, however, such measurements give few or no clues to the susceptible portion of interphase or the mechanism of the interference. One can, of course, indulge in some educated guessing if the antimitotic under consideration has a known physiological or biochemical effect. For example one might guess that the antimitotic action of chloramphenicol is associated with suppression of protein synthesis but it would be necessary to show a direct correlation between the two effects. Ideally one should use a system which allows a reasonable estimate of where in the mitotic cycle a cell was at the time

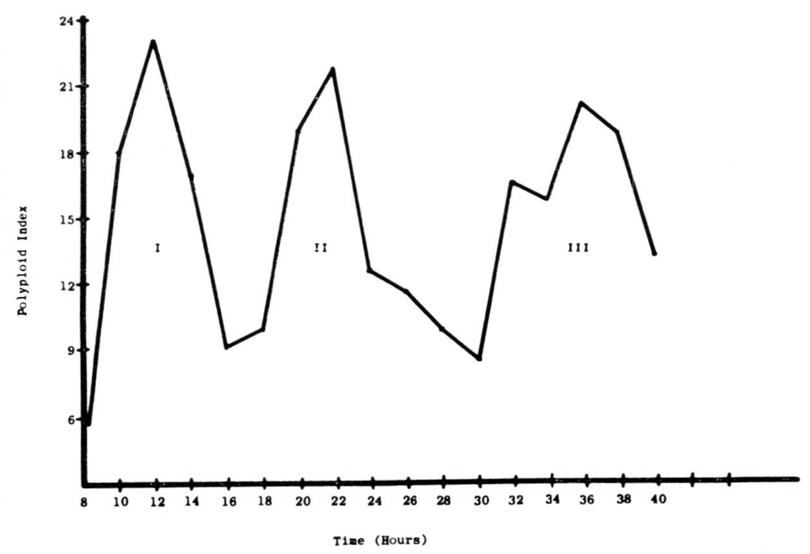

FIG. 3. Change in polyploid index in pea root with time. Each peak represents a cycle. Population marked with 30 minutes treatment with 3.76×10^{-4} *M* colchicine.

of treatment. This calls for some form of cyclic synchrony. In general one has the choice of attempting to induce synchrony in a population of cells as has been done in a variety of ways with microorganisms or taking advantage of natural synchrony such as may be found in early cleavage of amphibian eggs, or in meiotic (Stern, 1960; Sparrow and Sparrow, 1949), or first microspore divisions in suitable plants. If one can tag a group of cells at some particular point in the cycle in a rapidly growing tissue, it is expected that some portion of the tagged population will be reasonably synchronous for several mitotic cycles. Howard and Pelc (1951) made use of P^{32} incorporated into DNA in *Vicia* roots. Since then, similar studies mostly based on incorporation of tritiated-thymidine have been made by a number of workers; e.g., Quastler and Sherman

(1959) in mouse and Wimber (1960) in *Tradescantia*. Van't Hof, Wilson, and Colon (1960) described a method of inducing a limited (in both time and number) polyploid population of cells in pea meristems (Fig. 3). During the past 2 years this system has been explored quite extensively in our laboratory especially by Dr. Van't Hof. Within reasonable limits, this system may be used to determine such things as change in average cycle time and differential susceptibility of different segments of inter-

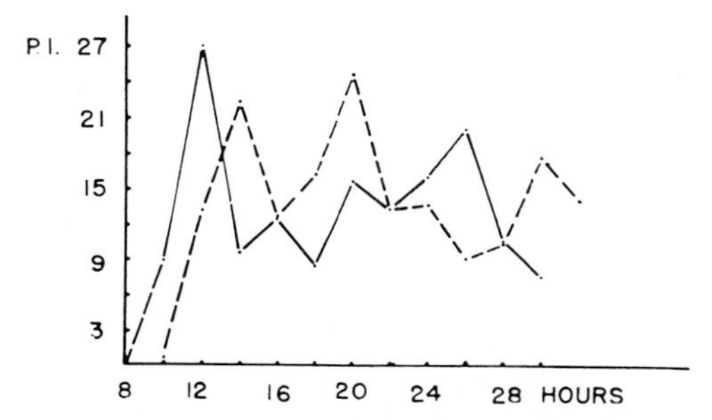

FIG. 4. Effect of 15-minute exposure to 4.3×10^{-5} M DNP 5 hours after tagging with colchicine. Solid line, control; broken line, treated.

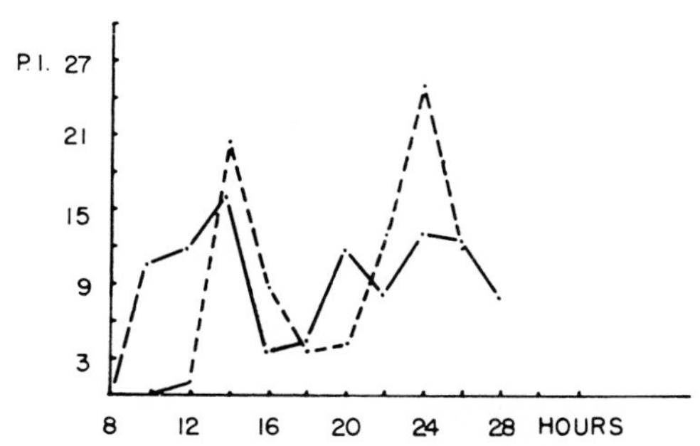

FIG. 5. Effect of 15-minute exposure to 4.3×10^{-5} M DNP plus 0.01 M KF 5 hours after tagging with colchicine. Solid line, control; broken line, treated.

phase. For example, treatment for 15 minutes with 4.3×10^{-5} M dinitrophenol (DNP) at anytime up to about 7 hours after tagging delays appearance of the marked population in mitosis about 2 hours and usually improves the synchrony as compared to control. Similar treatment after 7 hours also delays appearance and, in addition, destroys much of the synchrony. Unless the treatment has been given about 5 hours after tagging, the second cycle appears at the predicted "normal" time. If DNP is given at 5 hours, the second cycle is shortened by from 25–50% (Fig. 4). Identical results were obtained with 4.3×10^{-4} M KCN. This enhancement or accelerating effect is reversible with KF (Fig. 5). The exact reason for this enhancement is not known, but its reversal with fluoride does suggest involvement of glycolytic activity. In any event the segment of interphase between 4 and 6 hours (early midinterphase) does respond to respiratory disruption in a manner differing from that of any other segment. Presumably this time span coincides with rather rapid change in synthetic activities.

C. Some Sensitive Points in the Mitotic Cycle

As indicated earlier the worker concerned with the disruption of the mitotic cycle must be able ultimately to recognize characteristic responses of specific segments of the cycle. Figure 6 may serve to illustrate this point even though its inadequacies are numerous. These several segments may be characterized briefly as follows:

a. Segment O-1 represents about the first two-thirds of prophase. So far about the only measurable change induced is a change in relative frequency which may, with some caution, be interpreted as a rate change. For example, urethane (Cornman, 1954; L. V. Leak, unpublished) and adenine (G. Muhling and G. B. Wilson, unpublished) both increase the prophase index.

b. Point 2 represents the period of transformation from late prophase to prometaphase. The most conspicuous changes are breakdown of the nuclear region and "clumping" of the chromosomes into the center of the nuclear region. A number of substances (dubbed prophase poisons by D'Amato, 1948) will delay or prevent these series of changes. Among them are DNP, phenolic herbicides, and actidione. The term "prophase poison," of course, indicates the most conspicuous symptom not the disease.

c. Point 3 indicates the period of transition from prometaphase to metaphase to anaphase which transition is blocked by agents which prevent or destroy spindle organization (colchicine and other c-mitotic agents) or "immobilizes" the spindle (iodoacetic acid). These two kinds of spindle "poisons" can be distinguished by quite obvious differences in characteristic configurations (Wilson and Morrison, 1958).

d. Segment 4 corresponds more or less to G_1 of a number of authors (see Quastler, 1960). Treatment for short times during this period with a number of respiratory poisons including KCN and DNP at 10^{-5} *M* and lower delay the onset of mitosis by 2 to 3 hours. Similar treatments with the same molar range of actidione delays onset of mitosis by 5 to 7 hours. Whether the delay represents increase in G_1 duration or not is not clear from our data. Quastler, however, has noted (1960) that normal variation in cycle time seems to be largely attributable to variation in G_1 time.

e. Segment 5 corresponds more or less to the S or synthetic stage of a number of authors (Howard and Pelc, 1951; Howard and Dewey, 1959; Quastler and Sherman, 1959) during which DNA synthesis rises to a maximum. In our case it is the segment which responds to treatment with suitable respiratory poisons by decreasing the duration of the X_2 cycle as already noted.

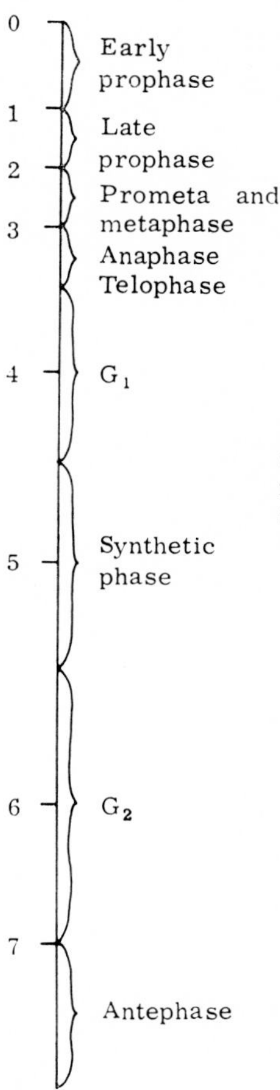

FIG. 6. Diagram illustrating theoretical segmentation of the mitotic cycle in terms of antimitotic action. See text for details.

f. Segments 6 and 7 may be considered to be equivalent to the G_2 period but for a variety of reasons we have elected to use Bullough's (1952) term, antephase,[2] for the latter portion. The reason for this dis-

[2] Since the term "antephase" as used in this and previous papers does not have the same significance as ascribed to it by Dr. Bullough, some other term should probably be substituted. One is tempted to use the term, "G_2," but this also does not really apply either in time or in its implication. It is with some trepidation, therefore, that I suggest "prekinetic phase" or "prekinesis" while expressing the hope that someone may offer a more constructive and suitable term.

tinction is simply the fact that, in peas at any rate, a population of cells in late interphase reacts somewhat differently from one at an earlier stage. For example, when pea root meristem containing a marked population of cells at this stage, is treated with 2,4-dinitrophenol, synchrony is largely lost as compared with controls. Similar treatment with actidione results in cleaving the marked population into two parts; one enters mitosis about the normal time and the other is delayed by some 5 hours.

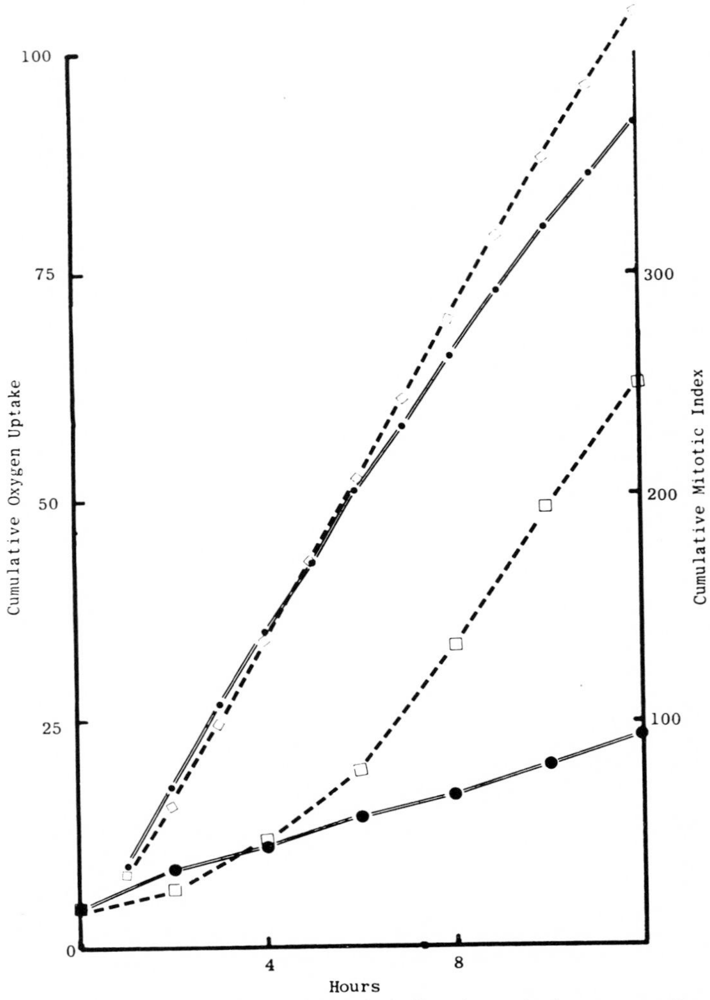

FIG. 7. Cumulative O_2 uptake and mitotic indices in excised pea roots. Dotted lines from roots treated with 1% glucose 8 hours after excision; solid lines from roots without glucose. O_2 uptake in microliters per hour.

Excision of the root also has a very similar effect; namely, part of the marked population enters mitosis at about the expected time while the other, and larger part does not begin to enter division for about 8 hours regardless of whether an exogenous carbon source is added or not. In recent years there has been considerable discussion concerning the energy requirements for passage from antephase to mitosis (see especially Bullough 1952, 1954, 1955; and Gelfant, 1960). I have no major contribution to add but, for what it is worth, I would like to mention two observations which have been repeated many times in our laboratory: (1) When excised pea roots are maintained in a salt solution for some 8 hours, the mitotic activity falls to a low level. The addition of a suitable carbon source results in a fairly abrupt but temporary rise in mitotic activity. Most of the cells entering mitosis during the rise must be assumed to have been in antephase at the time of addition of the carbon source. At the same time there is a marked increase in O_2 uptake as well as a definite increase in RQ from about 0.8–1.2 (Van Dreal, 1961). The effect of glucose versus no glucose on O_2 uptake and mitotic activity is shown in Fig. 7. That some of the exogenous glucose is respired is shown by the results of a radioactive tracer study (Table I). Why the C_1-labeled/C_6-labeled glucose ratio should be so different during the first 6 hours is not clear but seems likely to represent temporary preference for glycerol-fatty acid synthesis. (2) If the carbon source is provided at the time of excision or, indeed, any time prior to 8 hours, the changes in both mitotic and respiratory activity are the same as if no carbon source is added before the end of the "run-down" period. My only conclusion concerning the energy requirement for initiation of mitosis is that no very specific pronouncements are justified.

TABLE I

$C^{14}O_2$ COUNTED AS BARIUM CARBONATE AFTER TREATMENT OF EXCISED PEA ROOTS FOR 6 AND 12 HOURS AFTER 8 HOUR "RUN DOWN"

Compound	Counts/min. after 6 hours	Counts/min. after 12 hours
Glucose-1-C^{14}	360	2220
Glucose-6-C^{14}	1000	1913
Glucose-U-C^{14}		4080
Untagged control		65

II. Concluding Remarks

The study of antimitotic action may be undertaken, as indicated earlier, for one of two reasons: (*a*) on the premise that the ways in which a process may be disrupted provides some insight into the nature of the process itself or (*b*) on the basis that such studies should lead to control over the probability that a given cell will undergo mitosis at a given time and place. In practice it is not necessary to distinguish between the two motivations but it is essential to realize that "sledge hammer" tactics are not very suitable for either purpose.

REFERENCES

Biesele, J. J. (1958). "Mitotic Poisons and the Cancer Problem." Elsevier, Amsterdam.
Bullough, W. S. (1952). *Biol. Revs.* **27**, 133-168.
Bullough, W. S. (1954). *Exptl. Cell Research* **1**, 176-185.
Bullough, W. S. (1955). *Vitamins and Hormones* **13**, 261-292.
Cornman, I. (1954). *Intern. Rev. Cytol.* **3**, 113-130.
D'Amato, F. (1948). *Caryologia (Pisa)* **1**, 49.
Eigsti, O. J., and Dustin, P., Jr. (1955). "Colchicine." Iowa State College Press, Ames, Iowa.
Gelfant, S. (1960). *Ann. N. Y. Acad. Sci.* **90**, 536-549.
Guttman, R. (1952). *Am. J. Botany* **39**, 528-534.
Hadder, J. C., and Wilson, G. B. (1958). *Chromosoma* **9**, 91-104.
Howard, A., and Dewey, D. L. (1959). *In* "The Cell Nucleus" (The Faraday Society), pp. 155-162. Butterworths, London.
Howard, A., and Pelc, S. R. (1951). *Exptl. Cell Research* **2**, 178-187.
Prescott, D. M. (1960). *Exptl. Cell Research* **19**, 228-239.
Quastler, H. (1960). *Ann. N. Y. Acad. Sci.* **90**, 580-592.
Quastler, H., and Sherman, F. G. (1959). *Exptl. Cell Research* **17**, 420-438.
Scherbaum, O. H. (1960). *Ann. N. Y. Acad. Sci.* **90**, 565-579.
Schrader, F. (1953). "Mitosis," 2nd ed. Columbia Univ. Press, New York.
Sparrow, A. H., and Sparrow, R. (1949). *Stain Technol.* **24**, 47-55.
Stern, H. (1960). *Ann. N. Y. Acad. Sci.* **90**, 440-454.
Taylor, H. J. (1960). *Advances in Biol. and Med. Phys.* **7**, 107-130.
Van Dreal, P. (1961). Mitotic Activity and Respiration in the Pea Root Meristem. Ph.D. Thesis, Michigan State University.
Van't Hof, J., Wilson, G. B., and Colon, A. (1960). *Chromosoma* **11**, 313-321.
Wilson, E. B. (1925). "The Cell," 3rd ed. Macmillan, New York.
Wilson, G. B. (1959). *Can. J. Genet. Cytol.* **1**, 1-9.
Wilson, G. B., and Hyypio, P. A. (1955). *Cytologia* **20**, 177-184.
Wilson, G. B., and Morrison, J. H. (1958). *Nucleus (Calcutta)* **1**, 45-56.
Wimber, D. E. (1960). *Am. J. Botany* **47**, 828-834.

Discussion by John J. Biesele

Department of Zoology, The University of Texas, Austin, Texas

Dr. Wilson has presented an exposition of mitosis disruption with the general comment that the study of antimitotic activity is usually carried out for one of two reasons: either to learn something about the biological effects of possible controlling agents of interest, or to understand the cycle better by interfering with it. In the latter case, of course, the ideal would be to use agents whose mechanisms of action are well understood, in order that only the mitotic effects need to be interpreted. Unfortunately, as Dr. Wilson points out, the student of antimitotic activity is confronted at the same time with two interacting factors that are incompletely understood: one is the mitotic cycle and its aberrations, and the other is the mechanism of biological activity of the agent.

In this difficult situation, it obviously behooves us to set up experimental procedures that are as simple as possible and to use experimental material that is as little complicated as possible. It has been stated somewhere that about 95% of the research on mitotic poisons, especially of the "screening" variety, has been of questionable scientific merit; I do not consider my own work to be above this criticism. It is a hopeful sign to see the growing emphasis on synchronized systems, which are beginning to provide some of the more reliable data in this field, despite the warning by Mitchison (1957) that results with synchronized systems do not necessarily apply to unsynchronized growth. Dr. Wilson has used a colchicine-induced polyploid population in pea-seedling meristem (Van't Hof *et al.*, 1960). It has the virtue of maintaining a reasonable degree of mitotic synchrony for several cycles. It has a possible fault in not being absolutely distinctive, if later treatments of the meristem should involve use of polyploidizing agents, and if already naturally polyploid cells of other tissues should be brought to mitotic activity. As to the question of the induced polyploid population's not being representative of the total meristem, Van't Hof *et al.*, (1960) have replied that their experience does not bear out such a possibility.

Another system used by Dr. Wilson is that of the excised pea root, which exhibits a fall in activity to a minimum by 8 hours after excision, and thereafter can respond to provision of a carbon source by increase in both mitotic activity and oxygen consumption. It is because of this effect that Dr. Wilson prefers to think of the end of interphase as being a distinctive period, for which he uses Bullough's name of "antephase" (Bullough and Johnson, 1951). But is it not possible that the pea root simply requires 8 hours to recover from the shock of excision?

It is important to mention at this point that Gelfant (1960) has seriously disputed the validity of the concept of antephase, specifically in the excised mouse ear epidermis system used by Bullough and by Gelfant. Gelfant has concluded that his experiments show (1) no relationship between high oxygen tension and mitotic activity when glucose concentration is optimal, (2) no substitution of Krebs cycle intermediates for glucose in the support of epidermal mitosis, and (3) no evidence for the prevention of mitosis by specific and reversible action of inhibitors of carbohydrate metabolism, including iodoacetate, fluoride, malonate, azide, cyanide, and dinitrophenol, some of which made the skin necrotic. However, Gelfant is still faced with the problem of explaining the increase in epidermal mitotic activity when glucose is added to the culture medium.

In short-term experiments with a related system, the excised corneal epithelium of mouse eyes, Utkin (1961) has also dealt with the effects of inhibitors of carbohydrate metabolism on mitotic activity. Besides making the interesting observation that phosphate-buffered control medium appears to be more inhibitory to mitosis than Tris-buffered medium (Gelfant used phosphate-buffered saline), Utkin found that addition of potassium cyanide, dinitrophenol, or sodium fluoride slowed mitoses that were in progress. Anaerobic conditions also did so, and Utkin was led to suggest that an additional energy supply may be required by mitoses in progress, beyond the original "energy reservoir" of Swann (1953). Utkin (1961) made the further interesting observation that the initiation of mitosis *in vitro* was considerably greater in cultures to which glucose had been added than in control cultures, if the incubation had continued only 2 or 3 hours; after 4 hours incubation, the number of mitoses was about the same in the presence as well as in the absence of added glucose; but at 5 hours, corneas in saline alone might show more mitoses than those in saline plus glucose.

This suggests, of course, that the saline alone was acting as a mitosis-synchronizing agent by delaying initiation, while the glucose was preventing the piling up of cells ready for mitosis to the same extent, by permitting some of them to divide. In the pea meristem studied by Wilson, a synchronizing effect was also given by dinitrophenol, when this was added up to 7 hours after colchicine-tagging. The glucose effect was temporary, not only in Utkin's experiment, but also in Wilson's pea roots that had been allowed to "run down" for 8 hours after excision.

A general criticism may be leveled against short-term experiments, such as these with animal cells. Perhaps Dr. Wilson would tell us whether this criticism can be applied to meristematic plant cells. It is this: Paul (1961) has remarked on the "surprising observation . . . that the metabolic pattern dictated by a given environment persists for some considerable time after the stimulus is removed." He was speaking of mammalian cell strains subjected to changes in their media, such as variations in oxygen tension and glucose concentration, to which the cells eventually responded by altering their rates of glycolysis or respiration. In short, I wonder whether experiments involving only the first few hours after tissue excision might not be unduly complicated by adjustments of unknown extent going on in the metabolism of the cells under study.

The short-term experiments we have been considering dealt with explantations of excised tissue fragments, in which the cells maintain their original locations with respect to one another and thus may not suffer the abrupt change in their immediate environment and the attendant leakage of intermediary metabolites that is often a feature of the cultivation of cell strains (Paul, 1961). Gelfant (1960), indeed, found mitosis occurring only at the cut edge of his explanted mouse ears. This confirms a common observation made in the early days of tissue culture (Fischer, 1930; Ephrussi, 1932), when cutting a quiescent culture was found to be a way to renew mitotic activity. Harding and Srinivasan (1960) have recently remarked on the wave of mitotic activity that spreads outward over a period of days from the site of mechanical injury in rabbit lens epithelium, as though some stimulatory substance were diffusing from the site of injury (among other possibilities). In this case, however, a wave of deoxyribonucleic acid (DNA) synthesis precedes the mitotic wave.

There is an interesting parallel between certain data of Dr. Wilson's and some of the Copenhagen school. Dr. Wilson has told us that treatment of pea roots with dinitrophenol up to 7 hours after colchicine-tagging delayed the following mitosis and improved synchrony, while treatment after 7 hours also delayed mitosis but tended to destroy synchrony. Zeuthen (1958) summarized results for various metabolic inhibitors, including dinitrophenol (DNP), applied to synchronized populations of *Tetrahymena*. The maximum sensitivity to these inhibitors, in division cycles lasting about 1½ hours, was about ½ hour before division, and not immediately before, as with heat or cold. From this, Zeuthen concluded that the cell accumulates energy for division in an open store until about 25 minutes before division, when the energy is set aside in a form, perhaps a structure, no longer available for maintenance. Zeuthen suggested that after this time the realization of cell division might depend on energy currently made available. Could one then term the last 25 minutes before division in *Tetrahymena* antephase?

There is other work with anoxia and respiratory inhibitors that is closer to Dr. Wilson's research from the standpoint of material used. Lamardelle and Gavaudan (cited by Gavaudan *et al.*, 1960) examined root meristems of hyacinth and of wheat after 24 hours of anoxia, after 4 to 12 hours of exposure to potassium cyanide, after ½ hour exposure to hydrogen cyanide followed by 4 hours in normal medium, or after exposure to sodium azide. Respiration was diminished from one-fifth to three-fifths by these treatments. The findings included arrested metaphases (often star-metaphases) with short, thick chromosomes and with maintenance of the structure but not the function of spindle fibers. These figures seemed to represent many of the mitoses in progress when treatment was started. There was an accumulation of "preprophasic" nuclei, defined as those in which the interphasic perinucleolar aureole had disappeared and in which heteropycnotic masses had accumulated around the nucleolus; such nuclei were distinctly different from those in prophase. In the cytoplasm ,there was evidence of mitochondrial damage, for some mitochondria individually became swollen and vacuolated, while others gathered together in peculiar "scaly bodies." There were similar accumulations of preprophasic nuclei in roots treated with mercuric chloride, coumarin, ethylurethan, benzene, and other agents, but the mitostatic effect in these cases was not accompanied by visible changes in the mitochondria.

Although it is now known that adenosine triphosphate (ATP) synthesis in the nucleus, as well as in mitochondria, is inhibited by anaerobiosis, dinitrophenol, sodium azide, and sodium cyanide, which also block amino acid uptake into nuclear proteins (Allfrey and Mirsky, 1957) , other evidence concerning a relation of mitochondria to mitosis may be cited. In cultures of chick-embryo fibroblasts and myoblasts under a nitrogen atmosphere, Frederic and Chèvremont-Comhaire (1955a,b) noted that the filamentous mitochondria lengthened by end-to-end fusion of shorter elements and then formed networks with thickened nodes. Prophases and prometaphases slowed down, the chromosomes swelled, and the mitoses became abortive without metaphase plate formation (Frederic, 1958) . This work, as you see, suggests (but only suggests) a support of mitosis in progress by current respiratory activity of mitochondria. Of course, there is the alternative possibility that damage to mitochondria or perhaps to lysosomes might release or activate adenosine triphosphatase, which in turn might eliminate certain energy stores in the cell. Plesner (1958) , you will recall, found that nucleoside triphosphate in synchronized *Tetrahymena* reached a maximum concentration just when the division index started to move upward from zero.

There is the interesting observation of Agrell (1955) to the effect that mitochondrial numbers in *Psammechinus miliaris* eggs were at a maximum at some time between telophase and the next prophase. This mitochondrial rhythm showed agreement with the respiratory rhythm, such as those traced by Holter and Zeuthen (1957) with a Cartesian diver microrespirometer on marine eggs: respiration increased through interphase, reached a maximum in prophase, and then decreased through metaphase and anaphase to a minimum in telophase. The difference in oxygen consumption between the maximum and the minimum was not great, however, for the cyclically varying respiration that supports cell division is a minor part of the total respiration of the cell, including that which supports growth (Hamburger and Zeuthen, 1957) .

St. Amand *et al.,* (1960) have remarked that respiration and mitotic rate are separable phenomena, for mitotic inhibition from X-radiation is not accompanied by diminution of respiration. Their own findings with agmatine, or decarboxylated arginine, support this statement admirably: grasshopper neuroblasts carried out more mitoses and went through each more rapidly in the presence of agmatine than in control medium, but with no change in the level of respiration.

In the light of all this, Dr. Wilson may be said to exhibit wise restraint in concluding from his own work that no specific pronouncements on the energy requirement for initiating mitosis are justified. However, this is an area of inquiry in which continued work is certainly needed.

On perusing the literature, I find no precedent for the accelerating effect of dinitrophenol or cyanide (when applied at 5 hours) on the second division of the tetraploid pea cell population, nor on its reversal with fluoride. These are remarkable findings, and I trust Dr. Wilson and his collaborators will provide us with explanations.

The other major subject of my discussion is our *ad hoc* classification of mitotic poisons according to the cellular structures or the mitotic stages presumably most affected by them, e.g., "spindle poisons," "metaphase poisons," etc. It is evident, first of all, that an arrest in one mitotic stage may have been caused by injury or inhibition in some previous stage. This conclusion, in individual cases, would be most readily arrived at by the use of synchronized populations exposed to the mitotic poison for a brief treatment in various stages of the total cycle, and with samples of the treated population examined at various later times. Nonsynchronized populations would be expected to give more equivocal data with less "resolution" of the effect, submerged as it would be in the heterogeneous cell population.

One can envision the possibility that there may be several sorts of poisons of a given mitotic stage, depending on location of the sensitive preceding portion of the cycle, during which treatment leads to arrest in the stage concerned.

Then again, arrest in a given stage might have several different terminal mechanisms. Dr. Wilson, for example, has told us that colchicine and other c-mitotic agents, which do away with spindle organization, are to be distinguished from such agents as iodoacetic acid, which immobilizes the spindle, presumably leaving a recognizable structure. One might add that there may be an intermediate condition, exemplified by the star-metaphase, in which the achromatic apparatus may be developed to a small extent. According to the hypothesis of Lettré and Lettré (1959), the chromosomal fibers of the spindle still persist in such arrested figures; however, de Harven and Dustin (1960) saw only indications of diffuse, radially oriented spindle material between centrioles and chromosomes in star-metaphases induced by Colcemid and examined in the electron microscope. Gavaudan *et al.*, (1960) found a similar spindle-immobilizing effect to be produced by anoxia. One might even include the effect of deuterium oxide on marine eggs, described by Gross and Spindel (1960). Lehmann (1960) described a whole series of effects on the achromatic apparatus in *Tubifex* eggs, ranging from complete disappearance with colchicine through lessening degrees of destruction with various quinones.

Another point to be made about the mitotic stage classification of mitotic poisons is the lack of necessary exclusiveness. A given agent need not be purely a prophase poison, or purely a metaphase poison, for instance. Dr. Wilson gives us a good example in actidione, which appears to act at many points in the mitotic cycle (Hadder and Wilson, 1958). Deuterium oxide inhibits at all stages of the cycle (Gross and Spindel, 1960). Sentein (1961) has objected strongly to calling colchicine a metaphase poison, because it can produce stathmokinetic prophases, anaphases, or telophases as well as arrested metaphases in his amphibian blastomeres. It also seems that some agents may act most markedly at different points (or not at all) in different biological material, as a result of the diversity that tops off the underlying biochemical unity of living matter.

These generalizations have been made evident in several recent studies carried out in France and summarized below.

Deysson and Truhaut (1960) examined the mitotic effects on plant root tips of a variety of agents used in experimental cancer chemotherapy. These agents included mustards, with β-chloroethyl radicals, ethyleneimine compounds, urethans, certain antibiotics, and various purine and pyrimidine bases. In general, compounds of all these classes showed similar actions. Beginning with the weakest concentrations to show activity (near 10^{-5} molar for the urethan, and about 10^{-7} to 10^{-9} molar for the others) and progressing to higher concentrations, there was first a depression of mitotic incidence, then chromosomal breakage and bridge-formation. With higher concentrations, the so-called preprophasic inhibition became so strong as to prevent all entry into mitosis. Cell death followed prolonged treatment or use of still higher concentrations. Near the lethal concentrations, certain compounds among those tested inhibited spindle function or prevented cytokinesis. Deysson and Truhaut noted with interest that the "radiomimetic" agents and the presumed antimetabolites had no essential differences in effect. All were preprophasic inhibitors, and most were chromosomal poisons. A few were spindle poisons; very few were both chromosomal poisons and spindle poisons to any marked degree. Poussel *et al.*, (1960) have remarked that such mutagens as alkylating agents and antipurines have the same end effect, but the alkylating agents are chemically active, while the purine analogs act in a "physical" manner (that is, following the concepts of thermodynamic "activity" developed by Ferguson, 1939). I prefer to think of the purine and pyrimidine analogs as acting in a way expected for antimetabolites (Biesele, 1960).

The most enthusiastic classification of mitotic poisons of late is that of Sentein (1960, 1961). He used segmenting urodele eggs, in which mitoses succeed one another in rapid succession and in which the large chromosomes and well-developed achromatic apparatus are clearly visible. Sentein distinguished three large classes of antimitotic chemical agents: depolarizing agents, mutagens or agents of rupture, and agents of mixed or complex effect. Each of these classes was broken down further. Among those whose primary action was depolarization of the spindle (and destruction of the cytoskeleton) were (*a*) slowly reversible agents, including colchicine, its derivatives, and podophyllin and podophyllotoxin — these might also cause secondary breakage of chromosomes; (*b*) agents of medium reversibility, such as chloral hydrate, which evoked monocentric mitoses; (*c*) the more rapidly reversible urethanes, which also damaged chromosomes; (*d*) feebly acting agents such as sarcomycin and sodium cacodylate, which affected cleavage; and patulin, which also strongly damaged chromosomes. Agents whose principal effect was chromosomal breakage included (*a*) triethylene melamine and certain ethyleneimine quinones, (*b*) some related agents, with depolarizing effects in addition, including certain other ethyleneimine quinones and nitrogen mustard, and (*c*) some others with strong effects on polarity including the ethyleneimine quinone, E 39. The third class, with mixed effects on spindle, nucleus (prophasic inhibition), and chromatin, included (*a*) phenols, (*b*) derivatives of khellin, (*c*) para-aminosalicylate and isonicotinylhydrazine, (*d*) phenylmercuric borate, and (*e*) caffeine. Finally, Sentein's amphibian eggs gave no response to colchicoside, γ-hexachlorocyclohexane, actinomycin, Myleran, and a few others of known antimitotic effect in other systems.

It appears that classifying antimitotic agents by the cell structures chiefly affected is as poor a method of achieving a clear-cut classification as is classifying by inhibited mitotic stage. Lehmann (1960) even objects to calling colchicine a spindle poison, because according to his understanding it actually affects the centriole in stages prior to development of the spindle in his material, the eggs of *Tubifex*. He also finds colchicine to cause dissolution of chromatin in this material.

In sum, I can only agree with Dr. Wilson's statement that "perhaps the time has come to work out a usable taxonomy of antimitotics," because that is inevitably what one is tempted to try, when working in the field, but I fear that we are bound to flounder through several revisions of the taxonomy until mitosis is better understood.

References

Agrell, I. (1955). *Exptl. Cell Research* **8**, 232-234.

Allfrey, V. G., and Mirsky, A. E. (1957). *Proc. Natl. Acad. Sci. U. S.* **43**, 589-598.

Biesele, J. J. (1960). *In* "Fundamental Aspects of Normal and Malignant Growth" (W. W. Nowinski, ed.), pp. 926-951. Elsevier, Amsterdam.

Bullough, W. S., and Johnson, M. (1951). *Proc. Roy. Soc.* (*London*) **B138**, 562-575.

de Harven, E., and Dustin, P., Jr. (1960). "L'Action Antimitotique et Caryoclasique de Substances Chimiques," Colloq. Intern. C.N.R.S. No. 88, pp. 189-197. C.N.R.S., Paris.

Deysson, G., and Truhaut, R. (1960). L'Action Antimitotique et Caryoclasique de Substances Chimiques," Colloq. Intern. C.N.R.S. No. 88, pp. 223-234. C.N.R.S., Paris.

Ephrussi, B. (1932). "La Culture de Tissus." Gauthier-Villars, Paris.

Ferguson, J. (1939). *Proc. Roy. Soc.* (*London*) **B127**, 387-404.

Fischer, A. (1930). *Virchows Archiv.* **279**, 94.

Frederic, J. (1958). *Arch. biol.* (*Liége*) **69**, 167-342.

Frederic, J., and Chèvremont-Comhaire, S. (1955a). *Compt. rend. soc. biol.* **148**, 211-213.

Frederic, J., and Chèvremont-Comhaire, S. (1955b). *Compt. rend. soc. biol.* **149**, 2094-2096.

Gavaudan, P., Guyot, M., and Poussel, H. (1960). "L'Action Antimitotique et Caryoclasique de Substances Chimiques," Coloq. Intern. C.N.R.S. No. 88, pp. 73-100. C.N.R.S., Paris.

Gelfant, S. (1960). *Ann. N. Y. Acad. Sci.* **90**, 536-549.

Gross, P. R., and Spindle, W. (1960). *Ann. N. Y. Acad. Sci.* **90**, 500-522.

Hadder, J. C., and Wilson, G. B. (1958). *Chromosoma* **9**, 91-104.

Hamburger, K., and Zeuthen, E. (1957). *Exptl. Cell Research* **13**, 443-453.

Harding, C. V., and Srinivasan, B. D. (1960). *Ann. N. Y. Acad. Sci.* **90**, 610-613.

Holter, H., and Zeuthen, E. (1957). *Pubbl. staz. zool. Napoli* **29**, 285-306.

Lehmann, F. E. (1960). "L'Action Antimitotique et Caryoclasique de Substances Chimiques," Coloq. Intern. C.N.R.S. No. 88, pp. 125-142. C.N.R.S., Paris.

Lettré, H., and Lettré, R. (1959). *Nucleus* (*Calcutta*) **2**, 23-44.

Mitchison, J. M. (1957). *Exptl. Cell Research* **13**, 244-262.

Paul, J. (1961). *Pathol. et biol.* **9**, 529-532.

Plesner, P. E. (1958). *Biochim. et Biophys. Acta* **29**, 462-463.

Poussel, H., Gavaudan, P., and Guyot, M. (1960). "L'Action Antimitotique et Caryoclasique de Substances Chimiques," Colloq. Intern. C.N.R.S. No. 88, pp. 101-124. C.N.R.S., Paris.

St. Amand, G. A., Anderson, N. G., and Gaulden, M. E. (1960). *Exptl. Cell Research* **20**, 71-76.

Sentein, P. (1960). "L'Action Antimitotique et Caryoclasique de Substances Chimiques," Colloq. Intern. C.N.R.S. No. 88, pp. 143-166. C.N.R.S., Paris.

Sentein, P. (1961). *Pathol. et biol.* **9**, 445-466.

Swann, M. M. (1953). *Quart. J. Microscop. Sci.* **94**, 369-379.

Utkin, I. A. (1961). *Pathol. et biol.* **9**, 519-522.

Van't Hof, J., Wilson, G. B., and Colon, A. (1960). *Chromosoma* **11**, 313-321.

Zeuthen, E. (1958). *Advances in Biol. and Med. Phys.* **6**, 37-73.

General Discussion

Dr. S. Gelfant (Syracuse University) : I would like to make two additions to Dr. Biesele's excellent review of "Studies on the Disruption of the Mitotic Cycle."

The first point involves the relationship of injury to the stimulation of DNA synthesis and initiation of mitosis. I would like to confirm Dr. Biesele's prediction concerning the effects of wounding in mouse ear epidermis (Gelfant, 1960, *Ann. N. Y. Acad. Sci.* **90**, 536) as compared to the effects of injury in rabbit lens epithelium (Harding and Srinivason, 1960, *Ann. N. Y. Acad. Sci.* **90**, 610). It is true that in contrast to rabbit lens epithelial cells in which injury stimulates DNA synthesis before mitosis, mouse ear epidermal cells are stimulated from the G_2 period of the cell cycle (after DNA synthesis has occurred). We have just reported this observation last week at a meeting in Chicago (Gelfant, 1961, *1st Meeting Am. Soc. Cell Biol.* November, 1961, p. 67).

The second point I would like to add concerns your allusions to Dr. Utkin's work (Utkin, 1960, *Pathol. et biol.* **9**, 519) which suggests that an additional energy supply may be required during the actual process of visible mitosis — over and above the original supply presumably provided during interphase. There are several recent papers by Amoore at the University of Edinburgh (J. E. Amoore, 1961, *Proc. Roy. Soc. B* **154**, 95 and 109) showing that the visible stages of mitosis in excised pea-root tips also have a requirement for oxygen and energy metabolism. I think the observations of Utkin and of Amoore open up a new point for consideration in the energy requirements for cell division.

There are two other points I would like to comment on. In your paper you say that an arrest in one mitotic stage may be caused by injury or inhibition in some previous stage of cell division. I wonder if this is really the way it works, because I think there is a rather strict system of check points from one stage to the next and that if you inhibit any event in one stage the cell will not move to the next stage of mitosis. For example, a cell will not move from interphase into mitosis if any of the interphase events such as energy production, sulfhydryl conversions, or DNA synthesis are inhibited. Prevention of nuclear membrane breakdown in prophase by actidione (as Dr. Wilson has shown) stops the cell from entering metaphase. If the spindle is disoriented in metaphase, anaphase will not occur, and if anaphase movement is inhibited, telophase will not take place. And finally, if cytoplasmic cleavage is blocked during telophase the two daughter cells are not formed. So that I don't think you can really influence mitosis in one stage by a previous inhibition or injury because the cell never reaches the next stage.

The last point involves the question of multiple effects of inhibitors. I think your concept of multiple effects of inhibitors is an excellent one and should be considered in all experiments. However, with reference to deuterium oxide being a multiple inhibitor it should be remembered that its mechanism of action on the different stages of mitosis is the same. That is, its action on the cleavage furrow in telophase or on the spindle in metaphase is the same because D_2O has a direct influence on the individual hydrogen bonds that cross-link these macromolecular gel structures. So that in this case the multiple effects of deuterium oxide are due the same mechanism rather than to multiple biochemical effects.

The Histones: Syntheses, Transitions, and Functions[1]

David P. Bloch

Botany Department and The Plant Research Institute, University of Texas, Austin, Texas

I. Introduction

While a fair amount may be said about the nuclear histones, little relates directly to the problems of mitosis, the underlying theme of this symposium. Therefore I wish to bring this discussion into context with the central issue by considering first some general aspects of the problem of cellular heredity. The connection has not been contrived to fit the occasion. The idea that histones play a role in somatic heredity has served as a motivating hypothesis for some years. The role of histones may ultimately be shown to be trivial, and our introduction inappropriate. No matter – in the meanwhile we shall have had the fun of relating the work to an exciting biological problem, and at present the likelihood of such a gross error seems remote indeed.

A. Alternative Bases for Somatic Heredity

Two properties commonly attributed to the gene are control of cellular processes and replication. Mitosis guarantees progeny cells their equivalent complements of these physiological units. Although this formulation of division may suffice for many single-celled organisms in which gene expression appears to be similar for all members of a population in a uniform environment, it is inadequate for cell division in multicellular organisms, where stabilization of alternative patterns of cellular activity is essential for cell specialization. In a liver cell, the genes responsible for the production of bile are expressed. Those which produce visual purple are not. If the liver is induced to regenerate, the progeny cells continue to produce bile, and refrain from producing visual purple. Phenotypic characteristics of a differentiated cell may even be maintained during growth in culture (Schindler *et al.*, 1959; Sato and Buonassisi, 1961). Hence during

[1] This work has been supported in part by grants from the National Science Foundation, the United States Public Health Service, and the University of California Cancer Funds.

mitosis, information regarding which genes are to be expressed is transmitted as well as the genes themselves.

While it is generally conceded that change in gene expression arising through classic mutation is attributable to a physical or chemical change in the gene itself, there have been no compelling reasons for accepting a similar basis for changes in gene expression during development. On the contrary, trends in the development of genetics have conspired to prejudice against this idea. Gene mutations are described as unpredictable in their frequency and direction and uncontrolled by the system they affect, while the opposing qualities characterize developmental changes in the cell. Moreover, proof of mutation in higher organisms has depended upon the demonstration that the change is segregated as are genes during meiosis, excluding from this category any changes in the gene which may normally be restricted to somatic cells during the course of development.

As the bearer of heredity has been narrowed from the chromosome to deoxyribonucleic acid (DNA), and each in its turn described as constant during development, genes have come to be regarded as constant in substance, flexible in activity, responding on demand as their containing cell differentiates. As a corollary, the heredity of a differentiated state has been seen as resulting from continuity of a controlling cytoplasm during division.

1. *The Cytoplasmic Basis*

Many examples of cytoplasmic heredity can be cited which are consistent with the above view. One beautiful illustration of the clonal inheritance of phenotypic differences, among subclones of cells of similar genetic constitution grown in an identical environment, is given in the work of Novick and Weiner (1957). *E. coli* cells or clones once induced to form beta-galactosidase can be maintained in an induced state in concentrations of inducer which are too low to initiate induction. Induction is an all-or-none phenomenon, hence individual sublines in a mixed population of induced and uninduced cells can retain their respective states when grown in "maintenance" concentration of inducer. The "hereditary unit" immediately responsible for maintaining induction appears to be the cytoplasmic permease, itself inducible, which concentrates inducer within the already induced cells, and is passed on to the progeny as part of the cell wall. The genes may be pictured as reflexive in their action, and there is no need to postulate stable alternative gene configurations.

Observations similar to these persuaded Ephrussi 5 years earlier toward the view that "Unless development involves a rather unlikely process of orderly and directed gene mutation, the differential must have its seat in the cytoplasm" (1953).

2. *The Nuclear Basis*

That certain aspects of development may be founded upon cytoplasmic changes, there is little doubt. However examples may be cited of heritable developmental change in which the cytoplasm, albeit responsible for initiation, appears to be irrelevant to maintenance. The most well known are the experiments of King and Briggs (1956), in which it was shown that nuclei obtained from the gastrula stages of the frog, in contrast to nuclei from earlier stages, are unable to support complete development when transferred to enucleated eggs. Furthermore, the loss of nuclear potentiality is heritable, since such a nucleus transmits its change essentially unaltered when cloned during serial transplantation.

In Sonneborn's work with *Paramecium aurelia* we have an experiment which simulates the exchange of nuclei between differentiated cells (see Nanney, 1957). In this organism, conjugation occurs between two cells of opposite mating type. After conjugation, the mating types of each of the cells remains the same as before, apparently determined by the residual "differentiated" cytoplasms, for the micronuclei and the newly forming macro (somatic) nuclei of the two exconjugant cells are genetically similar. If cytoplasmic mixing occurs during conjugation, the exconjugants become identical. Either of the mating types may now be realized, or one of them experimentally elicited by controlling the environment. If the newly forming macronucleus is caused to abort, a functional macronucleus is regenerated from fragments of the old macronucleus. The latter is apparently differentiated, for where a conflict exists between the determinacies of the old macronucleus and the residual cytoplasm, the former overrides the latter. In short, a newly forming macronucleus is multipotential. Only one of these potentialities is normally realized, which one, determined by the cytoplasm. Once determined however, it remains fixed. The nucleus has differentiated.

3. *The Genic Basis*

The demonstration of stable hereditary changes in the nucleus does not constitute proof of the existence of stable hereditary changes in its genes. The possibility of nongenic controlling factors within the interphase nucleus which may be carried along with the chromosomes during, or retrieved from the cytoplasm after mitosis cannot be excluded. Rigorous proof of gene differentiation depends upon the demonstration of alternative potentialities of gene expression which persist when the gene is stripped of nongenic material.

There are indeed several examples in which these requirements are closely approximated. One of these is the transducibility of alternative states of a locus in *Salmonella* (Lederberg and Iino, 1956). A given species of *Salmonella* is defined in part by its two flagellar antigens which are alternatively expressed among individual cells. The expression of either antigen is clonally inherited with occasional cells showing the alternative phenotype. The antigens are under the control of two distinct loci, H-1 and H-2. Transduction experiments indicate the following: The H-1 locus, responsible for the production of its antigen, is potentially active in all cells. Whether or not this activity is expressed depends upon the H-2 locus, which fluctuates in its activity. When the H-2 locus is active, its own antigen will be produced, but not that of the H-1 locus. When the H-2 locus is inactive, the H-1 locus is expressed with the formation of its own antigen. Ultimate control of phenotype appears to lie with the H-2 locus. The basis for variability of the H-2 activity is unknown, other than that it would appear to be an intrinsic part of the locus itself, since this locus can be transduced without change of activity.

Certain classes of gene mutation exist in higher organisms, which segregate in a Mendelian fashion, yet which have attributes which might be expected of developmental mutations, if such, indeed, exist. Among these are the "paramutations" described by Brink (1958). Paramutation describes the change in expression of a gene which is induced by one of its alleles. The change is directed, occurring 100% of the time. In this respect, it is similar to change in expression during development. The change is self perpetuating, persisting indefinitely after the inducing allele is segregated out by crossing, yet it can be almost completely reversed by bringing the altered gene into a homozygous condition. The classical genetic techniques may be applied in the study of this change because unlike changes in expression which occur during the course of development, this one persists in the germ line. It follows that the change is a physical one, occurring at the level of the gene.

These phenomena clearly demonstrate that genes can exist in alternative metastable forms, and by analogy, bring more closely together, the kind of genic change bridging generations of organisms, and those occurring during development. The persistence of the paramutational change through the germ line denies its usefulness to the organism during development, it is true, for it does not permit selection of alternative states during development of a single organism. However, it does point up the question whether, just as the mutant gene is an aberration of the normal gene, some aspects of mutation in the classical sense might be aberrations of a process of gene change which normally occurs during development.

It may be but a short step from the classical mutation to the paramutation, from paramutation to McClintock's activator-dissociator system in which controlling genes similarly effect the timing of expression of controlled genes (1955), and from the resultant abnormal variegation to the normal variegation of differentiation.

B. Inheritance of Chromosomal Constituents

A physical basis for hereditable alternative states of the gene may lie in a variety of configurations or compositions of the DNA molecule which can be reproduced during DNA replication, or in alternative complexes between DNA and other substances which are faithfully reproduced during or reconstituted following chromosome replication. The chromosome consists primarily of four types of components which might transmit such information; DNA, ribonucleic acid (RNA), histone, and nonhistone proteins. We do not know to what extent constituents other than DNA are conserved during division.

In 1955, Bloch and Godman found closely parallel syntheses of DNA and its associated histone during the interphase period of chromosome synthesis leading to mitosis (1955a), suggesting that histone may be inherited as well as DNA, although the critical test of conservation of histone during division has not been carried out. Increases in dye-binding capacities of DNA and histone during the period of DNA synthesis, suggested in addition, that other substances bound to deoxyribonucleohistone during the presynthetic period became dissociated during chromosome synthesis. These substances were presumed to be protein in nature and bound to DNA by electrostatic linkages (Bloch and Godman, 1955b).

Studies of *Euplotes* appear to be particularly promising in making more detailed analyses of nuclear syntheses and conservation of nuclear components. The exact sites of synthesis within the elongated nucleus are marked by the presence of regeneration bands which travel from both ends toward the center as synthesis proceeds. Distal to these bands are areas in which chromosome synthesis has been completed, and proximal, areas which have not yet begun synthesis. Gall has shown that DNA and histone syntheses occur at the regeneration band (1959) thereby obtaining a much more precise spatial correlation of the two events than had heretofore been possible. Again, syntheses of substances other than deoxyribonucleohistone appear to be interrupted during DNA replication, for Prescott and Kimball (1961) have shown that while RNA synthesis occurs in regions both proximal and distal to the regeneration band, little if any occurs at the band itself. Interestingly, RNA already synthesized by the nucleus is removed as the band traverses a region. Very recently, Prescott had found that, in contrast to ribonucleotides, lysine is incorporated in the regeneration band during DNA and histone synthesis, and is conserved during a subsequent synthesis (unpublished).

These observations support the view of hereditary continuity of at least one chromosomal protein during division, discontinuity of some other chromosomal substances, and insofar as continuity may provide a hereditary basis for chromosome variability, gives some support to the idea proposed by the Stedmans over a dozen years ago that histones may play a role in differentiation (1951).

II. Histone Variability

A. General Considerations: Specification and Association; Two Aspects of the DNA-Histone Relationship

The Stedmans' views were based on the finding of slight differences among tissues, in (1) the prevalence of different electrophoretically separable histone fractions, and (2) the amino acid compositions of some histone fractions with similar mobilities. Subsequent investigations have confirmed the heterogeneity of histones, even when these proteins are obtained from a homogeneous population of cells (e.g., nucleated erythrocytes, see Neelin and Connell, 1959), suggesting that the histones within a single cell are heterogeneous. Granted heterogeneity, other comparisons of the spectrum of histones obtained from different adult tissues have stressed similarities rather than differences (Crampton *et al.,* 1957; Vendrely *et al.,* 1958). The question is left open, whether histones are truly cell specific (varying among different cells) or constant among cells, deriving their variation from the different genes with which they are associated (or of which they form a part).

This question can be partially resolved. The limits of histone variability fall far short of that necessary to provide a specific set of histones for each gene (or better, for each polynucleotide sequence). This can be deduced from considerations of some of the physical characteristics of the deoxyribonucleohistone complex, and some simple principles of information transfer (Bloch, 1962a).

Deoxyribonucleohistone consists of a molecule of DNA having a molecular weight upwards of 10 million, complexed over its length with many smaller molecules of histone whose molecular weights range from 3500 to 74,000 (Felix *et al.,* 1950; Ui, 1956; Cruft *et al.,* 1958). The two species exist in approximately equal weights (Mirsky and Ris, 1947; Zubay and Doty, 1959). Most of the phosphate groups of the DNA are neutralized by the basic groups of the protein (Davison and Butler, 1956), the combined arginine, lysine, and histidine residues of the protein and the phosphate group being approximately equal in number (Felix *et al.,* 1951; Vendrely *et al.,* 1959). The electron micrographs of Zubay and Doty

(1959) are in accord with the view that in deoxyribonucleohistone the DNA is almost completely covered by histone.

There are from three to four amino acids per nucleotide in deoxyribonucleohistone. Because of the relative simplicity of the nucleic acid code (four symbols) as compared with the protein code (twenty symbols) Crick and others have reasoned that at least three nucleotides in sequence are needed to specify the position of an amino acid (Crick *et al.*, 1957; Brenner, 1957; Sueoka, 1961). It follows that there is insufficient information in a polynucleotide sequence to specify the synthesis of its associated polypeptide sequence. To phrase this in another way; a polynucleotide

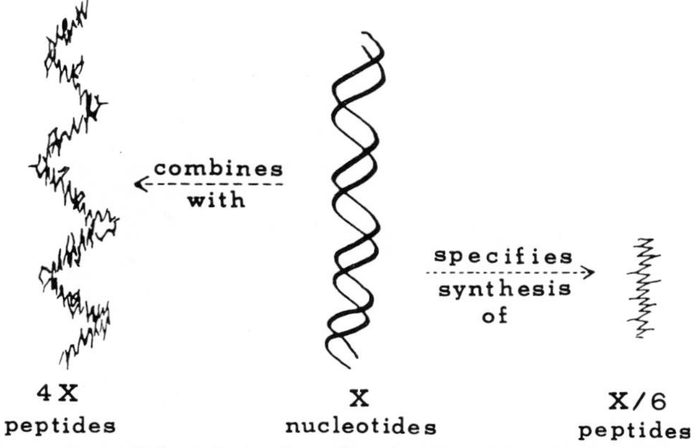

FIG. 1. Comparison of the relative lengths of polypeptides with which a polynucleotide becomes associated, and the synthesis of which it may specify.

can combine with more protein than it can specify (Fig. 1). If DNA provides the only ultimate source of information for protein synthesis, there must be fewer types of histone molecules than of their associated polynucleotides. Since DNA is nearly completely covered, different polynucleotide sequences must share common varieties of histones. Also, relatively few genes function in the syntheses of these proteins in spite of the fact that all genes appear to be associated with them.

The fact that alternative basic proteins (including the usual histones and those strange ones called protamines) successively occupy whole nuclei during spermiogenesis, fertilization, and development (Bloch and Hew, 1960a,b) indicates that alternative basic proteins may be combined with a given DNA: A gene will be associated with different histones during development. This aspect will be dealt with in more detail in the following sections.

These speculations do not prove, but they do accord with the idea that histones act in regulating some aspect of gene function, rather than in sharing with DNA the specification of gene activity, and that they may mediate in the influence of genes (histone producing) on other genes (histone associating), whatever this influence may mean.

B. Histone Variability during the Life Cycle

1. *Spermatogenesis*

During spermatogenesis the first of a series of changes occurs which extends through fertilization and early embryonic development. In these transitions the entire complement of chromosomal basic proteins is replaced by a succession of new proteins of differing composition. The replacement of a histone by a protamine in salmon, first described by Miescher (1897), was shown by Alfert (1956) to occur during sperm maturation. This alteration in the nuclear basic proteins during spermiogenesis is of common occurrence among both vertebrates and invertebrates. Its basic features in *Helix aspersa* and *Loligo opalescens* are described in Figs. 2 and 3. During the early stages of spermiogenesis there is a change from a histone which stains typically,[2] to one which stains very strongly with Alfert and Geschwind's method, and whose staining is indifferent to treatments which block or remove lysine epsilon amino groups. This change in staining properties reflects the replacement of the original histones in which lysine predominates; by arginine-rich histones which are intermediate between typical histones and protamines.

This first transition of the nucleoprotein is abrupt, being completed within 1 or 2 days, in contrast to the 2 or more months for the entire spermatogenic process. The change is accompanied by an incorporation of arginine into nuclear protein, indicative of a synthesis of new protein rather than conversion of a pre-existing one as had been proposed by Kossel (1928). Since this process occurs at least 3 weeks later than the previous (premeiotic) DNA synthesis, and is removed from the subsequent synthesis (in the fertilized egg) by at least two intervening protein syntheses, it seems unlikely that the histone synthesis in the intermediate spermatid is connected with the process of chromosome duplication, but probably serves some other function.

At a later stage in spermiogenesis, the nucleoprotein becomes much less resistant to the rigors of the technique used to demonstrate histones, and the nucleus of the mature sperm contained within the spermatophore can no longer be stained with Alfert and Geschwind's method.[3] It may be

[2] Most histones bind anionic dyes such as fast green at high pH's (Alfert and Geschwind, 1953) but lose this ability after removal of the basic groups of lysine by deamination.

[3] The nonstaining of many mature sperm provides a tentative cytochemical test for protamine.

visualized by a modification of this technique, utilizing picric acid for the hydrolysis of DNA, and eosin for staining of the protein. This procedure maintains the labile protamine as a precipitate, and stains this protein as well as histone.

Thus at least two steps occur during spermiogenesis in *Helix*. The first is the replacement of a typical histone by one which is very rich in arginine. This is in its turn replaced by a protamine.

A chemical analysis of these events has been carried out in the squid, *Loligo opalescens*. The same pattern of staining changes is seen in this organism. The mature sperm are contained within a spermatophore. The testis and spermatophore sac are large and yield sufficient quantities of nuclear basic proteins to permit extraction and chemical characterization. The testis contains the entire series of spermatogenic stages from gonial

TABLE I

MOLECULAR WEIGHTS OF SQUID PROTEINS DURING SPERMATOGENESIS[a]

Protein fraction[b]	Molecular weight		
	Ultracentrifugation		End group analysis (Phillips, 1955)
	(Archibald, 1947)	(Yphantis, 1960)	
Somatic histone (band 2)	46,500		—
Arginine-rich histone (band 3)	7,950	8020	
Testes protamine (band 4)	—	4600	3800
Spermatophore protamine (band 5)	—	5900	5800
Spermatophore protamine (band 6)	—	5000	5200

[a] Bloch (1962b).

[b] See Fig. 3.

cells to early morphologically mature sperm. Histones extracted from this organ contain fractions whose electrophoretic mobilities are similar to those obtained from the somatic tissues. These fractions, in the aggregate, have amino acid compositions similar to most histones. In addition, the testis contains two fractions with higher mobilities. The major and slower of the two fast fractions has all the amino acids usually found in histones. However, arginine accounts for approximately 60% by weight of the total, and outnumbers lysine by approximately 10 to 1 on a mole basis.[4] Accordingly this protein or class of proteins is designated arginine-rich histone. Its molecular weight is approximately 8000 as compared with 46,000 for one of the somatic histone fractions, and 4000 to 6000 for the protamines (Table I). The ability of the intermediate spermatids of *Loligo* and

[4] The author thanks Dr. W. A. Schroeder of the California Institute of Technology for his amino acid analyses of these proteins.

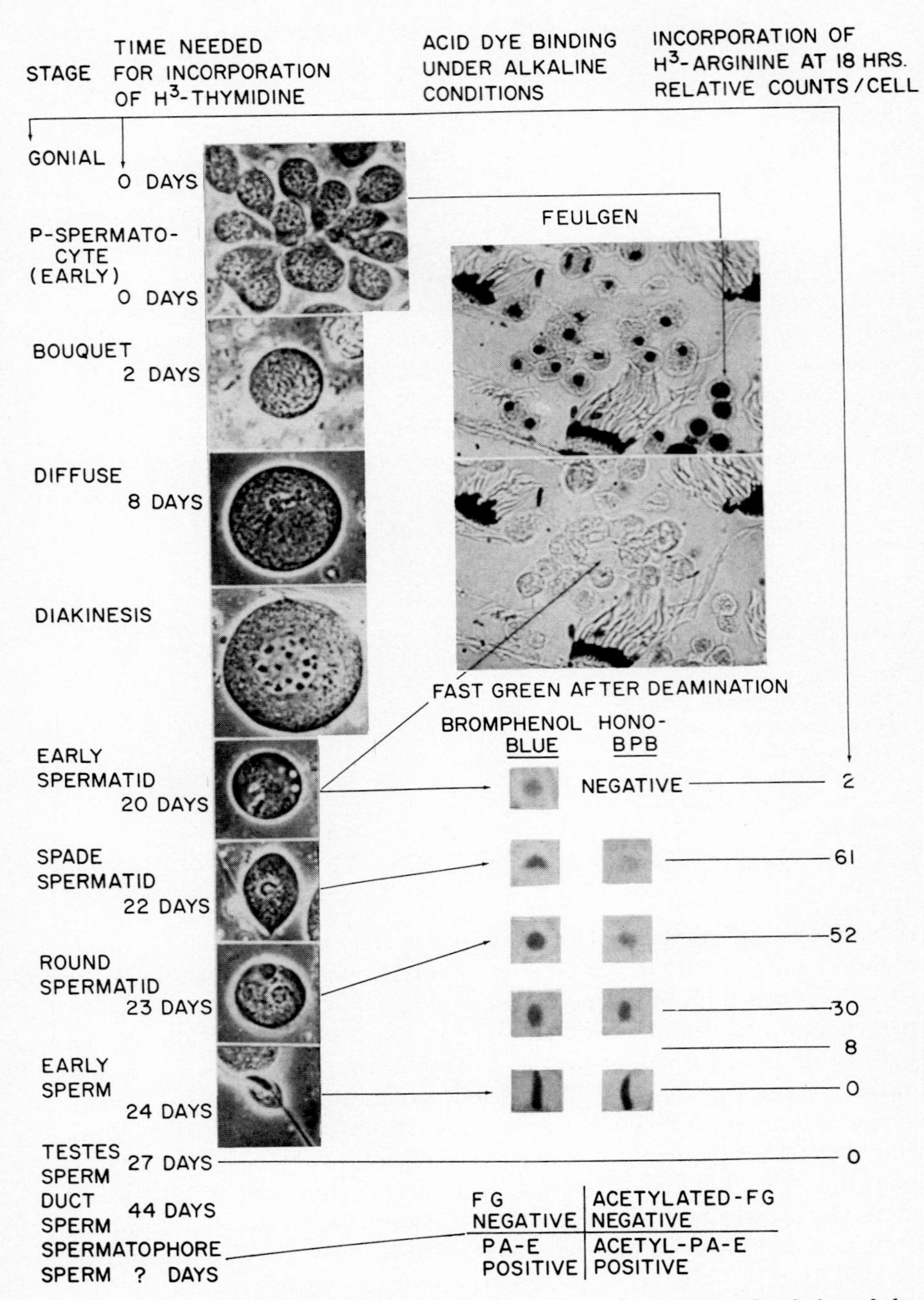

FIG. 2. Spermatogenesis in *Helix aspersa*. The first column shows the timing of the different stages depicted in the second column. These values were obtained by determining the times necessary for cells incorporating thymidine during premeiotic syn-

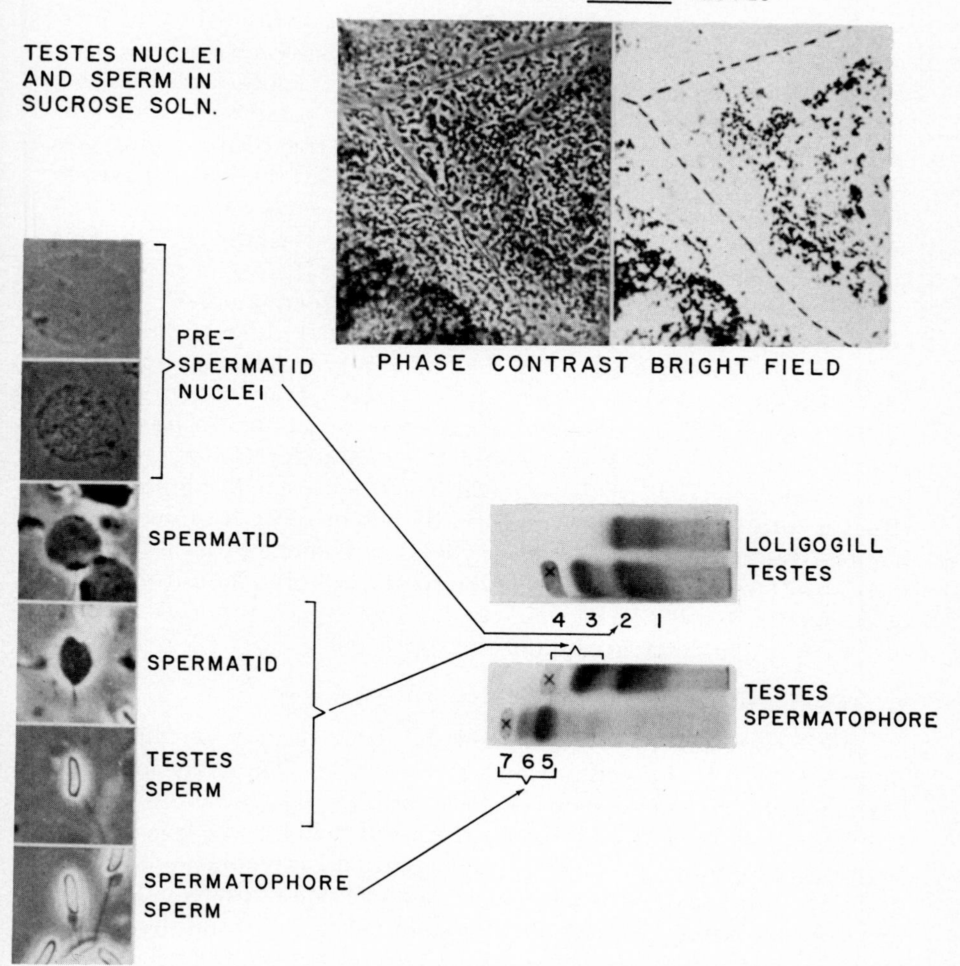

FIG. 3. Spermatogenesis in *Loligo opalescens.* The photographs in the upper right hand show the ability of the later spermatid and testes sperm to stain with acid dyes after deamination. The electropherograms on the lower right show a comparison of the banding patterns of the basic proteins extracted from several different organs. The shaded regions (x) indicate the positions of components which are present in variable amounts in these organs. The arrows show the sources of the component proteins which can be obtained from the separated preparations of nuclei and sperm. (Bloch, 1962b.)

---

thesis to reach these stages. The third column summarized the alterations in the staining behavior of the nuclear histones. HONO designates treatment with nitrous acid; BPB, staining with bromphenol blue; FG, fast green; PA-E, picric acid hydrolysis and eosin staining. The fourth column indicates the average number of silver grains overlying nuclei showing a positive autoradiograph 18 hours after administration of tritium-labeled arginine. (From Bloch and Hew, 1960a.)

Helix to stain with fast green after deamination is attributed to the presence of this protein. The fastest of the testes fractions has a molecular weight of approximately 4000; it contains only 8 amino acids. Its sole basic amino acid, arginine, constitutes 70% of the total, by weight. This protein can be classed as a protamine. Proteins extracted from a mixture of late spermatids and early sperms, separated from the earlier developmental stages by fractional centrifugation of the testes homogenate in a sucrose gradient, contain only the two fast fractions, i.e., the high arginine histone and protamine.

The two major fractions obtained from the spermatophore are also protamines, having amino acid compositions similar although not identical to the protamine of the testis. The molecular weights are approximately 6000 for the slower, and 5000 for the faster spermatophore fraction.

Thus at least three events mark the nuclear transitions during spermiogenesis in *Loligo*. In both *Loligo* and *Helix*, histones similar to somatic histones are replaced by histones which are very rich in arginine, and these in turn are successively replaced by one or more fractions of protamines. Interestingly, these changes occur after most of the other syntheses which result in formation of the specialized cytoplasmic organelles of the sperm. Hence it seems unlikely that they represent nuclear changes precedent to the differentiation of the sperm cell.

2. *Fertilization and Early Embryonic Development*

In *Helix aspersa* the protamine of the sperm head is lost during the process of fertilization (Bloch and Hew, 1960b). The early male pronucleus, still condensed, stains with neither the Sakaguchi test for protein-bound arginine, nor the picric-acid eosin method which is sensitive to protamine. Instead, a more sensitive picric acid bromphenol blue technique is needed to demonstrate the nucleic-acid bound protein. This protein differs substantially from the protamines it replaces, and the histones which it foreruns. So far it has been characterized only cytochemically, as being less reactive to acid dyes than protamines and histones, and has been presumed to be less basic than either. This protein is found in the male pronucleus, the female pronucleus, the polar body chromosomes, and all the nuclei during cleavage and morula stages.

At roughly 48 hours (at a stage corresponding to gastrulation in the frog) most if not all of the cells of the embryo revert to a state in which the histones can be stained in a typical manner with the fast green method of Alfert and Geschwind. The suggestion that such a histone change as this might provide a physical basis for Briggs and King's observed loss in totipotency in the frog nuclei prompted a spirited attempt to extend these studies to *Rana pipiens*.

3. *Variation in the Sequence of Transitions among Different Organisms*

Organisms differ in the number and the timing of the changes which occur during these processes, and also in the resultant products of the change. In many mammals, including the mouse, rat, guinea pig, and bull, the mature sperm contain an arginine-rich histone which has staining properties similar to that of the intermediate spermatid in *Helix*. Their acid-dye binding abilities are not affected by deamination. The proteins are presumed to be more complex than simple protamines because they are not leached from the cell during procedures which result in the loss of the latter class of substances. Indeed, Borenfreund and Bendich's finding that the DNA of mammalian sperm is extractable only after the use of procedures which reduce sulfhydryl bonds (1961) suggests that some of the DNA-associated proteins may be rich in sulfur-containing amino acids, although whether these include or accompany the associated basic proteins is not known for certain.

The formation of this arginine-rich histone occurs during spermiogenesis, however, in the mouse this process never "goes to completion," to form a simple protamine. Alfert did find that after fertilization the pronucleus loses its ability to bind fast green under alkaline conditions (1958). He attributed this loss to masking of the histone basic groups rather than loss of the histone itself. Whether it is truly a masking, or a replacement by a protamine, or by a protein similar to that of the cleaving snail egg, remains to be settled. The nuclei do stain typically within two or three divisions, and the process as described for the snail appears to be telescoped in the mouse, with the exclusion of several steps. In the frog the process is seen to be even simpler. There are no differences between the histones of the sperm and somatic cells which can be detected with presently available cytochemical techniques. The basic proteins of the sperm stain well with Alfert's method, indicating that they are not protamines, and their staining ability is abolished by blocking the basic groups of lysine, showing the histone to be typical (Hew and Bloch, unpublished). The amino acid composition is similar to that of somatic histones (Vendrely, 1957), and starch-gel electrophoresis shows the sperm histones and those obtained from somatic cells to have similar mobilities (Hew and Bloch, unpublished) (although the sperm proteins appear to be richer in those fractions having the higher mobilities, Fig. 4). The nuclei of cleavage and blastula stages stain with fast green and there are no obvious changes in the staining capacities at gastulation. The frog does not stand alone in this respect.[5] Rasch and Woodard found gametogenesis in *Tradescantia* to be similarly devoid of changes in histones (1959).

---

[5] Our findings contradict those of Horn (1960), who failed to find fast green staining

However, a study of nuclear histones in the frog is necessarily incomplete. The cytoplasm of the mature frog egg is known to contain DNA in sufficient amounts to account for all the nuclear DNA of the gastrula (Hoff-Jorgensen, 1954). Since no net synthesis of DNA accompanies the

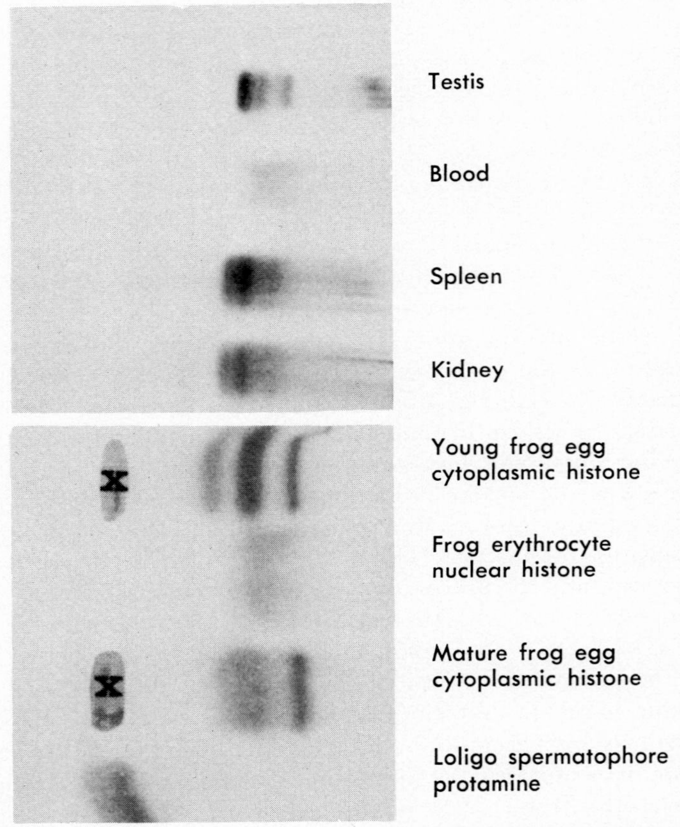

FIG. 4. Comparison of the electropherograms of histones and protamines extracted from different frog tissues, frog eggs, and squid sperm. The fastest fractions of the frog eggs (x) are present in substantial quantities, however, their inability to hold the stain during differentiation of the gel necessitated the marking of their position by shading. (Hew and Bloch, unpublished.)

nuclear proliferation of early embryonic development, it has been proposed that this DNA represents storage material to be utilized during very rapid cleavage. Furthermore, the nuclei have been thought to be

histones in the frog nuclei until gastrulation. The basis for these discrepant results may ultimately be found in quantitative measurements on the staining of histones as development proceeds, but at present, remain unexplained.

physiologically inactive during these stages except in the sense of proliferation of chromosomal material.

The cleavage nuclei of *Helix* differ from those of the frog. In *Helix* these nuclei are 50–100 times larger than adult diploid nuclei. The early cleavage nuclei contain a good number of nucleolarlike inclusions. While these "nucleoli" are atypical in some respects (Kersten, unpublished), their presence does suggest that the nuclei are active in carrying out syntheses in addition to those essential for chromosome replication. The nuclei of the cleaving frog's egg lack nucleoli (Colombo, 1949). Thus the question arose whether the cytoplasmic DNA of the frog embryo might be carrying out functions similar to the nuclear DNA in the snail embryo. A fair comparison of histone of *Helix* and *Rana* must include the basic proteins associated with the cytoplasmic DNA as well as the nuclear proteins.

Histones have been obtained from the cytoplasm of sea urchin eggs by Taleporos (1959), and frog eggs by Horn (1960). Starch-gel electrophoresis shows the major fraction of the cytoplasmic histones of the frog egg to consist of proteins whose mobilities are similar to the nuclear histones of somatic tissues, and also demonstrates a component similar to the protamines obtained from squid (Fig. 4). The amino acid composition of this latter component is indicative of a protamine. Thus protamines do arise in association with DNA, albeit in the cytoplasm, and the frog no longer appears to be so exceptional as we once thought it to be. Nevertheless, in their nuclear proteins, the frog and snail still offer an interesting contrast, and the frog provides a challenge to the idea that the extremity of the changes in nuclear histones so apparent in the snail and many other organisms, are of importance in the processes of cell differentiation common to all multicellular organisms.

III. Conclusion

Does a change in DNA-associated histone constitute gene differentiation? The answer to this question depends upon whether histone is indeed part of the gene. If histones are irrelevant to the process of gene activity, the changes in histones can be classed as differentiation only in the most trivial sense. For example, if histones act as protective agents, securing the DNA against the effects of intracellular enzymes, appropriate histone changes during development are of a similar order to those changes in other cell components during differentiation.

If changes in histones are merely the result of altered gene activity, as are changes in cytoplasmic proteins whose syntheses are governed by the gene, or perhaps changes in RNA associated with particular genes (puffs, etc.) during development (Beerman, 1952; Pavan and Breuer, 1952), but play no determining role, then histone changes, although perhaps reflecting gene differentiation, would not constitute gene differentiation.

If DNA and histone act in concert in specifying whether a particular type of activity can be realized under a given set of conditions, and if the information contained therein can survive DNA replication, then histone is part of the developmental gene and its changes would represent gene differentiation.

References

Alfert, M. (1956). *J. Biophys. Biochem. Cytol.* **2**, 109.
Alfert, M. (1958). *Ges. physiol. Chem. Colloq.* **9**, 73.
Alfert, M., and Geschwind, I. I. (1953). *Proc. Natl. Acad. Sci. U.S.* **39**, 991.
Archibald, W. J. (1947). *J. Phys. & Colloid Chem.* **51**, 1204.
Beerman, W. (1952). *Chromosoma* **5**, 139.
Bloch, D. P. (1962a). *Proc. Natl. Acad. Sci. U.S.* **48**, 324.
Bloch, D. P. (1962b). *J. Histochem. Cytochem.* **10**, 137.
Bloch, D. P., and Godman, G. C. (1955a). *J. Biophys. Biochem. Cytol.* **1**, 17.
Bloch, D. P., and Godman, G. C. (1955b). *J. Biophys. Biochem. Cytol.* **1**, 531.
Bloch, D. P., and Hew, H. Y. C. (1960a). *J. Biophys. Biochem. Cytol.* **7**, 515.
Bloch, D. P., and Hew, H. Y. C. (1960b). *J. Biophys. Biochem. Cytol.* **8**, 69.
Borenfreund, E., and Bendich, A. (1961). *Federation Proc.* **20**, 154.
Brenner, S. (1957). *Proc. Natl. Acad. Sci. U.S.* **43**, 687.
Brink, R. A. (1958). *Cold Spring Harbor Symposia Quant. Biol.* **23**, 379.
Colombo, G. (1949). *Caryologia* **1**, 297.
Crampton, C. F., Stein, W. H., and Moore, S. (1957). *J. Biol. Chem.* **225**, 263.
Crick, F. H. C., Griffith, J. S., and Orgel, L. E. (1957). *Proc. Natl. Acad. Sci. U.S.* **43**, 416.
Cruft, H. J., Mauritzen, C. M., and Stedman, E. (1958). *Proc. Roy. Soc. (London)* **B149**, 36.
Davison, P. F., and Butler, J. A. V. (1956). *Biochim. et Biophys. Acta* **21**, 568.
Ephrussi, B. (1953). "Nucleo-cytoplasmic Interactions in Micro-organisms," p. 4. Oxford Univ. Press, London and New York.
Felix, K., Fischer, H., Krekels, A., and Rauen, M. (1950). *Z. physiol. Chem., Hoppe-Seyler's* **286**, 67.
Felix, K., Fischer, H., Krekels, A., and Mohr, R. (1951). *Z. physiol. Chem., Hoppe-Seyler's* **287**, 224.
Gall, J. G. (1959). *J. Biophys. Biochem. Cytol.* **5**, 295.
Hoff-Jorgensen, E. (1954). *In* "Recent Developments in Cell Physiology" (J. A. Kitching, ed.), p. 79. Butterworths, London.
Horn, E. (1960). *Anat. Record* **137**, 365.
King, J. T., and Briggs, R. (1956). *Cold Spring Harbor Symposia Quant. Biol.* **21**, 217.
Kossel, A. (1928). "The Protamines and Histones." Longmans, Green, New York.
Lederberg, J., and Iino, T. (1956). *Genetics* **41**, 743.
McClintock, B. (1955). *Brookhaven Symposia in Biol.* **8**, 58.
Miescher, F. (1897). "Die histochemischen und physiologischen Arbeiten." Vogel, Leipzig.
Mirsky, A. E., and Ris, H. (1947). *J. Gen. Physiol.* **31**, 7.
Nanney, D. L. (1957). *In* "The Chemical Basis of Heredity" (W. D. McElroy and B. Glass, eds.), p. 134. Johns Hopkins Press, Baltimore, Maryland.
Neelin, J. M., and Connell, G. E. (1959). *Biochim. et Biophys. Acta* **31**, 539.
Novick, A., and Weiner, M. (1957). *Proc. Natl. Acad. Sci. U.S.* **43**, 553.

Pavan, C., and Breuer, M. E. (1952). *J. Heredity* **43**, 151.
Phillips, D. M. P. (1955). *Biochem. J.* **60**, 403.
Prescott, D. M., and Kimball, R. F. (1961). *Proc. Natl. Acad. Sci. U. S.* **47**, 686.
Rasch, E., and Woodard, J. W. (1959). *J. Biophys. Biochem. Cytol.* **6**, 263.
Sato, G., and Buonassisi, V. (1961). *Abstr. Am. Soc. Cell Biol.* **1**, 189.
Schindler, R., Day, M., and Fischer, G. A. (1959). *Cancer Research* **19**, 47.
Stedman, E., and Stedman, E. (1951). *Phil. Trans. Roy. Soc. London* **B235**, 565.
Sueoka, N. (1961). *Proc. Natl. Acad. Sci. U. S.* **47**, 1141.
Taleporos, P. (1959). *J. Histochem. Cytochem.* **7**, 332.
Vendrely, R. (1957). *Arch. Julius Klaus-Stift. Vererbungsforsch. Sozialanthropol. u. Rassenhyg.* **32**, 538.
Vendrely, R., Knobloch, A., and Matsudaira, H. (1958). *Nature* **181**, 343.
Vendrely, R., Knobloch-Mazen, A., and Vendrely, C. (1959). *In* "The Cell Nucleus" (J. S. Mitchell, ed.), p. 200. Butterworths, London.
Ui, N. (1956). *Biochim. et Biophys. Acta* **22**, 205.
Yphantis, D. (1960). *Ann. N. Y. Acad. Sci.* **88**, 586.
Zubay, G., and Doty, P. (1959). *J. Mol. Biol.* **1**, 1.

Discussion by Jerome S. Kaye

Department of Biology, University of Rochester, New York

I would like to present some preliminary results we have on the relation between chromatin structure and the nature of the basic protein in cricket (*Acheta domesticus*) spermatids. Application of the staining reactions used by Bloch to characterize histones and protamine, has indicated that some changes in structure (though not all) appear to be correlated with changes in the basic protein.

In the early spermatid nucleus, the basic protein shows the staining reactions typical of histone of somatic cells. It stains well with alkaline fast green, and its stainability is lost after deamination of free amino groups (by procedures used by Bloch). In electron micrographs these early nuclei have essentially the same appearance as typical interphase nuclei as shown in Fig. 1.

Somewhat later in spermiogenesis there is a large decrease in nuclear volume and a great deal of nonhistone protein is lost. The nuclear material reorganizes after this loss: electron micrographs show that a shell of very dense material has formed at the periphery of the nucleus, while the interior of the nucleus now has a very low density. Feulgen and alkaline fast green staining indicate that the DNA and the basic protein of the nucleus have become localized in the peripheral shell leaving only protein in the interior. Alkaline fast green staining of the shell is barely detectable after deamination, in spite of the very high concentration of basic protein there; hence the protein must still be of the typical histone type. An electron micrograph of a nucleus at the shell stage is shown in Fig. 2. The chromatin appears to be in the form of tightly coiled fibers. The fiber diameter ranges from 100 to 150 Å. The nuclear interior also appears to contain fibers.

Stages of nuclear elongation succeed the shell stage. In the earliest elongation stage (Fig. 3) the peripheral chromatin is formed of fairly straight fibers which are thicker than those of the shell stage. The fibers are circular in cross section, with diameters of 300 to 400 Å, and project from the membrane towards the interior for a distance of about 0.5 micron. Fibers in the interior of the nucleus are about 150 Å in diameter. In

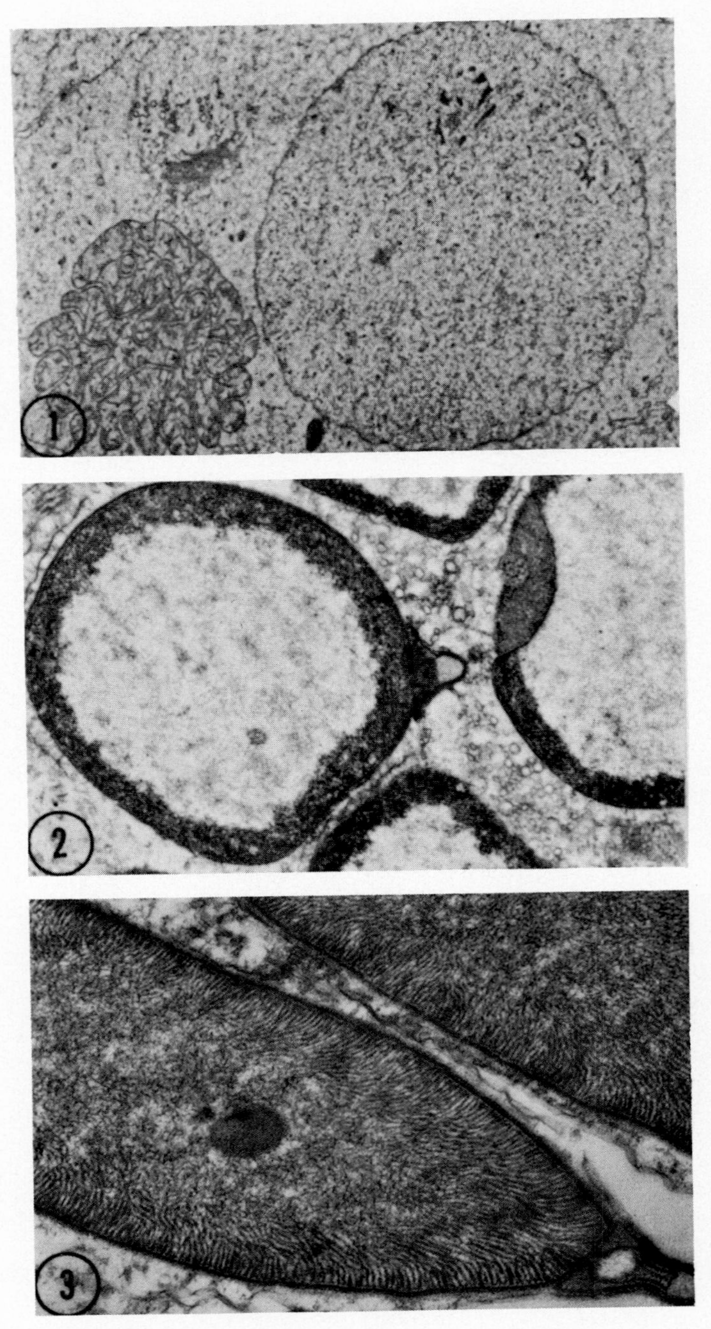

FIG. 1. An early spermatid of the house cricket, *Acheta domesticus*. The nucleus appears at the right. The chromatin has a granular appearance which is similar to that found in a typical somatic interphase nucleus. Magnification: 10,450 ×.

a later elongation stage the peripheral chromatin is formed of still thicker fibers, with diameters up to 2000 Å. They project interiorly for about half a micron as before. The interior region is now filled with lamellae, about 100 Å thick. In subsequent stages the peripheral material condenses into a solid shell. The lamellae remain visible in the interior for some time, but eventually the nucleus becomes completely solid in appearance.

Staining reactions indicate that a transition from typical histone to "arginine-rich histone" occurs during elongation. A considerable amount of alkaline fast green is bound by chromatin of the early elongation stage after deamination; however, deamination reduces its stainability. The chromatin of later elongation stages shows little, if any, decrease in stainability as a result of deamination. The arginine-rich protein of these elongated stages, including apparently mature sperm, is classified as histone rather than protamine, since the heads do not dissolve in hot TCA. The chromatin of mature sperm which are free in the ducts, does disintegrate in hot TCA, however, as would be expected of pure DNA-protamine.

In summary: the first observed rearrangement of chromatin, its condensation into a shell of coiled fibers, seems to be correlated with loss of nonhistone protein from the nucleus, rather than with changes in nature of the basic protein; the subsequent formation of straight 300 Å fibers is correlated with the appearance of some arginine-rich histone in the chromatin; the later arrangement in very thick (2000 Å) fibers is found when essentially all the histone is of the arginine-rich type; the final, completely solid structure is found for chromatin that consists of DNA-protamine, but may take form before the transition from arginine-rich histone to protamine is completed.

General Discussion

Dr. H. Ris (University of Wisconsin) : I would like to ask a question of Dr. Bloch. He cited work with bacteria to illustrate genic differentiation. As far as I know, bacteria do not contain histones in association with DNA. I wonder how he would reconcile this, with the view that histones play a role in differentiation. Various bacteria have been studied with the electron microscope, and have been shown to contain fibrils about 25 Å thick. These were called DNA fibers by Kleinschmidt. In their chemical studies, Zubay and Watson have shown DNA not to be associated with histone. Wilkins' studies of nuclear material from bacteria have indicated that it packs very differently from nucleo-histone. The DNA assumes a crystalline array in regions where the molecules are very closely packed, which would not be possible when DNA is associated with histone or protamine.

Dr. D. P. Bloch (University of Texas) : It is true that bacteria are supposed not to have histone, but it is also true that bacteria do not seem to differentiate in the sense that the cells of higher organisms differentiate. The only good example in which bacteria simulate the differentiation of higher forms is provided by the studies of Lederberg and Iino (1956, *Genetics* **41**, 743) on phase variation in *Salmonella*. Here is seen a relatively stable change in gene expression which is clonally inherited. Novick and Weiner's experiment (1957, *Proc. Natl. Acad. Sci.* **43**, 553) on the other hand, shows

---

Fig. 2. A later spermatid in the "shell" stage. The chromatin is in an electron-dense shell at the periphery of the nucleus. Magnification: 12,350 ×.

Fig. 3. A spermatid in early nuclear elongation stage. The chromatin has the form of numerous fibers about 300° in thickness at the periphery of the nucleus. Other fibers, about half that thickness, are in the interior. Magnification: 27,550 ×.

a much more contrived sort of "differentiation." Here is also seen a mixture of cells which differ in their gene expression, and which clonally propagate their state. But this situation is very unstable and depends upon constant control of the environment by the experimenter. Furthermore, this heredity appears to have a cytoplasmic basis. I think that if one were to draw any conclusions regarding a possible role of histones in differentiation in contrasting bacteria and higher organisms, it would be that bacteria do not have any demonstrable histones, and that differentiation among the cells of a bacterial colony does not occur to any high degree. The implication of this comparison is not very compelling. Chargaff has shown that bacterial DNA is associated with a protein, although he described this as a nonhistone protein.

Dr. S. Gelfant (Syracuse University) : Just a small question. I think you are selling yourself short, Dr. Bloch, regarding the concept that the DNA could not transfer information to its histone because the amount of histone is too great. In the past, you have proposed, as has Dr. Ris, that the complex should be considered, the total gene action resulting from the combination of DNA and histone. Why do you now "cleave" the two and consider only the DNA to act as the gene?

Dr. D. P. Bloch: I believe that certain DNA's of the complement have the job of making histones. Most DNA's do something else. What I have been proposing is that the integral gene which exhibits a certain effect in a cell is not only the DNA, but DNA plus the histone with which it is associated. There is no evidence for this, but it could provide a basis for differential gene activity. I would call the gene, DNA plus whatever it is that controls its activity, if "it plus whatever it is" can replicate as a unit. Let us be conservative for a moment and assume that gene potential is determined by DNA. Realization of the potential might be determined by the type of histone with which the DNA is associated. If this complex is able to replicate as a unit; that is, if when DNA replicates itself, the histone previously associated with DNA can specify which of the varieties of histone molecules already synthesized by the cell will associate with the replicated DNA, we would have a complex which is hereditary and which might be called the gene.

Dr. T. F. DeCaro (Villanova) : Is there any evidence that mutagenic agents have an effect on these basic proteins?

Dr. D. P. Bloch: None that I know of.

Dr. M. Bernstein (Wayne State University) : To the same point that Dr. Gelfant brought up, the dictation of histone synthesis by sequences of three nucleotides per amino acid residue. Assuming that each such sequence of three nucleotides specifies the placement of one amino acid, if, in fact, you consider all combinations, sets of three nucleotides can participate in directing the position of not one amino acid, but sixty-four. The other thing I wanted to ask you was this: From your talk, I'm left with the impression that it is not possible to draw a line, in a chemical sense, which separates protamines from histones. In functional terms, it becomes increasingly evident that no valid distinctions can be drawn; rather that one is dealing with basic proteins which exhibit quantitative differences in their degree of alkalinity. Furthermore, the expression or effective role of histones in terms of gene expression may conceivably be likened to what Crick has suggested, that at a given point of the double helix there may be an information suppressor, and that histones may be functioning as suppressors.

Dr. D. P. Bloch: We can't distinguish between histones and protamines on the basis of function because we don't know what the functions are. We can't separate them on the basis of any chemical attribute because protamines lie toward one end of a broad spectrum of basic proteins. As a matter of fact, the "histones" we find in the spermatid would have been called "triprotamines" by Kossel (1928), "intermediate histones" by

Hultin and Herne. We designate them "arginine-rich" histones. I would like to see the arbitrary classification of nuclear basic proteins into histones and protamines abandoned. I have reluctantly been calling "protamines" those proteins which everybody would agree to be protamines, but I believe that they might be better considered as an extreme form of histone, as are some other similarly simple, but very lysine rich "histones" obtained from somatic tissues.

Regarding the possible specification of sixty-four amino acids by all possible sets of three nucleotides: Crick's point was that to remove the problem of overlap, a large proportion of sequences must be without sense (or sensible for nonexistent amino acids). Crick supposes that of the sixty-four different combinations, two of three are nonsense, with the result that the remaining codes do not overlap, but specify the twenty or so naturally occurring amino acids.

DR. S. GELFANT: Just for the records, Dr. Bloch, considering the problem of gene expression at the somatic level, where would you place the nonhistone protein of the chromosome?

DR. D. P. BLOCH: This protein is present in small amounts. I've usually considered residual protein, as Mirsky described it, to be what is left when you extract DNA, RNA, and histone. It is relatively nondescript. Because of its high turnover I have been inclined to regard it as a gene product, rather than part of the gene itself. This is my particular prejudice. It isn't a firm one.

DR. O. SCHERBAUM (University of California at Los Angeles): I would like to ask a question concerning Alfert's finding that under certain growth conditions in Tetrahymena, for example, during the exponential phase, but not during the stationary phase, fast green staining material was found in the cytoplasm.

DR. D. P. BLOCH: At present we are involved in a study in which I believe we may find nuclear histones synthesized in the cytoplasm. In the grasshopper spermatid, the transition of the nuclear basic proteins occurs during a stage when there is no detectable RNA in the nucleus.

When cytoplasmic basic proteins are found, I would defer judgment on what to call them. If they can be shown to have their origin in the nucleus, or have a nuclear destination, they might correctly be called histones.

As you probably realize, our ability to selectively stain basic proteins and find them localized within the nucleus is sheer luck. Many basic proteins will stain with Alfert's fast green method, for example, lysozyme, and cytochrome. Fortunately such proteins are usually not present in the cell in sufficient amounts to cause confusion. Occasionally, one encounters cells such as eosinophilic leucocytes with large quantities of cytoplasmic basic proteins which can be characterized as nonhistone. When an unknown basic protein is encountered in the cytoplasm, we must wait and see. They may be histones.

DR. H. RIS: A word about the residual protein. You have given the usual definition, which is what is left after nuclei are extracted in strong salt in which nucleohistone dissociates and goes in solution. This fraction is by no means always a trifle. In some nuclei, such as those of liver, kidney, pancreas, it's a large part of the nucleus, 50% or more. Furthermore it is not in the nuclear sap but part of the chromosome. It is nondescript only in the sense that one doesn't know just what is does except that it turns over as much as any cytoplasmic protein. I don't think we should dismiss it when thinking about nuclear differentiation. It is true, we don't know yet just how it fits into the chromosome and how it is connected to the nucleohistone. In calf thymus nuclei it appears as more coarsely fibrous material between the DNA containing "chromatin" which consists of 100 Å fibrils. It does not change in appearance if isolated nuclei are extracted in 2 *M* NaCl, while the 100 Å fibrils are extracted and disappear.

With regard to the nucleohistone, I would like to say that the diagram from the work of Wilkin's group is no longer accepted by them. Recent X-ray diffraction studies on nucleohistone indicate that the histone is not wrapped about the DNA, but dispersed more randomly with concentration at 34 Å and 70 Å intervals. They think that the histone binds adjacent DNA strands together. My electronmicroscope studies on nucleohistone fibers from calf thymus indicate that always two DNA double helices are tied together by histone to compose a 100 Å fibril. In sperm where protamine-like proteins replace histone one finds that the 100 Å fibrils fall apart into two thinner ones.

Dr. J. S. Kaye (University of Rochester) : Our observations don't agree with Dr. Ris' scheme that chromatin fibers separate into two 40 Å fibers when protamine appears in the nucleus. The finest chromatin fibrils are found in early cricket spermatids when there is at most, a very low concentration of protamine. In the later spermatids, where arginine-rich protein is in high concentration, the chromatin fibers have increased from the thinnest dimensions, to thicker dimensions.

Dr. H. Ris: You have to distinguish two processes occurring during spermiogenesis. One is the appearance of the 35 Å thick fibrils (two for each 100 Å fibril) , and the second is the lateral association of the fibrils into larger complexes, which may be bundles of increasing thickness of membranelike structures such as are found in the grasshopper and certain snails.

I think what you are referring to are the thicker bundles of 35 Å fibrils which appear later in spermiogenesis during the gradual compacting of the sperm nucleus.

Dr. D. P. Bloch: I wish I had a picture with me of the electron micrographs of deoxyribonucleohistone published by Zubay and Doty (1959) . These strongly suggest that histone covers the whole length of the DNA molecules. The strands of the complex show occasional regions where the thickness decreases to 20 Å, but adjacent to these regions can usually be seen a clump, as though some material had peeled off and doubled back. Zubay and Doty, and others, have stressed the importance of using procedures for extracting deoxyribonucleohistone, which will not degrade the complex. Degradation occurs readily, and once degraded, it is virtually impossible to bring together a structure which is strictly linear rather than a network.

Wilkins may again change his mind. He need not have had to change it in the first place. Their original conception of the deoxyribonucleoprotamine complex was one in which a sequence of adjacent arginines alternated with two nonbasic amino acids which looped out, so that the electrostatic interaction between the arginines of the protamine and the phosphate groups of the DNA could continue without interruption. Now one of the main differences between a protamine and a histone is the relative prevalence of nonbasic amino acids in the latter. One might visualize the loops of histones being larger than those of protamines. This would be necessary to accommodate all the nonbasic residues which exist in histones. If these loops are sufficiently large, and if any degree of periodicity exists in their disposition one might see this in X-ray diffraction data. The reflections mentioned by Dr. Ris may not be due so much to a periodic association of histone along the DNA molecule as to periodicity of regions which are relatively rich and poor in loops.

Whatever the real structure, I don't think it matters too much as far as concerns the conceptual device of relating histones and DNA in showing there to be too little information in the DNA to specify the histone with which it is associated. The question of the exact physical configuration of deoxyribonucleoprotein is irrelevant.

Contractility

Andrew G. Szent-Györgyi[1]

Institute for Muscle Research at the Marine Biological Laboratory, Woods Hole, Massachusetts

---

I WILL ATTEMPT TO DISCUSS THREE PROBLEMS. Contraction of long chain polymers; polymerization of actin as an example of reversible fiber formation; and, finally, contraction of muscle. I hope such a description may help to find an answer to some of the questions which are raised in this symposium concerning cellular motility, spindle fiber formation, and directed movement of particles in the cell.

I. Contraction of Long Chain Polymers

Shape change, and mechanical work at the expense of chemical reactions is not a specific property of muscle, or components of muscle. Polymer molecules which have a structure different from a random coil will contract when conditions are changed, leading to the disruption of this structure. Viscosity changes accompanying denaturation of proteins or deoxyribonucleic acid (DNA) are examples of such a shape change. If, in addition, continuity is achieved between the molecules by condensing them into fibers and the two ends of the fibers are connected to a suitable apparatus, the tension and work which result from the folding of the molecules in the polymer system can be readily measured. Folding can be achieved in various ways which leads to the disruption of the forces maintaining the three dimensional structure of the polymers, such as temperature, urea, pH changes, "plasticizers" such as KI, etc. If the structure is regular, the forces maintaining the structure are crystalline-like forces and one may talk of the melting of crystals or phase transition. The criteria of such a phase transition have been discussed and treated thermodynamically by Flory (1957) who also proposed that this is the molecular event responsible for contraction of muscle.

---

[1] Established Investigator of the American Heart Association. This work was sponsored by Grant No. 54 F 46 E of the American Heart Association, Inc., Grant No. H-2042 R of the National Heart Institute and the Muscular Dystrophy Associations of America, Inc.

Present address: Department of Cytology, Dartmouth Medical School, Hanover, New Hampshire.

A more general treatment of the contraction of long chain polymers has been made by Katchalsky (1954) and by Kuhn *et al.* (1960). Katchalsky considered the contributions of electrostatic, elastic, and osmotic forces and developed a Carnot-like cycle in which the chemical energy of neutralization was reversibly converted into mechanical work. Kuhn *et al.* have shown experimentally that quantitative conversion of chemical energy into mechanical work can be achieved in long chain polymers. Moreover, mechanical work exerted on the system could be fully stored as chemical energy by reversing the processes initially used to convert chemical energy into mechanical work. If the long chain molecules are arranged in a three dimensional network in which the molecules cannot creep, and a reagent can cause reversible folding of the individual molecules, then the work performed will be equivalent to the change in the free energy which is derived from the change of the chemical activity of the reagent. Conversely, stretching the system will reverse the process. This is Kuhn's "teinochemical" principle, an expression he coined to signify that this is a separate and new way of producing work from chemical energy. According to Kuhn's explanation of these experimental results, the actual driving force of the system is a Donnan osmotic force and is due to the dilution tendency of the mobile ions within the gel. Great changes in the water content accompany the shape changes of the system.

The interconversion of mechanical and chemical energy was measured experimentally using copolymers of polyacrylate and polyvinyl alcohol prepared in such a way that three dimensional swelling was prevented and volume change was reflected by length changes only. The fibers prepared in such a way will shorten and lift weight if the pH is changed or the concentration of sodium ions is changed. The increase in the activity of these ions was also measured at various extensions. Knowing the force needed for stretching the system, the work performed, the activity of the ions, the quantitative interconversion of mechanical and chemical energy was demonstrated.

Of particular interest are systems in which the polymers precipitate. Among these are the polyacrylate-polyvinyl alcohol with calcium and the polyvinyl alcohol polyallyl-alloxan which in the presence of a redox system is oxidized to insoluble polydialuric acid. In both cases contraction is due to insolubilization of the complex and stretching will lead to the increased solubilization.

We do not have, as yet, any direct evidence that the molecular change in muscle contraction is folding in the above sense. Still, some movements in the cells may have very similar mechanisms. For example, the extracted preparations of Vorticella stalk (Levine, 1956; Hoffman-Berling, 1958) show features which are strikingly similar to Kuhn's system.

Let me summarize some significant features of the shortening of this system:

(*a*) It is a reversible equilibrium reaction, not a cycle. The length of the chain depends on the activity of the reagent causing contraction, and conversely, activity will depend on length.

(*b*) There is no need for a particular phosphagen as energy donor. The system is not ATP driven.

(*c*) There can be a variety of reagents leading to contraction, e.g., reagents which neutralize charges, or reagents which lead to the precipitation of the polymer.

(*d*) The long chain system is essentially a one-component polymer system. (The polyvinyl alcohol was used to serve as a mechanical support, but does not participate in the reaction. The reactive polymer is polyacrylic acid or polyallyl alloxan.)

To summarize: if the mechanism of some cellular mobility is analogous to the system described here, there is no compelling reason to believe that the movement itself requires the integrity of the whole cell. Once the structure is isolated one would expect that contraction and elongation could be reproduced with isolated material using substances present in the cell. An example of this is the observations on Vorticella (Hoffman-Berling, 1958). Indeed, such a mechanism is ideally suited for such a type of study because of the simplicity of the reaction and the variety of simple substances present in the cell which may serve as active agents.

II. Actin Polymerization

I discuss this subject for obvious reasons. Spindle fibers are formed at a certain phase in the cell's life and disappear at the anaphase of cell division. In fact, there have been attempts to find out whether or not spindle fibers may be considered to consist of an actin-like material. I do not know how far this concept is valid, but, nevertheless, polymerization of actin is one of the few examples of how a protein fiber can be formed from globular monomers. The work in the field is very active at present and many important factors have been brought forward.

In the reaction, a monomer with a molecular weight of about 57,000 is converted into filaments of several micra in length. The sharpness of the small angle X-ray reflection is an indication of the atomic precision with which the monomers are linked together.

The conversion of globular actin (G-actin) into fibrous actin (F-actin) is, in general, governed by ionic conditions (Straub, 1943). G-actin is stable in the absence of ions, or at high concentrations of strongly electronegative ions such as iodide or thiocyanate.

Adenosine triphosphate (ATP) is intimately involved in polymerization (Straub and Feuer, 1950). Both forms of actin have bound nucleotides in a mole per mole ratio based on the molecular weights of the actin monomer (Mommaerts, 1952). The bound nucleotide of G-actin is ATP, the bound nucleotide of F-actin is adenosine diphosphate (ADP). Actin itself is not an ATPase. Still, during polymerization and only during polymerization the ATP associated with actin is dephosphorylated. The structural change involved in the G-F transformation is intimately connected with the hydrolysis of the P-O-P linkage. In conditions where polymerization and depolymerization can proceed simultaneously, actin will simulate ATPase action and there can be a steady liberation of phosphate (Asakura and Oosawa, 1960).

The availability of nucleotide depends also on the state of actin. The ADP on F-actin is inaccessible to enzymes which interact with ADP free in solution (Laki *et al.*, 1950; Laki and Clark, 1951; Perry, 1954); it does not exchange with carbon labeled ADP and is not recovered from actin by extensive dialysis. The attachment of ATP to G-actin is less strong. It will exchange with ATP added to the medium (Martonosi *et al.*, 1960). It also can be dephosphorylated by apyrases and can participate in the reaction catalyzed by creatine kinase (Strohman, 1959). Ion exchange resins which readily adsorb free ATP from solution do not remove ATP from G-actin (Hayashi and Tsuboi, 1960). The nucleotides have also a protective action on the protein component. In general, removal of ATP from G-actin quickly leads to the loss of polymerizability.

Binding of adenine nucleotides appears to be mediated by divalent cations. There is one mole of calcium in a mole of actin (Chrambach *et al.*, 1961). Removal of the Ca by ethylenediaminetetraacetic acid, which, in itself, does not interact with ATP, leads to the dissociation of the nucleotide from actin. It is likely that the binding of ATP to actin is mediated by chelate formation via calcium (Strohman and Sanorodin, 1962). It is possible, though, that calcium is not directly interacting with ATP but is maintaining a certain conformation on the protein needed for binding of ATP.

What ATP does and what happens as a result of the dephosphorylation is not known. It may be somehow connected with the very specific way in which the monomers are linked together, and be correlated in exposing active sites for specific protein-protein interaction. Dephosphorylation may be the price we have to pay to have a specific interaction which is reversible in the sense that the starting components can be recovered unaltered. The price is still a small one if we compare it with the expense in energy required for the synthesis of monomers which would be necessary if fibers were synthetized *de novo*. It should be pointed out that actin is not the only example of formation of protein filaments which is characterized by the high precision of the interacting sites. The formation of fibrin from fibrinogen is also a reaction of great specificity. Here the uncovering of active sites is performed by the removal of peptide material, fibrinopeptide A and B, from fibrinogen through the proteolytic action of thrombin (Bailey *et al.,* 1951). The altered fibrinogen molecules associate to form fibrin. In this reaction the change is essentially irreversible, and one cannot recover the starting material in significant quantities.

I do not know how far the analogy between spindle fiber formation and actin polymerization is valid, whether it is valid at all. It seems that in certain cells the spindle fibers occupy a considerable mass of the cell. If it is formed from precursors, it would appear to be an economic way not to lose these precursors or monomers entirely by an irreversible change during the formation and disappearance of the spindle and to preserve them for the division of daughter cells. It would be of interest to see how far the material of spindle fibers of one division is conserved during subsequent divisions.

If it can be demonstrated that spindle formation has some degree of reversibility as indicated from the temperature dependence of birefringence (Inoué, 1959), the reactions of actin may be of help in the study of formation and disappearance of the spindle.

III. Contraction in Muscle in General

At first I will try to describe those general features of the chemistry of muscle contraction which I consider relevant, then I will discuss some problems relating to the structure and contraction of striated muscle.

Treatment of muscle tissues with solvents of low ionic strength leaves the main structural features characterizing the particular muscle relatively unaltered and does not extract the components responsible for contraction. This relative insolubility of the proteins participating in contraction is the reason why simplified muscle preparations like glycerol-extracted psoas, etc., can be used in studying some of the reactions in contraction (Szent-Györgyi, 1949), and allows us to devise experiments to account for the structural features in terms of the component proteins. The contractile proteins can be extracted, a simplified process of contraction reproduced, and their properties and reactions studied (Szent-Györgyi, 1951).

Table I shows the protein composition and some of the parameters of the main proteins of washed myofibrils. It is easily seen that about 80–90% of the total protein content of the myofibril can be accounted for by three relatively well studied proteins; myosin, actin, and tropomyosin. Of these, actin and myosin are directly involved in contraction and, as shown by Albert Szent-Györgyi (1951), the simplest contractile system consists of actin, myosin, and ATP. At low ionic strength, in the presence of small concentrations of Ca^{++} and Mg^{++}, contraction will take place when actin, myosin, and ATP are mixed. The presence of both proteins are necessary for contraction. Both actin and myosin specifically interact with ATP, neither of them are contractile alone. I have discussed, previously, the role of ATP on actin polymerization. The interaction of myosin with ATP is an enzymatic one and myosin is capable of hydrolyzing the last phosphate of ATP (Engelhardt and Ljubimova, 1939) and related triphosphorylated nucleotides. Since the energy for contraction is derived directly or possibly somewhat indirectly from the high group transfer energy potential of the P-O-P linkage, this means that the protein expending energy in contraction is somehow capable of regulating its supply. In fact, inactivation of ATPase leads to loss of contractility in actomyosin or extracted muscle preparations. The reactions involved

TABLE I
PROTEINS OF THE MYOFIBRIL

	Molecular weight	Estimated shape in A	Function	Per cent of protein in rabbit myofibril
Actin			Contraction	15–20%
Globular	60,000	25 × 290?		
Fibrous	∼ 4,000,000	∼ 10,000		
Myosin	450,000	25 × 1600	ATPase contraction	∼ 55%
Tropomyosin	53,000	15 × 385	?	∼ 10%

in the energization are not known. The hydrolytic step itself may be an artifact which occurs in the *in vitro* preparations only. But, nevertheless, it may be taken as an indication of a necessary property of myosin involved in the energization process. At low ionic strength and presence of Mg^{++} ions, the ATPase activity of actomyosin is many times greater than that of myosin, and there is a parallel between contractility and enzymatic activity (Weber and Portzehl, 1954). Contraction proceeds only at an ionic strength at which myosin and actomyosin are precipitated. This fact restricts the use of tools of physicochemistry designed to study size and shape changes in solution and is a reason for our present ignorance of how far such changes are involved in contraction.

Myosin itself is a complex molecule. Short treatment with proteolytic enzymes breaks it up into essentially two kinds of components; heavy and light meromyosin (Gergely, 1953; Mihalyi and A. G. Szent-Györgyi, 1953; A. G. Szent-Györgyi, 1953). The heavy meromyosin contains the active centers necessary for ATPase activity and combination with actin. It is readily soluble at low ionic strength. The light meromyosin has a highly ordered structure. It is one of the few proteins which appear to be fully helical, has a structural regularity with a 430 Å repeat period (A. G. Szent-Györgyi *et al.*, 1960). It is not an ATPase and does not interact with actin. Myosin thus can be considered as having a head and a tail, the head performing specific activities, and the tail giving rigidity and structural regularity.

As mentioned above, when ATP is added to actomyosin at low ionic strength, contraction occurs. Action of ATP is specific and can be simulated only by the other triphosphorylated nucleotides in the presence of higher $MgCl_2$ concentrations. At higher ionic strength, in which actomyosin is in solution, ATP has a different effect. It will dissociate actomyosin into its component proteins, actin and myosin. The dissociating action of ATP is less specific. Inorganic pyrophosphate in the presence of Mg^{++} ions has a similar action.

We can picture the changes in contraction in the following fashion. At rest there is no interaction between actin and myosin. The conditions are such that the presence of ATP in muscle will prevent interaction between these two proteins. The main result of excitation is to bring about combination of actin and myosin. Once actomyosin is formed it contracts since there is ATP present. During contraction, the energy of the last phosphate of ATP, in possibly more than one step, is utilized to pay for the energy expenditure in contraction, resulting in the liberation of inorganic phosphate. After the disturbance accompanying excitation has passed and contraction has been performed, interaction between actin and myosin will be suspended and the resting state restored.

$$\begin{array}{ccc} A + My - ATP & \rightleftharpoons AMy - ATP \rightarrow & AMy^* - ATP \\ \uparrow & & \downarrow \\ (AMy^* + ATP) & \leftarrow \leftarrow & AMy^* - ADP + Pi \end{array}$$

This scheme is a rather superficial one in that it does not describe any of the reactions in a precise way, nor does it give an account of the structural changes during contraction. Even on the basis of such a crude description, one arrives at certain conclusions and contrasts it with the relatively simple contraction of the long chain polymers discussed at the beginning:

1. The contractile system of muscle is a two-component system. This allows the control over rest and activity to be more precise and sharp without the necessity of a sequestering mechanism for the removal of the small molecule causing contraction. The mechanism is also a more specific one.

2. The reaction is an ATP-driven reaction. The energy balance is not taken care of by changes in activity of ions, but by some mechanism utilizing the high energy phosphate bond of ATP.

3. The contraction of muscle is not a simple equilibrium. It involves a number of steps. While these individual steps may be in equilibrium, the over-all process is a cycle. The restitution of the resting state is not simply a reversal of steps which lead to contraction.

It appears thus that motility which resembles muscle contraction will require the presence of ATP or possibly other phosphagens and will involve enzymatic processes. Such a system seems to be operating in contraction of fibroblasts, Jensen sarcoma cells (Hoffman-Berling, 1954, 1956), and the retraction of the fibrin clot (Bettex-Galland and Luscher, 1961). Systems of this kind appear also to be amenable to study in simplified form, and it is likely that the structures containing the contractile material can be isolated and rudimentary processes of motion reproduced in cell-free conditions.

IV. Structure and Contraction of Striated Muscle

A. The Sliding Theory of Contraction

In the past it has been generally assumed that contraction is the result of the folding of the contractile material. One would have thought that our increased knowledge of the behavior of polymers would help in the understanding of muscle contraction. Indeed, specific proposals have been put forward recently that contraction involves a phase change, melting of crystalline regions of the contractile material (Flory, 1957). The advantage of these theories is that the mechanism of such reactions is fairly well understood. On the other hand, there is no direct evidence as yet indicating that the contractile proteins undergo such change during contraction.

The studies of Hanson and Huxley (1955; Huxley, 1960) on the structure of striated muscle at various lengths led them to propose that contraction does not involve shape change of the contractile molecules, but rather that shortening is the result of the sliding of rigid rods which are out of register at rest length.

The theory is based on the following type of observations:

(*a*) The filamentous organization of striated vertebrate muscle shows a double array of filaments in a hexagonal pattern. Thick filaments (100 Å) form a hexagonal pattern with thin filaments (50 Å) at the trigonal points (Huxley, 1957). While it may well be that electronmicroscopic studies do not show all the material which is present, at least as a skeleton structure the double hexagonal array of filaments represent the basic filamentous organization of these muscles. From the thick filaments, cross bridges extend towards the thin filaments connecting the two. Between each thick and thin filament there is a cross bridge for every 420 Å. Since a thin filament is shared by three thick filaments, there is a cross bridge at every 140 Å around a thin filament.

(*b*) The relative position of the thick and thin filaments correspond to the band pattern of striated muscle (Huxley and Hanson, 1954, 1957). The thick filaments are about 1.5μ long and extend from one edge of the A band to the other edge. The thin filaments are 1μ long and protrude from the Z membrane to the edge of the H band. The anisotropy of the A band is due to the presence of the thick filaments. The extra density at the lateral portions of the A band is reflection of the presence of thin and thick filaments there. This is the overlap area. The H band is less dense since only the thick filaments are there. The density of the I band is even lower, containing only thin filaments. Its birefringence is also low. According to this interpretation, the length of the bands are then correlated with the length and position of the filaments.

(*c*) Identification of the filaments with actin and myosin. The thick filaments consist of myosin, the thin filaments of actin (possibly with tropomyosin). This is based mostly on the observation that extraction of myofibrils with solvents which remove mostly myosin, leads to the disappearance of the extra density in the A band, and also to the disappearance of birefringence (Huxley and Hanson, 1954, 1957).

(*d*) Band pattern changes during stretch and contraction. During stretch, the A band stays constant while both the I band and H band elongate in proportion to the stretch. This has been shown for living single fibers of frog (A. F. Huxley and Niedergerke, 1954) and for psoas muscles glycerinated at various lengths (Huxley and Hanson, 1954, 1957). In the initial stages of contraction, the A band stays constant, the length of the I bands reduces proportionally to the shortening. The H band will eventually disappear, but it is difficult to decide, at present, how this is correlated with the shortening itself. At more extensive shortening, contraction bands, regions of higher density, appear in the center of the A band and at the Z membrane. These are called the Cm and Cz bands. It may be noted here that contraction is not the reversal of stretching and the changes in the H band are not clear yet.

The explanation of these changes, then, is that in stretch the actin filaments are pulled out and the overlap between the myosin and actin filaments decreases. In contraction, the actin filaments are sliding further in between the myosin filaments, their movement mediated somehow by the cross bridges which would be the interacting sites between actin and myosin filaments. At later stages of shortening the filaments fold up and this is how the contraction bands are explained.

Thus contraction, at least at the beginning, does not involve length changes of molecules, rather the driving force of contraction is due to the interaction of the two types of filaments containing the different proteins which leads to the directed sliding.

These are very beautiful experiments which not only yield relevant information of the structure and changes of structure of striated muscle, but through their exactness and quantitative predictions, establish criteria for any theory which is devised to explain contraction in terms of the structural components of muscle.

At present, the sliding model of contraction is perhaps the most seriously and widely considered theory of contraction. There are a number of features, though, which disturb me personally and make me wonder whether this is the only mechanism which fits the observations.

It does not seem possible to explain the movement of huge molecular aggregates in one direction, mediated by the repeated and additive action of equivalent sites, based on known physicochemical principles.

Another general objection is that this mechanism of contraction hardly explains the *in vitro* contraction of extracted unoriented actomyosin. In such systems the sliding of one filament is compensated by the sliding of another in the opposite direction and no visible change should be observed. In fact, ATP addition to unoriented actomyosin leads to a drastic contraction. Neither is this theory in agreement with some of the observations when fluorescent labeled antibodies were used to localize myosin at rest and during contraction.

B. Immune Studies on the Localization of Myosin

I would like to speculate on the possibility of a contraction mechanism where contraction is driven by a structural change in myosin but would still be in agreement with the observed band pattern changes and basic filamentous organization. I will also describe some experiments using the fluorescent labeled antibodies. This work was initiated by Marshall *et al.* (1959) who prepared antibodies against actin, myosin, and meromyosins and measured in chicken muscle where the antibodies were bound, that is, where the antigens are localized. I will concern myself only with myosin here. In myofibrils at rest length, anti-myosin is deposited in the A band only (Fig. 1). This result agrees with the studies of Huxley and Hanson on the localization of myosin. Antibody against the light meromyosin component of myosin stains preferentially the lateral portions of the A band. Antibody against heavy meromyosin gives a strong staining at the M band region and the staining of the rest of the A band is weaker and more diffuse.

The differential staining of the myofibril by the meromyosins is difficult to interpret in a simple way. The distance from the edge of the A band is about 7500 Å which is five times or so greater than the value of

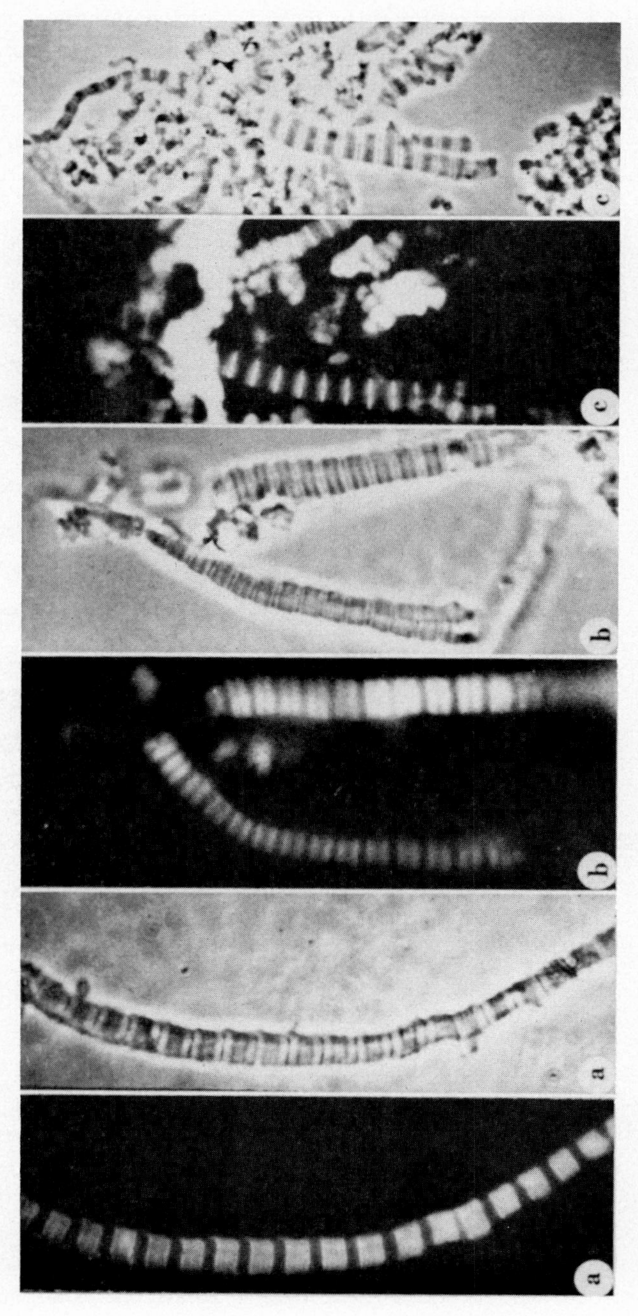

FIG. 1. Fluorescent band pattern of myofibrils: a, myosin antibody; b, light meromyosin antibody (LMM); c, heavy meromyosin antibody (HMM). (Marshall *et al.*, 1959).

1600 Å found for the length of myosin from hydrodynamic measurements (Holtzer and Lowey, 1956). If the staining represents the position of the meromyosins within the A band, the myosin molecule must be in a more extended configuration in the myofibril than in solution (Marshall *et al.*, 1959) or the meromyosins are separated from each other and myosin is formed during extraction and during activity. Some of the differential

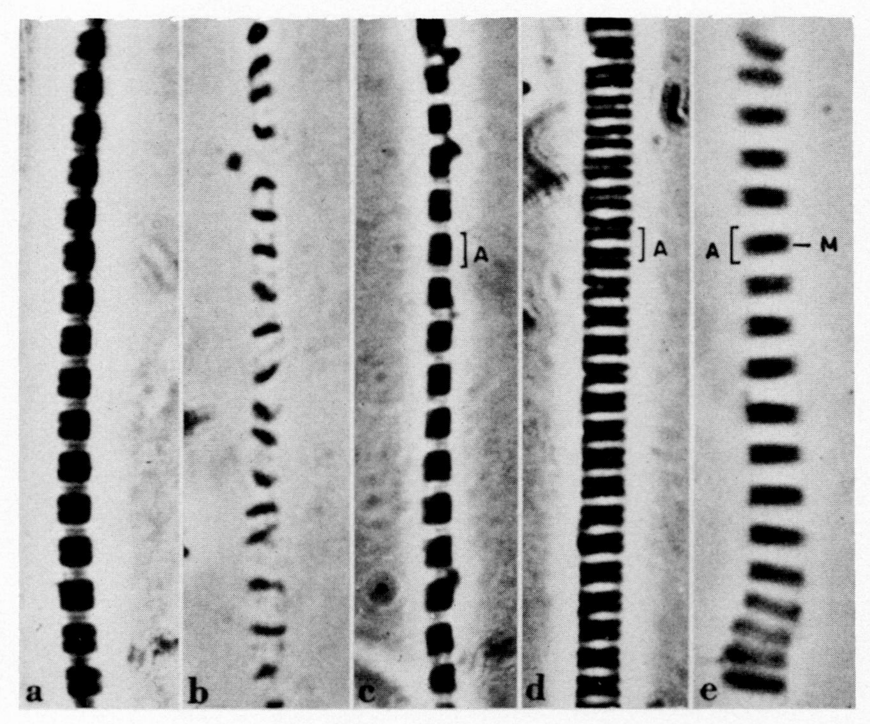

FIG. 2. Fixation of myofibrils with antibodies: a, control unextracted; b, control extracted with 0.6 *M* KI; c, extracted after myosin antibody treatment; d, extracted after light meromyosin antibody treatment; e, extracted after heavy meromyosin antibody treatment. (A. G. Szent-Györgyi, and Holtzer, 1960.)

staining may also arise from impurities, i.e., antigens other than myosin. There is also the possibility of steric hindrance and of blocked antigenic sites which influence staining differently when using myosin antibody rather than the individual meromyosins.

It was found (A. G. Szent-Györgyi and H. Holtzer, 1960) that the local precipitin reaction in the myofibril which results from the antibody-antigen interaction preserves the stained portions from the action of solvents which, in the absence of antibody treatment, removes these structures (Fig. 2). Thus anti-myosin treatment renders the whole A band insoluble in 0.6 *M* KI. Treatment against anti-light meromyosin protects the lateral portions of the A band, treatment against anti-heavy meromyosin protects the central portions of the A band. This finding is in agreement with the fluorescent band pattern. The fixation studies may also help in the interpretation of that pattern. By determining the

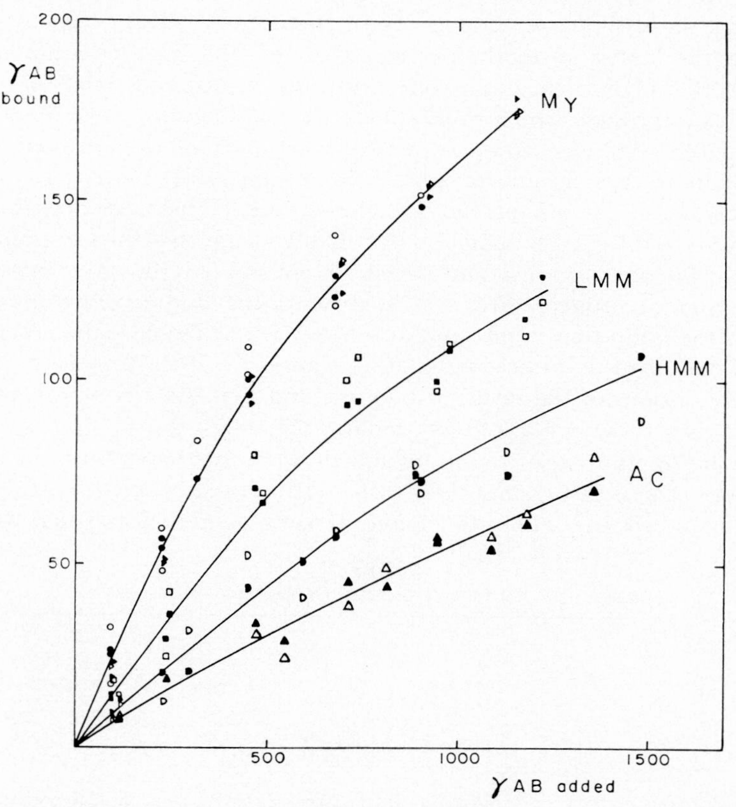

FIG. 3. Antibody uptake of myofibrils; 80γ myofibril. Key: open symbols, protein determination; solid symbols, fluorescence determinations.

amount of myosin fixed, questions concerning the availability of antigenic sites, the role of impurities in the antibody preparations, and the state of meromyosins in muscle are easier to answer. Dr. Holtzer and I performed these types of experiments (A. G. Szent-Györgyi and H. Holtzer, in preparation). Since the ATPase activity of myosin is not changed by reaction with the antibodies used (Samuels, 1961), and myosin represents most of the ATPase of washed myofibrils, the reaction of myosin with antibodies could be followed by measuring the distribution of ATPase in the soluble and insoluble fractions. The distribution of myosin was also obtained from protein balance, estimating the antibody concentration independently from fluorescence measurements. This check was necessary to show whether ATPase with greatly different molecular weight from myosin was solubilized or not.

Figure 3 shows the antibody uptake of myofibrils when increasing amounts of antibody are added. With the increase of antibody concentration, the bound antibody increases although the binding is not linear and at the higher antibody concentrations the uptake falls off. The amount of antibody which can be bound is considerable. The myofibrils are capable of binding their own weight or more of the antibody. One can calculate that 1 mole of myosin in the myofibril reacts with 10 or more moles of myosin antibody. The different antibody preparations are bound to different extent. This probably means that the total number of antigenic sites against anti-myosin is greater in the myofibril than the number of antigenic sites against either of the anti-meromyosins. The sum of the bound anti-light and anti-heavy is not much higher than the bound anti-myosin indicating that the antigenic sites for these are at different regions of the myosin molecule, and that there is no great overlap between these two antibody preparations.

In the fixation studies, the myofibrils were incubated with antibody not more than 8 to 10 times their own weight, an antibody concentration far from saturating amounts. Table II summarizes the results. It can

TABLE II
EFFECT OF ANTIBODIES ON THE EXTRACTION OF MYOSIN

	γAB[a]			ATPase	ATPase	Myosin calculated from protein
	Added	Bound	Extracted	Soluble (%)	Insoluble (%)	Soluble (%)
Control				87		
Myosin 10-12-14[I]	2,340	416	70	47		45
Myosin 8-11	8,780	1,760	247	23	79	24
Myosin 10-12-14[II]	7,900	1,752	137	20	78	17
LMM[I]	7,200	1,144	66	15	82	23
LMM[II]	8,840	1,424	117	21	80	27
HMM[I]	7,270	789	125	35	60	40
HMM[II]	8,420	812	159	37	60	38
HMM[III]	12,000	940	96	15	72	6

[a] Antibody was added to 1000γ Myofibril. Antibody Bound was also calculated for 1000γ Myofibril and corrected for the 10–30% loss incurred during the washing.

be seen that antibodies against myosin, light meromyosin, and heavy meromyosin are capable of extensive fixation of myosin. At least 80% or so of the myosin molecules are reacting with the antibody. There is no extensive steric hindrance or blocked antigenic sites operating under these conditions. The protein balance checks well with the ATPase runs. There is no low molecular weight ATPase released in these experiments. This and the fact that light meromyosin antibodies can fix ATPase indicate that myosin is not dissociated into meromyosins in the myofibril.

These antibody preparations all contain antibody against myosin. While not all the staining is necessarily due to the myosin antigen in the myofibril, these findings mean that those portions of the myofibril which are not stained do not contain myosin or those portions of myosin which have the antigenic sites against the particular antibody. The lack of staining in the center of the A band, using antibody against light meromyosin, indicates that the center of the A band either does not contain myosin or does not contain the light meromyosin portion of myosin. That means that as far as myosin is concerned, the A band, or the thick filaments, have a center of symmetry and the myosin-containing structures extend from the edge of the A band to near the center of the A band.

The solvent used here was somewhat milder than KI in order to preserve ATPase activity better. The sarcomere band pattern changes, upon extraction, were not as excessive as with KI although they followed similar tendencies. Extraction under conditions which left most of the myosin in the myofibril resulted in changes in band pattern which differed depending on the type of antibody used. Extraction which followed myosin antibody treatment did not lead to a significant alteration (Table III). Extraction after anti-light treatment caused an apparent decrease in the density at the center of the A band as if material moved toward the lateral regions of the A band. Sometimes the A band seemed to be compressed toward the cleared central zone and its width decreased somewhat. Extraction after anti-heavy treatment led to an apparently greater contrast of the M band region. The width of the A band reduced significantly. Sometimes the edges of the remaining A band were stronger, the pattern gave a triple banded appearance. The fluorescent pattern agreed with the phase observations, except for being somewhat wider in the unextracted myosin and LMM antibody-treated fibrils.

These findings may be explained by assuming that the myosin-containing structures moved either laterally or towards the center. These two regions may represent anchoring points of myosin, and movement will depend upon which end is released or fixed. The movement of myosin in a different direction using heavy and light antibody depends in this case upon which of these anchoring points is reinforced. The continuity of myosin structures may be achieved in either of two ways: (*a*) each myosin molecule is extended from the edge to the center of the A band, or (*b*) continuity is the result of the interaction between myosin molecules within the filaments.

TABLE III

SARCOMERE LENGTH OF MYOFIBRILS[a]

	No. of Fibrils	Sarcomere (μ)		A-Band (μ)	
		Phase	Fluorescence	Phase	Fluorescence
CONTROL					
Unextracted	9	2.22 ± 0.16		1.54 ± 0.03	
MYOSIN 8-10-12, II					
Unextracted	12	2.14 ± 0.12	2.16 ± 0.15	1.57 ± 0.05	1.68 ± 0.04
Extracted	14	2.03 ± 0.09	2.02 ± 0.11	1.60 ± 0.07	1.60 ± 0.07
LMM II					
Unextracted	13	2.08 ± 0.17	2.07 ± 0.14	1.61 ± 0.04	1.67 ± 0.07
Extracted	14	2.00 ± 0.19	1.99 ± 0.19	1.48 ± 0.09	1.49 ± 0.09
HMM II					
Unextracted	16	2.18 ± 0.17	2.18 ± 0.16	1.56 ± 0.03	1.62 ± 0.05
Extracted	17	1.87 ± 0.11	1.85 ± 0.11	1.17 ± 0.08	1.12 ± 0.10
HMM III					
Unextracted	24	2.28 ± 0.19	2.28 ± 0.19	1.54 ± 0.06	1.55 ± 0.07
Extracted	30	2.12 ± 0.12	2.11 ± 0.12	1.31 ± 0.07	1.31 ± 0.08

[a] Same myofibrils as shown in Table II.

The movement of myosin occurs here under experimental conditions which are artificial and cannot be equated with changes in contraction. Important observations were made by Tunik and H. Holtzer (1961) who showed that in contracted myofibrils the staining pattern against myosin and light meromyosin antibody is different from the staining pattern of myofibrils at rest length. In contracted fibrils these antibodies stained only the lateral portions of the A band. In myofibrils showing more and more advanced stages of contraction, the central region of the A band, devoid of staining, became wider, the deposition of antimyosin and anti-light was more and more restricted to the lateral edges of the A band. Finally, the Cm band did not stain and the antibodies deposited on the Cz band only. Assuming that steric hindrance does not play a significant role in the antibody-antigen reactions of contracted myofibrils, this observation means that, in contraction, myosin moves away from the center towards the lateral edges of the A band. Such a finding is incompatible with the sliding theory of contraction which demands rigid rods within which actin and myosin molecules stay put.

C. An Alternative Theory of Contraction of Striated Muscle

I will attempt to propose a scheme which, to my mind, satisfies most of the observations of Huxley and Hanson on muscle structure and also accounts for the antibody findings. This is based on the following assumptions for which I have mentioned some of the experimental evidence.

1. Myosin or myosin-containing structures have a continuity and extend from the edges of the A band to near the center of the A band.

2. Interaction with actin, which may be localized in the thin filaments, leads to a folding or to a change in the lattice arrangement of myosin molecules resulting in the accumulation of myosin at the edges and an increase in the width of the myosin-free region at the center of the A band. The driving force of contraction is this rearrangement of myosin.

3. The myosin containing structures are connected with the contralateral thin filaments of the same sarcomere. Such an arrangement will preserve the over-all constancy of A band length and lead to a change in the length of the I band at the initial stages of contraction, even though structural changes are taking place within the A band.

Figure 4 shows the changes in the myosin-containing structures of various sarcomere lengths.

In stretch, the connection between the thick and contralateral thin filaments, which is highly extensible in muscle at rest, is responsible for the sarcomere length change. As a result, the thin filaments are pulled out from between the thick filaments. Increase in sarcomere length is accompanied by a proportional increase in the width of the I band, increase in the H band, the A band staying constant. The arrangement is very similar to the scheme of Huxley and Hanson. The real difference is in the position of the elastic connection. In their picture it was supposed to be between the thin filaments, in ours between thin and contralateral thick ones. It should be mentioned that none of these elastic connections have been observed, as yet, in the electronmicroscope with the exception that the M band in the center of the A band shows the remainder of additional material. Since muscle which is greatly extended, even beyond the 3μ sarcomere length returns elastically to equilibrium length when released, their presence is indicated.

In contraction, myosin "contracts" toward the lateral edges of the A band, the connection between contralateral thin and thick filaments is rigid and the movement of myosin carries along the contralateral thin filaments, the change in the I band length is proportional initially to shortening of the sarcomere while the distance between the edges of the A band stays constant. With more extensive shortening, myosin will be found only at the Cz band but not at the Cm band and not between the regions of the Cm and Cz bands, contrary to a localization one would expect from the sliding filament theory.

A few remarks may be useful before we examine the consequences of such a theory. The schematic diagram shows only the position of myosin

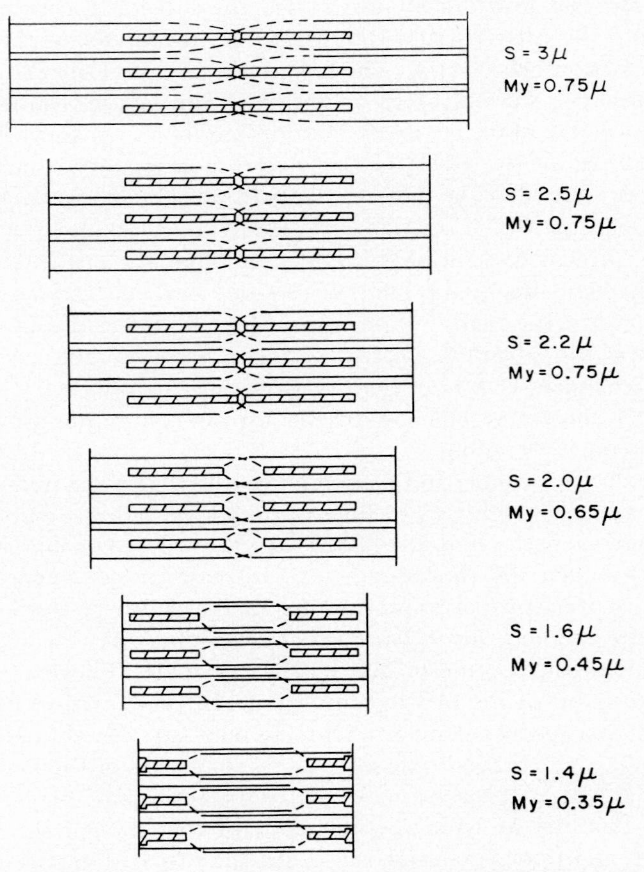

FIG. 4. Position of myosin and actin and the connections in the sarcomere at different lengths. Key: bars, myosin; uninterrupted lines, thin filaments (actin); broken lines, connections. S, sarcomere length; My, length of myosin lattice.

in the thick filaments and the thin filaments with the contralateral connections. Recently, it was proposed independently by Lowey and Cohen (1962) and by J. Marshall (personal communication) from structural reasons, derived mainly from an efficient packing arrangement, that myosin occupies only the surface of the thick filaments, and the core of the filaments is something else, possibly tropomyosin. Such a proposal would fit well in the scheme described here and the region from which myosin moved away is not empty, but still may contain the core.

The fixation of myosin-containing structures at the edge of the A band may be achieved in two ways. Either the core extends to the Z band, in which case the core should be incorporated and accumulate at the center of the A band as the myosin shell contracts. An alternative mechanism on which Dr. W. H. Johnson and I are working at the present, is that anchoring of myosin-containing structures is achieved by the cross links between myosin and the actin component of the ipsilateral thin filaments. This cross linkage, the interaction between actin and myosin, is actually the event which initiates the myosin "contraction," thus making contraction a myosin-actin interaction. This linkage would be broken as the reaction at this site terminates leaving behind altered myosin. A mechanism of this type will also anchor the lateral portions of the A band, and by assuming that movement leads to an increased probability of interaction between ipsilateral actin and myosin structures many of the data of heat production and of kinetics of contraction may be simply derived (A. G. Szent-Györgyi and W. H. Johnson, in preparation).

What are the structural consequences of such a mechanism which I have proposed? On the filamentous organizational level, a contralateral attachment of filaments demands that in contracted muscle more thin filaments are present at the center than in the rest of the sarcomere. Their number in a cross section should be twice as many as at the lateral parts of the A band and in the I band. H. E. Huxley (1962) has reported that this actually happens, that the Cm contraction band is not due to the folding of the thin filaments but rather that they overrun and bypass each other and in more contracted muscle this area of "overrunning" can be quite extensive. There is no pressing reason why contraction considered from the point of view of a simple sliding theory should lead to this phenomenon. On the other hand, doubling up of thin filaments is a necessary consequence of a contralateral attachment and movement of myosin.

The structure of the thick filament should be different in resting and in contracted muscle. Thick filaments in contracted muscles should be denser at their ends than at their middle. The sliding theory of contraction does not allow such change. Thick filaments dissociated from the thin filaments were obtained by Huxley (1962) from resting muscle. We do not have, as yet, similar material from contracted muscles, and do not have definitive information concerning the shape of the thick filaments in contracted muscles.

With proper fixation, electronmicroscopic techniques may reveal the presence and the connections of the elastic material. The predictions as to where these connections should be, differ in the two theories. I may mention here, also, that for the mechanism I have outlined to work, one has to assume that at rest the connection must somewhere be highly extensible and during activity the extensible region must become rigid. No such an assumption is required in the sliding theory where the elastic connections are between the contralateral thin filaments. On the other hand, the sliding theory gets into trouble in explaining the overrunning of thin filaments, especially if the ends are connected with an elastic material. Neither of the theories is in a very good position to explain the elastic behavior of resting muscle during stretch. Actually, the information of the behavior of whole muscle during stretch, its effect on activation, and reversibility, is scanty and the problem needs further extensive study.

On the sarcomere level, both theories explain the changes in A and I band width in stretch and contraction. There are important differences, though, in how the density of the A band should change in contraction. The appearance of the Cm band does not simply follow from the theory outlined here, rather a density increase of the lateral regions of the A band would be expected. It is not clear, at present, how far the Cm band arises from the position of the filaments or what its relation is to the filaments. Neither have the density changes been followed, as yet, quantitatively enough to make a very definite statement. The Cm band does not appear to follow the overrunning of thin filaments which would be expected from the sliding theory. Better and more quantitative information is needed as to how the density and birefringence of the A band changes in contraction before one could attempt to explain these changes adequately with any theory of contraction.

There are some other observations which are simpler to explain, on the basis of the picture I presented here, than the sliding theory which I will mention here only briefly. Among these is the decrease of the overlap area between thick and thin filaments at isometric tension (Carlson, *et al.,* 1961). This could be explained by a slight stretching of the connecting piece. The sliding theory would rather demand an increase in the overlap area in isometric tension. The length-tension diagram of intact muscle does not show a plateau between 2.0μ and 1.5μ sarcomere length as expected from the sliding theory. According to the theory proposed here, the interacting sites will start to decrease in number below equilibrium length as the concentration of unreacted myosin falls off which corresponds better with the observed length-tension curve.

It should be stressed that the contralateral attachment is proposed only for those muscles which shorten, at least partially, with no change in the length of the A band. This type of band pattern change would be the result of such a specific arrangement. The driving force of contraction is the structural change in myosin. Striated muscles in which the A band shortens (de Villafranca, 1961) when the muscle contracts could be explained by different organization and connections of the filaments. Once a change in myosin structure, or in its organization, is the cause of contraction, contractility of smooth muscle, or extracted actomyosin, poses no special problem. In contrast to this, in the sliding theory the essence and driving force of contraction is the sliding itself; any folding is a secondary consequence of sliding. On such a basis, it is difficult to explain contraction with a change in A band length, contractility of smooth muscle, and contractility of extracted actomyosin.

I do not claim that muscle contracts the way I have described. But it is a theory which, to my mind, describes available information as well and as quantitatively, in some cases better, than the sliding theory of contraction. It certainly points to an alternative way of explaining the observations. Most importantly, its consequences are different from the sliding theory. These can be tested and may thus lead to new experimental approaches.

References

Asakura, S., and Oosawa, F. (1960). *Arch. Biochem. Biophys.* **87**, 273.

Bailey, K., Bettelheim, F. R., Lorand, L., and Middlebrook, W. R. (1951). *Nature* **167**, 233.

Bettex-Galland, M., and Luscher, E. F. (1961). *Biochem. et Biophys. Acta* **49**, 536.

Carlsen, F., Knappeis, G. G., and Buchtal, F. (1961). *Biophys. and Biochem. Cytol.* **11**, 95.

Chrambach, A., Barany, M., and Finkelman, F. (1961). *Arch. Biochem. Biophys.* **93**, 456.

Engelhardt, W. A., and Ljubimova, M. N. (1939). *Nature* **144**, 668.

Flory, P. J. (1957). *J. Cellular Comp. Physiol.* **49**, Suppl. 1, 175.

Gergely, J. (1953). *J. Biol. Chem.* **200**, 543.

Hanson, J., and Huxley, H. E. (1955). *Symposia Soc. Exptl. Biol.* **9**, 228.

Hayashi, T., and Tsuboi, K. K. (1960). *Federation Proc.* **19**, 256.

Hoffman-Berling, H. (1954). *Biochim. et Biophys. Acta* **14**, 182.

Hoffman-Berling, H. (1956). *Biochim. et Biophys. Acta* **19**, 453.

Hoffman-Berling, H. (1958). *Biochim. et Biophys. Acta* **27**, 247.

Holtzer, A., and Lowey, S. (1956). *J. Am. Chem. Soc.* **78**, 5955.

Huxley, A. F., and Niedergerke, R. (1954). *Nature* **173**, 971.

Huxley, H. E. (1957). *J. Biophys. Biochem. Cytol.* **3**, 631.

Huxley, H. E. (1962). *In* "The Cell" (J. Brachet and A. E. Mirsky, eds.), Vol. 4, p. 365. Academic Press, New York.

Huxley, H. E. (1962). *In* "Muscle as a Tissue" (K. Rodahl and S. M. Horvath, eds.), pp. 64, 153–155. McGraw-Hill, New York.

Huxley, H. E., and Hanson, J. (1954). *Nature* **173**, 973.

Huxley, H. E., and Hanson, J. (1957). *Biochim. et Biophys. Acta* **23**, 229.

Inoué, S. (1959). *In* "Biophysical Science," A Study Program (J. L. Oncley, ed.), p. 402. Wiley, New York.

Katchalsky, A. (1954). *Prog. in Biophys. and Biophys. Chem.* **4**, 19.

Kuhn, W., Ramel, A., Walters, D. H., Ebner, G., and Kuhn, H. J. (1960). *Fortschr. Hochpolymer-Forsch.* **1**, 540.
Laki, K., Bowen, W. J., and Clark, A. (1950). *J. Gen. Physiol.* **33**, 437.
Laki, K., and Clark, A. (1951). *J. Biol. Chem.* **191**, 599.
Levine, L. (1956). *Biol. Bull.* **111**, 319.
Lowey, S., and Cohen, C. (1962). *J. Mol. Biol.* **4**, 293.
Marshall, J., Holtzer, H., Finck, H., and Pepe, F. (1959). *Exptl. Cell Research* Suppl. **7**, 219.
Martonosi, A., Gouvea, M. A., and Gergely, J. (1960). *J. Biol. Chem.* **235**, 1700.
Mihalyi, E., and Szent-Györgyi, A. G. (1953). *J. Biol. Chem.* **201**, 189.
Mommaerts, W. F. H. M. (1952). *J. Biol. Chem.* **198**, 469.
Perry, S. V. (1954). *Biochem. J.* **57**, 427.
Samuels, A. (1961). *Arch. Biochem. Biophys.* **92**, 497.
Straub, F. B. (1943). *Studies Inst. Med. Chem. Univ. Szeged* **3**, 23.
Straub, F. B., and Feuer, G. (1950). *Biochim. et Biophys. Acta* **4**, 455.
Strohman, R. C. (1959). *Biochim. et Biophys. Acta* **32**, 436.
Strohman, R. C., and Sanorodin, A. J. (1962). *J. Biol. Chem.* **237**, 363.
Szent-Györgyi, A. G. (1949). *Biol. Bull.* **96**, 140.
Szent-Györgyi, A. G. (1951). "Chemistry of Muscular Contraction." Academic Press, New York.
Szent-Györgyi, A. G. (1953). *Arch. Biochem. Biophys.* **42**, 305.
Szent-Györgyi, A. G., Cohen, C., and Philpott, D. E. (1960). *J. Mol. Biol.* **2**, 133.
Szent-Györgyi, A. G., and Holtzer, H. (1960). *Biochim. et Biophys. Acta* **41**, 14.
Tunik, B., and Holtzer, H. (1961). *J. Biophys. Biochem. Cytol.* **11**, 67.
de Villafranca, G. W. (1961). *J. Ultrastruct. Research* **5**, 109.
Weber, H. H., and Portzehl, H. (1954). *Progr. in Biophys. and Biophys. Chem.* **4**, 60.

Discussion by Laurence Levine

Wayne State University, Detroit, Michigan

There have been many examples presented here of an underlying motility supporting the cell in mitosis. The most obvious of these is Dr. Rehbun's saltations, and among the most subtle are the changing conformities of pattern profiles of the endoplasmic reticulum.

A contraction hypothesis in mitosis is among the most venerable being originally suggested in 1878 by Klein and further elaborated upon in subsequent years (van Beneden, 1887; Boveri, 1888). Such an hypothesis has oscillated in and out of favor ever since (see reviews by Schrader, 1953; Mazia, 1961). The most recent resurgence was made by Cornman in 1944. One approach to this hypothesis was treated by Inoué (1959), who on the basis of birefringence data, concluded that the basal granules of cilia, kinetochores, chromosomes, and centrioles all act as centers of fibrous organization in cells, and further implied that these elements were in fact identical, taking on different functions at their various stations in the cell. I think it is interesting that these bodies would organize structures (cilia, flagella, spindle fibers) which appear as tubules or some organization of tubules in the electron microscope. Another and more recent approach having rather firm implications for a "muscle-spindle analogy" is the discovery of ATPase associated with isolated spindles (Mazia *et al.*, 1961). I will have more to say about this later.

The anaphase movement has received the most directed attention. The usually tight synchrony of chromosomes moving toward the poles has all the elements of a good mystery. There are some clues but the final chapter is yet to be written. It is unfortunate that some of the events of prophase and premetaphase (e.g., kinetochore orientations to centers, premetaphase stretch) have received less attention. They could have important bearing on the over-all mechanism of chromosome movement (cf. Ostergren, 1960).

Dr. Hans Ris (1943, 1949) made quantitative studies on anaphase movement in insect spermatocytes. He concluded, from direct measurements, that chromosomes are moved by chromosomal fibers in two ways. These fibers may shorten and thereby pull the chromosome directly. They may also act passively by transmitting the elongation of the continuous fibers to the chromosomes via the separating centers. Thus chromosomes in anaphase may be pulled directly or pushed indirectly. Separation by shortening of chromosomal fibers may be superimposed upon or separated from disjunction by elongation of continuous fibers. The former applies in cases with diffuse kinetochore attachment of chromosomal fibers (Ris, 1943), and the latter with localized connection (Ris, 1949). In one case (*Tamalia,* first spermatocytes) anaphase is executed by elongation only. Anaphase may also be brought about in this manner in hypermastigote flagellates, as Dr. Cleveland has stated here and elsewhere.

Dr. Szent-Györgyi has provided us with a very lucid account of the current status of knowledge on contractility. His paper presents some rich resources for guidance in the design of experiments to direct attack on the issue of contractility in mitosis or embellish some of those that already have been made. Indeed, Dr. Szent-Györgyi has emphasized that we may, at present, invoke two models for cell motility. These are the one and two component systems, polyelectrolyte and actomyosin, respectively. It should be possible to distinguish between these systems on the basis of the kinds of substances which drive them, namely relatively simple molecules for the polyelectrolyte and ATP or closely related molecules for actomyosin systems. It would be of extreme value to know which system, if either, applies to the spindle. To find out, suitable models would have to be produced. These may be made by extraction of control systems under conditions which lend stability and maintain integrity in the motile substructures. This may be managed with cold glycerol or other suitable solvents; exactly how would have to depend upon the cell. The work of Hoffman-Berling, a master model maker, directly applies here. He showed that glycerated models of cultured subcutis fibroblast could be reactivated to complete cytokinesis (Hoffman-Berling, 1954a) with the addition of ATP in a suitable ionic millieu. In addition, it was also demonstrated that anaphase elongation could be similarly produced (Hoffman-Berling, 1954b). However, this latter required "supra-optimal" concentrations of nucleotide, imposing conditions of cell relaxation (Weber, 1958). It must be emphasized that sodium pyrophosphate could also induce this movement, although at higher concentrations. Furthermore the spindle movement was not inhibited by such sulfhydryl inhibitors as Mersalyl, which are extremely effective in most other models. So there is a suspicion that the continuous fibers in these cells may behave in whole or in part as polyelectrolyte. It would be of extreme value to repeat these experiments on such cells as: hypermastigotes of termites, where anaphase disjunction is solely by continuous fiber elongation; on glycerated insect spermatocytes with diffuse kinetochore attachments, where elongation is separated from a shortening of chromosomal fibers. These experiments would also bear repetition upon cells whose chromosomal fibers were inhibited by chloral hydrate (Ris, 1949).

The recent discovery of ATPase in isolated spindles of the sea urchin (Mazia, 1961) needs special attention. It is further demonstration of the power of isolated spindles for the elucidation of the biochemistry of the mitotic apparatus (cf. Zimmerman, this volume). Although encouraging, I think it is in order to be somewhat conservative with regard to rationalizing a contraction mechanism between ATP and a two-component system. Sulfhydryl dependence for the enzyme should be demonstrated in view of its low order of activity, but even this may be deceiving. In Vorticellids, where the stalk is most probably of the polyelectrolyte variety, there is also ATPase (Levine, 1960) which is sulfhydryl linked (Levine, 1962). While the ATPase is extracted by solvents for myosin (Hanson and Huxley, 1957) neither calcium-induced coiling nor EDTA-induced uncoiling is affected (Levine, 1962). Now let it be understood that I am not making a direct analogy, but attempting to establish some of the parameters for naturally occurring polyelectrolyte systems where the presence of ATPase does not necessarily imply direct association with contractile phenomena. It would, therefore, be necessary to establish whether the spindle of the sea urchin behaves as the one or two component or some other model before its ATPase could be functionally categorized as being engaged in motility or elsewhere.

References

van Beneden, E. (1883). *Arch. Biol. (Liege)* **4**, 265.

Boveri, T. (1888). *Zellenstudien* **II**. Fischer, Jena.

Cornman, I. (1944). *Am. Naturalist* **78**, 410.

Hanson, J., and Huxley, H. E. (1957). *Biochim. et Biophys. Acta* **23**, 250.

Hoffman-Berling, H. (1954a). *Biochim. et Biophys. Acta* **15**, 332.

Hoffman-Berling, H. (1954b). *Biochim. et Biophys. Acta* **15**, 226.

Inoué, S. (1959). *In* "Biophysical Science — A Study Program" (J. L. Oncley, ed.), p. 402. Wiley, New York.

Klein, E. (1878). *Quart. J. Microscop. Sci.* **18**, 315.

Levine, L. (1960). *Science* **131**, 1377.

Levine, L. (1962). Cytochemical correlates of contractile function in *Vorticella*. In preparation.

Mazia, D. (1961). *In* "The Cell, Biochemistry, Physiology, Morphology" (J. Brachet and A. E. Mirsky, eds.), Vol. III, pp. 77-412. Academic Press, New York.

Mazia, D., Chaffee, R., and Iverson, R. (1961). *Proc. Natl. Acad. Sci. U.S.* **47**, 788.

Ostergren, G., Molè-Bajer, J., and Bajer, A. (1960). *Ann. N. Y. Acad. Sci.* **90**, 381.

Ris, H. (1943). *Biol. Bull.* **85**, 164.

Ris, H. (1949). *Biol. Bull.* **96**, 90.

Schrader, F. (1953). "Mitosis, the Movement of Chromosomes in Cell Division," 2nd ed. Columbia Univ. Press, New York.

Weber, H. W. (1958). "The Motility of Muscle and Cells," p. 52. Harvard Univ. Press, Cambridge, Massachusetts.

Summation

Dr. L. Levine (Wayne State University, Detroit, Michigan): This is very much in the way of an experiment, the purpose of which is to see whether we could emerge from our deliberation with crystals, perhaps, tactoids, or some form of structural organization. The discussion will open on Dr. Szent-Gyorgyi's paper.

Dr. S. Gelfant (Syracuse University, New York): Dr. Szent-Gyorgyi, can you dis-associate the activity of the individual thick and thin fibers by reciprocal inhibitors? If this were possible would it help in resolving the thesis you presented?

Dr. Andrew G. Szent-Gyorgyi (Dartmouth Medical School, New Hampshire): As shown by Holtzer *et al.*, contraction of glycerol-extracted muscle is blocked after treatment with antibodies prepared against myosin, the meromyosins, or actin. It is hard, then, to dissociate the activities of thick and thin filaments in the myofibril using such a functional test.

If the muscle is disintegrated to the extent that the thick and thin filaments can be observed lying separated from each other, then as demonstrated by H. E. Huxley and Pepe, myosin antibody is deposited on the thick filament. This is a further evidence that the thick filaments contain myosin. Extension of such an approach may yield more direct answers to your question.

Dr. L. Rebhun (Princeton University, Princeton, New Jersey): With respect to the blocking of contractility by the antibody, do you suppose that the antibody is actually blocking an active site on the thick filaments by binding to it or is the blocking of contractility some form of steric hinderance due to the fact that the antibody molecules clog up the spaces between filaments.

Dr. A. G. Szent-Gyorgyi: Reaction with antibodies do not interfere with ATPase activity. This was shown by Samuels, using isolated myosin and myosin antibody. We have found the same, treating myofibrils with antibodies prepared against myosin, meromyosins, and actin. The antigenic sites are thus not near to the enzymatic centers, or at least do not block them. Whether the inhibition of contraction is a result of blocking some other active sites, or is due to some form of steric hindrance, is not known. The fact that ATPase activity remained unaltered made it possible to do the experiments I have described and allowed us to follow the solubilization of myosin by measuring the distribution of ATPase between the soluble and insoluble fraction.

Dr. A. W. Burke (Rhode Island Hospital, Rhode Island): I would like for Dr. Szent-Gyorgyi to repeat for the record the statement that he made in a private conversation when discussing the relationship between products of the centrioles such as flagella, cilia, central spindle, etc., and the muscle fiber or fibril. The question was put to him: Had he thought much about the statement of Wolbach that the centriole or centriole-like bodies produce the primitive myofibrils in rhabdomyosarcomata? His answer showed marvelous restraint in approaching a unified hypothesis for the origin of fibrils in living cells.

Dr. A. G. Szent-Gyorgyi: I really do not know how and where the myofibrillar proteins are synthesized.

Dr. A. W. Burke: The restraint that I am trying to get you to reiterate is: "I haven't seen it yet, and I wonder if anyone has seen it."

Dr. H. Ris (University of Wisconsin, Madison, Wisconsin): In connection with this I would like to correct a statement Dr. Levine made about spindle fibers coming from centrioles like the filaments of cilia coming from centrioles. This certainly is not so. As seen in electron micrographs, the spindle fibers never go up to the centriole. They stop quite a way from the centriole and so do the astal rays. It is certainly puzzling that a centriole should have these two functions, both of which are connected with the formation of protein fibers. There is no justification to homologize these two kinds of fibers.

DR. L. LEVINE: Thank you. It's a very interesting question whether apparently homologous structures organize homologous fibers. However, I think it is impressive that both the centriole and basal granule should organize fibers which appear as tubules in the electron microscope.

DR. H. RIS: Now which tubular structures are you referring to? In cilia, the outer nine and the inner two are quite different and neither look like spindle fibers at higher resolution.

DR. L. LEVINE: With respect to the seemingly structureless region between the ends of spindle fibers and centriole, how would one visualize anchors in the centriolar region. I think that negative electron microscopic evidence does not necessarily mean that there is no structure.

DR. A. W. BURKE: Your comment, Dr. Ris, about the spindle fibers or astral rays not actually going to the centriole in the electron micrograph may be the result of the uniqueness of the material which you have studied. The possibility that these structures may be continuous with the centriole has been shown by Dr. Cleveland in those forms (hypermastigote flagellates) which do not have a centrosome, where the spindle fiber is right up on and abutts the centriole. These observations, however, were made at the level of resolution of the light microscope.

DR. RIS: I'm talking about a level of resolution where you can decide whether it touches the centriole or not.

DR. BURKE: Yes, but, in the corresponding light microscope studies of your same material, as I understand it, the centrioles are not seen to be continuous with the spindle fibers either, so that the final answer would lie in investigating with the electron microscope, the spindle and astral ray and chromosome fiber connections to what we are calling the posterior elongation of the centriole in hypermastigotes, to evaluate this physical connection, and to see whether it does exist or not, as we see it on the light microscope. This hasn't been done yet.

DR. H. RIS: In addition we shouldn't forget that chromosomal fibers can be present in the absence of a centriole. This happens of course in higher plants and in many animal oöcytes, during meiosis.

DR. A. W. BURKE: There, Daniel Mazia's word of "euphemism" for the area as, "a centriole, lacking body, but functioning as a center for the spindle," would certainly be applicable.

DR. G. B. WILSON (Michigan State University, East Lansing, Michigan): I am sure Dr. Ris will agree with me that there is nothing in the higher plant cell that you could call centrosome or centriole. We have some pretty weird ones in some of the fungi. We've looked at a lot of them recently with electron microscope. The fibers fan out around the area, but the area is not really differentiated from anything else by any kind of fixation we've ever been able to manage.

DR. L. REBHUN: I want to change the subject back and make some comments on the applications of the kinds of contractile systems one finds in muscle and related systems to the mitotic spindle. There are indications of biochemical similarities between contractile systems and the spindle in terms of the size of the molecules involved, the shape of these molecules, ATPase, and so forth. One of the pieces of evidence used in support of this analogy is the work of Hoffman-Berling on glycerol-extracted fibroblasts. Specifically, it is claimed that anaphase movements of chromosomes can be induced by the addition of ATP, under certain conditions, to glycerinated fibroblasts. I would like to say that the kind of anaphase movement which one, at least I, would be interested in as a true anaphase movement is a motion of chromosomes relative to a fixed spindle or at least a spindle which is not moving at the same rate as the chromosomes. From these movies (after looking at them several times), I would say that there is no relative motion between the spindle and the chromosomes and that the phenomenon seen was probably spindle elongation which, as some of you may remember was easily induced in the anaphase spermatocyte spindle, by Belar simply by the addition of distilled water. That is, elongation of the spindle is relatively nonspecific and I think this elongation is probably what's seen in Hoffman-Berling's films of the so-called anaphase separation. If so, this would be dubious evidence to use in support of applying so-called contractility theories to anything that goes into the mechanism of spindle separation of chromosomes. I don't doubt that there are analogies and that parts of the molecular biology of muscle will lead to hypotheses for the mechanism of spindle action but I don't think one can use the term contractility and immediately put both muscle and spindle phenomena under it.

DR. ANDREW G. SZENT-GYORGYI: I fully agree with what you have just said. When Hoffman-Berling describes the apparent anaphase movement of chromosomes, this is accompanied with a considerable elongation and swelling of the whole cell. It is not clear to me how much this swelling and elongation contributes to the "movement" of chromosomes. The observed chromosome separation may entirely be the reflection of the changes induced by ATP on fibroblasts and is not a direct demonstration of an inherent contractility of spindle fibers.

In a way, the lesson is that if you must kill the cell in order to demonstrate the activity and analyze the behavior of some of its components you had better go all the way. Before assigning any contractile properties to the spindle, I would like to have some motion reproduced using isolated spindle or mitotic apparatus. Such a demonstration would be the first step in analyzing what the mechanism of that motion is and how similar it is to what is happening in the intact cell.

DR. R. C. RUSTAD (Florida State University, Tallahassee, Florida): I would not like to be accused of supporting Belar's stemmkorper hypothesis, but chromosome movements suggest some interesting possibilities. As Dr. Levine pointed out, in many types of cells there are two distinct chromosome movements: a movement toward the centrioles and an elongation of the whole spindle. In the sea urchin egg I found that these two movements are separated in time and that the elongation seems to be correlated with movement of RNA into the interzonal region. Dr. Porter's electron micrographs of rather different types of cells showed a definite invasion of rough endoplasmic reticulum during anaphase. These varied observations suggest that we might look to another possible site of mechanical action, not just the chromosomal and continuous fibers, but also the invading endoplasmic reticulum.

DR. L. REBHUN: I wonder if Dr. Szent-Gyorgyi could possibly straighten out some of the things I may get wrong in this, but if I remember Hoffman-Berling's ideas correctly, it might be possible, if there are contractile processes in the spindle, for both contraction and elongation to be due to the same mechanism. That is, the optimal ATP concentration for contraction might be lower for one of the spindle components, shall we say continuous fibers than for chromosomal fibers. In this case you might have a supra-optimal concentration of ATP for continuous fibers and actually get elongation due to relaxation in these fibers at the same time that contraction occurred in the chromosomal fibers. I think he thought that at least there's the possibility that the same system is involved in both effects but because of ion effects or because of other processes, some of the fibers might be modulated so that at a given ATP concentration they would contract while other ones at this concentration would elongate. Just parenthetically, I would like to say that there are cases known in which at times the chromosomes are separating and the entire spindle is shortening. This is something that happens for instance in first meiotic divisions in many marine eggs.

Dr. Andrew G. Szent-Gyorgyi: Yes. In muscle there is the relaxing system present which operates as one of the control mechanisms. This system is operative in the presence of a relatively higher ATP concentration but is inactive if ATP concentration falls below a critical level. Thus contraction and relaxation of the component may depend on the ATP level of the surroundings. Conversely, if we assume that a system, similar to the relaxing factor system, is absent at certain areas of the cell and is present in others a simultaneous contraction and relaxation is possible in different regions of the cell at constant ATP level. This was the analogy which Hoffman-Berling used to explain an apparent cleavage furrow formation in glycerol-extracted fibroblasts upon addition of a certain amount of ATP. The area where the cleavage furrow is formed is devoid or has a lower level of the "relaxing factor system" than the rest of the cell. To me this is a very reasonable analogy. I do not think a great deal could be said, at present, concerning his anaphase movement of the chromosomes. In the case of the rest of his cell models there is a definite dimensional change. Some of them behave quite like actomyosin, others are quite different. The evaluation of the analogies and differences from muscle-like action require further investigations. Does that answer your question?

Dr. H. Ris: Comparing spindle fibers with muscles we should remember that the spindle fibers disappear as they contract. Secondly the fibers in muscle are anchored. We don't know just how the chromosomal fibers are anchored. In the pictures of Belar and Huth of a monaster in urechis eggs, the chromosomes don't form a metaphase plate but they line up in the monaster, one chromosomal fiber going to the center and the other away from it and the two daughter chromosomes move away from each other, one to the pole of the monaster and one away from it. Apparently, they are pulled by chromosomal fibers the same way as they are in a bipolar spindle. Now where are these fibers anchored and how can they change the anchorage as the chromosomes move?

Dr. Andrew G. Szent-Gyorgyi: I hope I was clear enough. By no means did I want to imply that wherever there is motion it must be similar to muscle. The reason I described the behavior of synthetic polymers was to illustrate some other mechanisms. What we learned about muscle may be useful if for nothing else than to restrict our generalizations. If there is an analogy perhaps it is rather in the approach to the problem. What I would state again, in an optimistic mood, is that any directed motion should be connected with a structure and there is a good chance that such a structure can be isolated and the rudimentary reaction reproduced, using such simplified systems. Such an approach may help to identify the components of the system and may be the first step toward studying what the basic change is which is responsible for that particular motion. Its control and its ramification may be an entirely different story and will have to be studied separately with each cell. The isolation of the responsible components and the *in vitro* demonstration of some aspects of the movement may not be a hopeless adventure.

Dr. L. Rebhun: With respect to these chromosomal fibers I think it would be hard to pass up the subject without mentioning some of the beautiful films of Bajer and Inoue in Haemanthus in which one sees the chromosomal doublets before they get onto the metaphase plate, which appear to have a conical region of birefringence with a wide base at the centromere. These can point to any portion of the cell during the period when a continuous spindle is formed. In metaphase they align themselves onto this spindle (which has a much lower birefringence than the actual fibers); I guess you'd call the cones chromosomal fibers themselves. These chromosomal fibers do not seem to decrease in birefringence during early anaphase and are structures more birefringent than the spindle itself. They originate at the chromosome and appear to be involved in the chromosomal motion.

Dr. S. Gelfant: I would like to ask Dr. Scherbaum to make a general comment about the experimental system he used for studying cell division. What answers do you think the synchronized *Tetrahymena* system can eventually provide with regard to problems of cell division? Secondly, would you consider devising any other methods for studying the biochemistry and physiology of cell division?

Dr. O. Scherbaum (University of California, Los Angeles, California): Predictions of this nature are difficult to make. Generally speaking a synchronized system can be looked upon as a "magnified" system. In the ideal situation all cells do the same thing at the same time. Sampling such a mass culture between two subsequent divisions should offer a possibility for chemical analyses which are not possible on a single cell. Result: direct information on the normal cycle. In most cases such a synchrony has to be induced in the mass population. Potent synchronizing agents are temperature, light, or specific chemicals. Obviously we disrupt the normal metabolism occurring in a random manner in the culture. For example, the cells continue to grow, but do not divide. Here I feel we have a unique opportunity to learn something about vulnerable sites in the cellular metabolism connected with mitosis. In other words we can study the function of the sites which "decide" whether compound X shall be channeled to pathway A, leading to division, or pathway B, leading to increased cell size. The result of this approach then is indirect information on the normal cell cycle.

Synchronization procedures have been described which involve mere selection of cells in the same stage of their life cycle. For example, by filtration of a bacterial culture only the small cells after division can be pooled and grown together. Biochemical studies in this system can be directly related to the normal life cycle of the individual cell.

The reason we work with *Tetrahymena* is simply that this protozoon has a short generation time of about 2½ hours and can be grown on synthetic medium. Whether the complex organization of these cells is an advantage or disadvantage for these studies is a matter of attitude. Certainly the fact that all processes necessary for maintenance of the cell as an organism go on side by side in this microscopical unit and can be a disadvantage: for example powerful hydrolytic enzymes of the digestive system can make a biochemical approach very difficult. The elaborate cortical structures in *Tetrahymena* can be of advantage and looked upon as a "biological marker" signaling distinct changes in cellular morphology in connection with mitosis.

It would certainly be extremely interesting to learn how lightly specialized cells of multicellular organisms behave, once they are synchronized. Before this can be done however, quite formidable problems concerning culture technique and induction of synchrony have to be solved.

Dr. L. Levine: I thought Dr. Gelfant made a thought-provoking comment earlier, with regard to the status of injury in stimulating or otherwise affecting the initiation of cell division. I wonder whether you would care to further elaborate.

Dr. S. Gelfant: At the moment I consider it as an experimental tool. I think that the question of injury as a general stimulus has not been considered in many *in vitro* studies in which people feel they are dealing with physiologically normal material comparable to the same material *in vivo*. This is particularly true with regard to the process of cell division and the initiation of mitosis *in vitro*. I have no idea of what the chemical ramifications of injury might be nor how to even define injury in an experimental system. For example, in Dr. Scherbaum's system of synchronized *Tetrahymena* one might even consider the heat treatment as a type of injury which in this case blocks cell division.

DR. O. SCHERBAUM: This certainly is an interesting comparison. Whether the heat treatment is an injury depends, I think, on the definition of injury. Injury in the sense of disruption of structure certainly involves disruption of function, release and mixing of spatially separated material within a cell. If we look at the molecular level and assume shifted equilibria and blocked pathways with the result that a phenomenon such as mitosis is delayed or blocked this surely can be classified as "injury." In Weisz' grafting experiment which I had mentioned we have an interesting example: A *Stentor* which would divide within let's say half an hour normally is delayed in its division when it is grafted together with a young post fission cell. This ideally is about 10 hours. Now if one takes the same mature cell and just injures it in the same way as during the grafting experiment then it is also delayed about 10 hours. I think this is a very interesting aspect, that simple injury of cortical structures might affect the whole system very drastically and cause this division delay of 10 hours.

DR. L. REBHUN: With respect to naturally synchronized nuclei, one case which is always brought up is that of marine invertebrate eggs. However, here we seem to have a set of cells which are primed for the first few divisions so that they go off synchronously and I think this has nothing to do with correlation between the cells. There are cases of the synchronous divisions in organisms such as the noncellular slime molds in which large numbers of nuclei go through divisions simultaneously. This would seem to be reasonably ideal material because one can separate the process of cell division in some sense from the triggers. I think that's what Dr. Gelfant was talking about. There are a number of triggers which can be pulled in various metazoan tissues, which are under the control of the tissue as a whole and of the organism and which determine the rate of cell division. I think that would be one of the reasons for avoiding such a system. You don't know just what the triggers are. If you could synchronize all the cells in some sense you could balance out the trigger and in a naturally synchronized system like the noncellular slime molds you wouldn't have much possibility of injury due to external synchronizing influences.

DR. O. SCHERBAUM: Where *Tetrahymena* is concerned I am not afraid of the injury imposed by the heat treatment. If we want to refer to the temperature effect as injury it may be a matter of convenience. But the question is should the line be drawn. A poikilothermic system such as free living protozoa has emerged from evolution with adaptive mechanisms to survive temperature changes within physiological limits. As we approach these limits "damage" can be done to the systems resulting in morphological changes. Maybe it would be better to reserve the term injury for such cases. In answer to Dr. Rebhun's comment on naturally occurring synchronous systems, marine invertebrate eggs and slime molds are used extensively by many investigators interested in mitosis. With the synchronous sea urchin eggs many aspects of preparation for division can be followed, but it is unfortunate that we cannot study the subtle interaction between growth and division processes. We do not have to induce synchrony experimentally, because nature has done it already by separating growth and division processes; the eggs grow in the ovary and upon simultaneous fertilization go through the first three to four mitoses synchronously.

As I mentioned earlier bacteria can be induced to divide synchronously by fractional filtration and collection of small cells after fission. An important question is then how long the culture remains synchronized. In the mass culture the average generation time is rather constant under controlled conditions. If we record the generation times of isolated individuals from the same culture however a considerable variability is observed (Powell E. O., *Biometrica* **42**, 16, 1955). Let's assume that in an ideal experiment we were able by some means to pick from such as mass culture all cells right after division and culture them together. We assume further that the metabolism in these cells is not influenced by our method of selection. If we wait for one generation only and see how well the cells are synchronized in respect to division, we would very likely be surprised how poorly the synchrony is. This prediction is based on the observed variability of individual generation times I just mentioned. Of course this is a simplified example: we have deliberately omitted the possibility of interactions between the cells and the question whether the cells release metabolically important compounds into the culture fluid.

DR. L. REBHUN: I think this is the case in marine eggs. The first cleavages, at least in eggs I've worked with, are rarely, if you do the things right and don't crowd the eggs, more than a minute apart which means asynchrony of 1 or 2% of the total cycle. The second cleavage is usually synchronous too, but by the third and fourth and certainly by around fourth or fifth division, there is a little synchrony. This indicates events which have been timed for the first couple of cleavages independently but subsequently get out of synchrony due to lack of communication.

DR. O. SCHERBAUM: I might add that the synchronous sea urchins had inspired us (Scherbaum and Zeuthen, *Exptl. Cell Research* **6**, 221, 1954) to establish such a system experimentally for having synchrony available all year round independent of what nature has to offer us at a marine station.

DR. G. B. WILSON: I might say a few words on behalf of a plant meristem. It's always been treated as though it were a bag of dividing cells, which it is not. It is a delightfully and beautifully organized tissue set up presumably for the purpose of providing new cells for growth. The control on it is amazing. In the first place it is impossible to produce a tumor in a meristem. You can produce tumors in other parts of the plants with the greatest of ease, even when you don't want them. But not in meristem. The control over the rate of reproduction of cells is considerable even where you are dealing with only 2-mm section of the pea root containing nothing but the meristem. Half of each generation of cells goes somewhere else. We've calculated that only about 28 or 30% of the cells are actually in a mitotic cycle. We can use 10 to 15% as actually differentiating farther up. This leaves us with quite a balance of cells whose purpose we don't know at all, and which we presume exercise a good deal of control. Plant physiologists always thought of the control as coming from some apical stem, or apical cell way up the stem somewhere. But the root itself has a built-in control system which is simply tremendous.

DR. L. REBHUN: I did want to make one comment relative to Dr. Buck's paper. Unfortunately he isn't here, however I have spoken to him about this. It would be possible to get the impression from his paper that the mechanism for cytokinesis which he discussed, the formation of a series of vacuoles in a region in which a cleavage furrow is to appear, is a relatively universal one. I think that one has to have considerable caution here. Again if we go to egg material there are a variety of types of cleavage which may occur. In very rapidly cleaving eggs where the cleavage furrow is completed, there is considerable evidence, that it does not grow inwards but actually the membrane stretches, or if you think of the expanding membrane theory, it expands. At any rate, the point simply is that making experiments (putting particles on the cortex or using naturally included particles such as echinochrome granules or various refractile granules as Hiramoto, Dan and others have shown), all indicate that the surface actually moves into the furrow with not much new material added. Subsequently, these particles move out of the furrow which indicates that there has been addition of new cell membrane material and probably cortical material after the cleavage furrow is completed. On the other hand there are some very beautiful studies by Selman and Waddington, on cleavage in a European species of Newt in which cleavage takes place in about 40 minutes. They are able to verify an older hypothesis of Schechtman that there actually is a plate of materials that can be seen in the light microscope (they didn't use electron microscopy) which appears to grow considerably in advance of the actual furrow and that surface material does not move significantly into the cleavage furrow. New material appears to be laid down in this plate which then splits to form the furrow. This suggests that the second process, the formation of a large number of vacuoles in a plate, which then fuse to open up, would be a mechanism of cytokinesis in these amphibian eggs. Thus, there probably are at least two mechanisms, one involving rapidly cleaving eggs in which either a contraction process, which I think most of the evidence would indicate, or a surface expansion process such as Mitchison and Swann support, takes place followed by synthesis of furrow and cortex material, and in other cells an ingrowing cell plate may occur to form the furrow followed by fussion at vacuoles and splitting.

Dr. A. M. Elliot (University of Michigan, Ann Arbor, Michigan): There's just absolutely nothing in the area of furrowing of *Tetrahymena,* as far as I can tell, and I just wondered whether anyone here . . . .

Dr. H. Ris: In certain marine eggs, a dense material inside the plasma membrane and characteristic for the furrow area has been described.

Dr. L. Rebhun: That appears to be a set of microtubules of some sort which form a network that outlines the actual cleavage furrow. Right above that Mercer and Wolpert saw the plasma membrane apparently extended in a series of rather random looking extensions, which were possible expanded microvilli.

Dr. H. Ris: In some insect spermatocytes I have seen a ring of denser material, restricted to the furrow itself, which is retained around open connections between primary and secondary spermatocytes.

Dr. L. Levine: Thank you very much. With our sincere appreciation for your coming here and with a thanks to the members of our local committee, Dr. H. Rossmoore, Dr. W. Heim, Dr. M. Bernstein, and Dr. T. Tchen, we close these meetings.

Author Index

Numbers in italics indicate the pages on which the full references are listed.

Subject Index

A

B

D

N

P

R

S